# The Farmer's Benevolent Trust

STUDIES IN LEGAL HISTORY

*Published by the*

University of North Carolina Press

*in association with the*

American Society for Legal History

*Thomas A. Green and Hendrik Hartog, editors*

VICTORIA SAKER WOESTE

# The Farmer's Benevolent Trust

Law and Agricultural Cooperation

in Industrial America,

1865–1945

*The University of North Carolina Press / Chapel Hill and London*

© 1998 The University of North Carolina Press

All rights reserved

Manufactured in the United States of America

The paper in this book meets the guidelines for permanence and durability
of the Committee on Production Guidelines for Book Longevity of the
Council on Library Resources.

Library of Congress Cataloging-in-Publication Data

Woeste, Victoria Saker.

The farmer's benevolent trust: law and agricultural cooperation in
industrial America, 1865–1945 / Victoria Saker Woeste.

    p.   cm. — (Studies in legal history)

Includes bibliographical references and index.

ISBN 0-8078-2421-6 (cloth: alk. paper). —

ISBN 0-8078-4731-3 (pbk.: alk. paper)

1. Agriculture, Cooperative — Law and legislation — United States —
History.   2. Raisin industry — Law and legislation — California — History.

3. California Associated Raisin Company.   I. Title.   II. Series.

KF1715.W64   1998

346.73'0668 — dc21                                  97-32599

CIP

02 01 00 99 98  5 4 3 2 1

12168900

*For Keith and Helen*

The welfare of this whole community is so bound up in the

prosperity of the raisin business, and the progress of that business

is so dependent on organized and more or less public action,

that raisin affairs have always been treated, and properly treated,

as public affairs. . . . The raisin industry, in this community,

can not be treated, and never has been treated,

as "business" in that narrowly private sense.

—CHESTER H. ROWELL,

EDITOR, *FRESNO REPUBLICAN*, 1909

ॐ

# Contents

# Illustrations, Maps, and Figures

# Tables

# Acknowledgments

I take great pleasure in performing the author's ritual declaration of intellectual and personal debts.

The American Bar Foundation in Chicago, where I have the good fortune to work, has provided me with generous support and a convivial environment in which to write. Bryant Garth, director of the ABF, has encouraged my interests and projects, supplied wise insights into the academic world, and supported my career in important ways. My ABF colleagues have created an academic community that is without peer; I look forward to resuming my participation in the lively conversations at bagel time. I especially thank Carol Nielsen and Roz Caldwell of the ABF support staff for their efficiency and dedication.

I would be remiss if I did not acknowledge the members of my dissertation committee who did so much to shape the book's eventual direction. The topic was originally the idea of Morton Rothstein; his continued friendship and support over the years keep me in his debt. Jim Gregory and Martin Shapiro raised questions that I am still thinking about years later. My adviser, Harry Scheiber, has supplied advice, criticism, and mentoring over the years; I am proud to count myself among his students.

The manuscript began the arduous process of becoming a book under the steady guidance of Thomas Green, co-editor of the Studies in Legal History Series. Very early on Tom expressed interest in publishing my book and waited patiently as I thought through the conceptual and historical issues he challenged me to address. When Dirk Hartog joined the editorial team, he spent hours talking with me about the book's possibilities and shepherded the manuscript through formal review. Comments from the outside reviewer, Dan Rodgers, helped me to sharpen the focus of the first part of the book. Lewis Bateman of the University of North Carolina Press has been unstinting in his enthusiasm for the book; I have benefited from his many years of editorial experience and his vantage on the profession.

I imposed on many friends and colleagues to read chapter and manuscript drafts. My thanks to Ann Fidler, David Hounshell, Laura Kalman, David Morrison, Janice Okoomian, John Sayer, Art Stinchcombe, Christopher Tomlins, and Barbara Leibhardt Wester for their close readings and vital help. I wish as well to acknowledge the contribution of participants at seminars where I have presented parts of the book: the Five College Social History Seminar at Smith College; the Boston University Law School Legal History Seminar; the Legal History Workshop sponsored by the Center for Comparative Legal History at the University of Chicago; the Law, Deviance, and Society Research Colloquium of the Department of Sociology, University of Wisconsin, Madison; and the Economic History Workshop at Northwestern University.

In addition to the ABF, I have benefited from financial assistance supplied by the University of California, the Business History Conference, and the Agricultural Cooperative Section of the U.S. Department of Agriculture, which provided the grant that supported the dissertation.

Research was greatly assisted by the staffs at the National Archives in Washington, D.C.; Waltham, Massachusetts; San Bruno, California; Chicago, Illinois; and the Census Microfilm Rental Program in Annapolis Junction, Maryland. The research librarians at the Bancroft Library at the University of California at Berkeley put their legendary expertise at my disposal and promptly tracked down obscure documents. The archivists at the Fresno Historical Society made my visits there productive and responded efficiently to phone and mail requests. Edward Ayers answered my queries about the ICPSR electronic census files, and John Manley smoothly transferred the large data sets for me to work with. Heartfelt thanks go to Jennifer Culbert, Benjamin Forest, Llezlie Green, Sandy Pittman, Amy Toro, and Samuel Weinstein; their help was vital in the final stages. Caroline Goddard, my research assistant, chased down many California sources from Chicago and performed the particularly tedious task of reading manuscript census schedules on microfilm.

Sun-Maid's association with this book should be clarified at the outset. When I began my research in 1988, the cooperative's secretary, Gary Marshburn, gave me open and unrestricted access to all extant records. In 1993, I asked to borrow photographs from the cooperative's impressive collection, but by then the attitude of Sun-Maid officials had changed. My published articles had brought to light the incidents of violence associated with early Sun-Maid membership campaigns, and Sun-Maid's president, Barry Kriebel, was concerned about the impact of these publications on the cooperative's reputation and sales. Kriebel made permission to use photographs and other Sun-Maid art contingent on the publication in this book of a foreword written by him. Naturally, neither the Press nor I could agree to such a condition, and my working relationship with Sun-Maid ended at that time.

I am particularly indebted to the friends and teachers who have made a difference in my life. Dave Farrer was present literally from the moment I stepped off the plane to begin graduate school at Berkeley. In many ways he is family to me. Tekla Harms, James Hughes, and Debra Barbezat made life as a beginning assistant professor not only bearable but fun. In Chicago, I discovered again how rich in talent and generosity the academic community is when Carol Heimer and Wendy Espeland befriended me. Anita and Bob Rieder helped make Chicago home with their gifts of music and hospitality. William and Barbara Wester have been constant friends and companions even when careers and miles separated us. To Barbara, my graduate school buddy, I owe more than I can detail here. Let it suffice to say she wins the prize for reading the most drafts of the manuscript. Our long and frequent conversations about history, the profession, and publishing have been an essential part of my intellectual formation; her patience and enthusiasm renewed me when mine flagged. My teachers at the University of Virginia bear no responsibility for my choice to become a historian, but I know that without their example my life would have taken a different course. Robert Rutland, Martin Havran, and especially Charles McCurdy have shaped my life and career through their impeccable commitments to teaching and research.

Debts to family can never be completely discharged. The interest and encouragement of all my in-laws have bolstered me throughout my career. My father-in-law, John T. Woeste, steered me toward the USDA grant and to contacts in the agricultural extension system.

My parents, Patricia and Theodore Saker, and my family deserve special thanks for their love and support over the years. I have known for a long time that my intellectual training began in my parents' home, where a vast library of history books and biographies and my father's ravenous appetite for daily newspapers taught me the importance of keeping up with one's reading. I have my sisters and brothers to thank for early lessons in cooperation. They and their spouses cheered me on as the finish drew near. None of my grandparents lived to see me complete graduate school, but their belief in the power of education has remained with me.

I dedicate this book to my husband, Keith, and our daughter, Helen. Keith's relationship with this work has been nearly as intimate as my own; he not only read everything I wrote but also provided insight into horticulture and methodology. It is a tremendous advantage to live with one's statistics consultant, and I am sure I abused the privilege. Helen is old enough to register objections to the impact of my work schedule on her daily life and young enough to grant prompt and complete forgiveness. I am grateful to both of them for their forbearance and for daily reminders of life's larger purposes.

Earlier versions of parts of this book have been previously published as articles in the *Law and History Review* 10 (Spring 1992), 93–129, and *Audacity* 2 (Summer 1994), 48–61, and are adapted here with the permission of the copyright holders.

# The Farmer's
# Benevolent Trust

# Introduction

*When co-operation becomes general, there will be no great private fortunes,*

*no involuntary poverty, no international trade rivalry, and, therefore, no war.*

*Co-operation will turn the economic and social interests of the world into the*

*channels of peace and good-will.* — Sun-Maid Herald, 1918

This book is about the struggle of farmers to adapt to industrialization. It is a story that has never been told before, though that is not its chief appeal. Rather, it is significant because it revisits familiar historical issues in an unexpected context. This is the story of how farmers combined the traditional ideology of cooperation with the economic and legal powers of the business corporation, used this new creation to organize their industries and redefine their relationship to the market, and in the process simultaneously exploited and undermined the myth of the Jeffersonian agrarian farmer. Told chiefly through a chronicle of the producers in California's raisin industry and their influence on national law and administrative policy, this story reveals ironic, even paradoxical dimensions of the changes that restructured the U.S. economy and law between 1865 and 1945.

I became interested in issues associated with agriculture's relationship to the state during my undergraduate days at the University of Virginia. At Mr. Jefferson's university, courses in American legal history and American studies introduced me to the clash between the enduring image of the yeoman agrarian and the shifting economic and regulatory contexts of the late nineteenth and early twentieth centuries. To study industrialization and its impact on American law and society using industry as a conceptual framework, it seemed to me, was to take the straightforward path. And that approach dominated the economic and legal history literature, for entirely understandable reasons. Reading J. Willard Hurst and Morton Horwitz taught me that the rise of the national market and the transformation of American law during the nineteenth century came chiefly at the expense of nonindustrial economic activities and nonurban land uses. Developments such as the railroad and the transportation revolution, the rise of urban manufacturing centers and the labor movement, and the courts' willingness to endorse the dedication of property to its most valuable economic uses all were valued for their contributions to urban society, not balanced against their cost to rural society.[1] Studying agriculture's adaptation to the new economy, it seemed to me, presented the opportunity to explore the relationship between industrialization and legal change from the perspective of the sector industrialization supposedly left behind. By focusing on the period before World War II, I could treat agriculture's economic transformation during a time when contemporary observers and historians defined farming as nonindustrial.

Except for historians who specifically examined agriculture, scholars have long treated industrialism and agriculture as distinct. Certainly ample historical evidence suggests that agriculture's gains between 1870 and 1945 were sporadic. The agricultural sector began a steady decline after industrialization quickened its pace after the Civil War, a decline measured by population loss, a decrease in rural purchasing power relative to the price of retail goods, and the gradual disappearance of that most hallowed of icons, the family farm. Early twentieth-century magazines and newspapers sounded a drumbeat warning that farmers were losing ground, literally and figuratively, as urbanization transformed American society. Agricultural historians have been centrally concerned with understanding the reasons for these developments. But with few exceptions, they have paid little attention to agriculture's relationship to and impact on law and the pre–New Deal administrative state.[2] The history of the cooperative movement suggests that the elements of what we recognize as modern American agriculture—large-scale corporate ownership of farmland, government price supports, and marketing quotas—were rooted in legal issues and controversies that antedate the turn of the twentieth century.

The legal history of agriculture may seem like a mere extension of the work of J. Willard Hurst, Harry Scheiber, Lawrence Friedman, and the Wisconsin school of legal history. This project certainly began as such. The Wisconsin school essentially created the legal history of economic change. Hurst's study of the lumber industry in Wisconsin stands as the cornerstone of this new approach to legal history. In calling attention to the ways in which courts and legislatures made and remade law in dynamic interaction with the economic enterprise of logging, he shifted the methodology—as well as the conceptual framework—of the field from the doctrinal vagaries of U.S. Supreme Court decisions to the things that more immediately constitute the sociolegal order of everyday life. The way people dynamically deployed law to organize their polities and secure their commercial transactions, as well as the way the legal system actively promoted the "release of creative entrepreneurial energy" of the American people, Hurst argued, shaped the course of the nation in ways every bit as fundamental as Supreme Court decisions.[3] It was Morton Rothstein, the longtime agricultural historian at the University of Wisconsin, who suggested to me that it might be interesting for a legal historian to look at marketing cooperatives; my first morning's research in the law library more than confirmed his supposition. I discovered that the cooperative movement generally and the California organizations in particular played essential, heretofore unacknowledged roles in shaping some of the fundamental constitutional and regulatory issues of the Progressive Era. Telling the story of the raisin growers from the vantage of the irony and paradox created by their use of law became the goal of this book.

To write about historical actors and their "use" of law invites association with the functionalist/pluralist mode of history most often linked to the Progressive historians Charles Beard, Richard Hofstadter, and Hurst himself. Hurst in particular has been roundly criticized for arguing that resort to the law was an unproblematic extension of the democratic system, the implication being that anyone, regardless of personal status, could claim access to the law and its promise of economic and social advancement.[4] Not surprisingly, this criticism came during the 1970s and beyond from social historians of law, whose interests in women, racial and ethnic minorities, and such economically disadvantaged groups as labor led them to less sanguine conclusions about law's capacity for achieving social change.[5] Hurst's thesis, we should remember, was not that American law worked to the advantage of every marginalized individual or group but rather that its capacity for promoting economic growth prevented vested interests from retarding the development of society.[6] What limits the relevance of this thesis to the social history of law is Hurst's relative disengagement from the specific social contexts within which individuals and government interacted. The release of energy paradigm does not recognize the restraints that law imposes on those who seek to use it.

Many historians have argued that when women, African Americans, Native Americans, or immigrants used the courts to defend their rights, the results were mixed. Legal victories on specific claims often had fairly short shelf lives, with implications that participants could not anticipate.[7] For example, when courts awarded Chinese aliens due process rights in deportation proceedings, the administrative state responded with even more restrictive policies. The political priority of keeping out undesirable aliens withstood the sustained assault the Chinese mounted in the courts. And when Native Americans sought to obtain enforcement of their rights to fish, hunt, and gather at their traditional places, rights that had been ostensibly reserved to them by federal treaty and on occasion vindicated by the federal courts, the states redefined those rights to be consistent with the interests of white property owners. Individuals and groups used the law to pursue specific goals, but they did not always attain those goals in their imagined form, and even when they did the legal system often responded in ways that confounded their victories. Legal structures can be and were used to resubordinate those seeking to attack power relationships in society that law held in place. In short, the use of law often led to unpredictable outcomes, characterized by neither the democratic access of Hurst's dynamic law nor the entirely repressive hegemonic law of his critics.[8]

The history of the cooperative movement supplies a case in point of this idea. Cooperation embodied a series of puzzling incongruities. It combined the ideologies of self-help and agrarian self-sufficiency, even as it embraced the trend toward combination and consolidation. The movement borrowed from corporations and labor unions alike, yet it managed to evade the public enmity heaped on trusts and ultimately to avoid the legal pitfalls that courts carved out for unions. The movement proclaimed the unity of interests among farmers and the importance of agricultural solidarity, yet it resorted to zealotry and violence to reinforce its economic combinations. The movement was born out of farmers' protests against monopolies and trusts, yet it staked its own existence on a privileged use of monopoly and the trust. The movement invoked the myth of Jeffersonian agrarianism, even as it demonstrated the incongruity of that myth in an industrial economy. In short, cooperation was a movement of contradictions. While cooperation's traditional image kept it at the forefront of official agricultural policy for decades, the changes that farmers made in order to apply cooperation to the industrial market undermined the movement's claim to the legal privileges that farmers so doggedly sought.

The raisin growers learned this lesson the hard way. The extraordinary legislative gains they won altered the legal status of cooperatives, but they also transformed the cooperative ideal itself and made recourse to that ideal—as justification for the granting of still more legal privileges—increasingly problematic over time. And as farmers discovered during the critical decade of the 1920s, it was impossible at once to make predictive claims about how cooperatives would work in the market and then make cooperatives perform according to those claims. Formal legal victories had less effect on what happened in the economy than cooperation's proponents recognized. As I conclude, the cooperative movement succeeded in challenging the legal and social constraints that bound its development, but it could not change the underlying dynamics of an industrial economy. Ultimately, the downfall of the California Associated Raisin Company (CARC)—now known as the Sun-Maid Raisin Growers—occurred because law restrained cooperatives even as it endowed them with freedoms it did not grant to other entrepreneurs. Because cooperation failed to live up to its promise of saving agriculture and because the legal powers cooperatives attained were unequal to that task, agriculture's fate ended up more tightly linked to the state after the Great Depression and farmers no less subordinate to the law and the market than they had been before.[9]

This book tests this revised model of legal functionalism by examining the cooperative movement's intersection with three related areas of regulatory law.[10] Antitrust, corporation, and equal protection law furnished the central legal issues the cooperative movement faced between 1865 and 1945. During this period, cooperatives sought to do for farmers what labor unions were attempting to do for wage workers: provide the economic and organizational benefits of collective bargaining to individuals otherwise left on their own.[11] To bring farmers more profit for what they produced, cooperatives used price-fixing, monopoly, and anticompetitive trading to dislodge the "middlemen," thus putting farmers smack in the middle of the burning legal and economic issues of their day. Yet farmers' involvement in these issues and the historical significance of their contribution to regulatory culture have been overlooked. Economic and legal historians studying antitrust have concerned themselves with defining restraint of trade, understanding the web of legal rules governing market behavior after the Sherman Act of 1890, and explaining how the system of dual federalism created regulatory vacuums during the Progressive Era.[12] Because the proposal to exempt farmers from the Sherman Act failed to make it out of committee, it has been easy to conclude that farmers had little to say about federal antitrust law or enforcement. In fact, they said plenty, but they said it in arenas on the periphery of regulatory action after 1900. While the nation's businesses and consumers looked to Congress to set policy on the permissible degree of combination and monopoly in the national economy, agriculture was only partly affected by national legal and political developments during this time. Most of the antitrust and corporation law controversies involving farmers' organizations took place at the state level. Consequently, the antitrust conflicts generated by the cooperative movement continued to enliven state antitrust enforcement long after similar industrial conflicts had moved to the federal arena. Federalism thus provided an essential, if at times convoluted, regulatory context for agriculture and cooperatives.[13] The formal distinction between state and federal authority so prominent in Supreme Court jurisprudence during this era did not preclude farmers from conducting lively, ongoing experiments with the rules governing market behavior — or from pushing beyond those rules to see how far they could go.

Farmers pushed especially hard when it came to employing the business corporation to achieve the purposes of cooperation. Again, the legal and constitutional tensions of federalism underlay farmers' struggle with the question of the legal form of their enterprise. Legal historians interested in the use of the corporate form as a means of organizing enterprise and promoting the accumulation of capital have looked closely at state laws creating corporations and governing their powers. As historians have shown, states engaged in a virtual bidding war at the turn of the century, competing to attract corporations by offering tax breaks, subsidies, and other public benefits.[14] This history is important for understanding both the lack of antitrust enforcement at the state level and the pattern of corporate domiciling after 1900. Because this story has focused on industrial manufacturers and their strategic legal choices, however, it reinforces the general impression that the corporation was a more potent economic weapon in the hands of urban industrialists and had little relevance for the unsophisticated entrepreneurs on the farms. Hurst's emphasis on legal forms and their availability to individuals should teach us to be skeptical of such conclusions. When farmers created corporate cooperatives to reorganize their industries and redefine their relationship to the market, they were not merely copying the strategies and organizational innovations of big business.[15] Rather, they were remaking the corporation so it would be more useful to the agricultural sector. The enduring appeal of the Jeffersonian image and the continuing connection in the public's mind between national prosperity and a thriving agriculture played directly into farmers' hands. Thus farmers could draw on the corporation, the trust, and the labor union as institutional models and descriptive metaphors, while seeming to maintain a dignified distance from the specter of conspiracy, collusion, and corruption that these institutions evoked in the public mind.

During the period from 1890 to 1930, most groups in the American economy sought antitrust exemptions. The courts generally refused to permit one class of entrepreneurs to enjoy privileges not granted to others, on the grounds that no constitutionally recognizable disadvantage lay with the favored class. Farmers, however, thought that they constituted a special class of entrepreneurs because their occupation required them to sacrifice personal gain for the nation's good. They contended in state and federal courts that the cooperative constituted a distinct form of organization. They believed that they suffered from an impaired ability to compete in the industrial market, and they worked for decades to overturn a Supreme Court decision that insisted otherwise. The equal protection issues raised by farmers—whether farmers were a distinct class of entrepreneur, and if so whether they deserved special treatment at law—are not so different from those raised by organized labor, the women's movement, or even the struggles of Native Americans or African

Americans seeking their full rights under the Constitution, if only in the sense that farmers believed that they as a group bore a disproportionate burden of the costs of economic growth. By framing their response in legal terms and forms made legitimate by powerful interests, they staked a claim to the economic justice that eluded them in the market.[16]

Antitrust, corporate identity, and equal protection were all framed by broader institutional and intellectual developments that shaped legal change between 1870 and 1945. Because the cooperative movement espoused political and economic values that were consonant with Progressive Era Americans' vision of themselves and their society, farmers succeeded in gaining legal recognition for their special status long before these other groups did. The result was that relatively early in the twentieth century farmers established the basis for a special constitutional status that translated directly into long-standing policies of governmental privileges, subsidies, and favoritism. I argue that the equal protection doctrine affected economic status and regulatory culture in ways that standard legal and constitutional histories have missed.

The Progressive Era administrative state and agriculture had an unsettled relationship. Secretaries of agriculture under Theodore Roosevelt, William Howard Taft, and Woodrow Wilson consistently promoted the goal of getting American farmers to produce more as the number of farmers decreased. At the same time, the advent of bureaus of marketing in both the U.S. Department of Agriculture and the states provided the administrative capacity for addressing farmers' economic and legal complaints about the distribution system. Both administrators and agencies sang the praises of cooperation as the salvation of agriculture, but governments refrained from organizing cooperatives or supplying any other assistance that directly involved the state in mediating market relations. This result was driven less by party politics than by the contemporary constitutional and intellectual limitations on the government's involvement in the economy. The 1920s produced a wave of federal and state laws establishing cooperation as the pillar of agricultural policy, and these laws were direct precursors of the expanded administrative regulation of agricultural production, prices, and marketing that came during the New Deal. Here I argue that rather than standing as an example of the failed associationalist movement of the 1920s, cooperation established a regulatory model that led directly to the New Deal's vision of an orderly, centrally controlled agricultural economy.[17]

Recently, historians have begun to focus on the impact of ideology on legal discourse to sharpen and expand the Hurstian paradigm. By the 1980s, legal historians seeking to understand the impact and relevance of legal rules and doctrine on economic change had turned to the role of ideology and ideas, first to reveal the hidden fabric of the constitutional thought of important judges, then to explain the institutional and political contexts of organizations and movements.[18] Ideology plays no less essential a part in the history of the cooperative movement.

The cooperative movement had its own distinctive ideological heritage, drawn from European socialism, experimental utopianism, and a burgeoning of worker-owned purchasing collectives following the experiment of the Rochdale weavers in England. Although I will not undertake much examination of the European and socialist experiences with cooperative ideology and practice in this book, I notice the implications of those roots for cooperatives struggling to adapt to market conditions after 1900. The radicalism of their European and socialist origins sat side by side with the increasingly conservative emphasis of the American cooperative movement. When the movement pursued changes in antitrust and corporation laws, it sought only to make farmers better equipped to deal with other agents in the market, not to transform the structures and processes that defined free and competitive bargaining in the industrial economy. In short, in this country, agricultural cooperation reinforced the ideological and political assumptions underlying corporate capitalism. Indeed, such a position was necessary for the movement to gain legal legitimacy.[19]

Although Americans never embraced the European movement's more radical implications, the emphasis of cooperation on self-help meshed with widespread concerns about the impact on society of industrialization, the concentration of bargaining power, and the sheer bigness of big business. Cooperation represented an alternative to capitalism that did not force society to give up the benefits of capitalism, instead reinforcing the belief that people could, through self-help mechanisms, share fully in the fruits of industrialization. Cooperation suggested that an "equitable capitalism" was still possible after 1900, as long as people adhered to the organizational principles of Rochdale cooperatives. Indeed, its promoters boldly maintained that cooperation could end all economic war by taming competition and ensuring basic fairness.

The cooperative movement continued to make this appealing argument as it moved to legitimize collective marketing in agriculture. It offered the striking possibility that agriculture could solve its own problems without rocking the ship of industrialization. The innovative organizations that arose in the California fruit industries, which claimed to fulfill the cooperative vision of a producers' collective while building airtight monopolies over supply and marketing, stabilized industrial horticulture and integrated California agriculture into the national economy. In the process, however, California growers rejected the Rochdale model and its emphasis on decentralized decision making and noncapital operations in favor of a more corporate being with the commercial and legal powers necessary for conducting trade in the national market.

Putting this model into practice necessarily transformed the cooperative ideal. Farmers' success in obtaining legislative recognition of their new form of corporation pushed the movement toward conservative economic goals and away from the image of the agrarian farmer it purported to preserve. The cooperative movement's reorganization of horticultural marketing provoked heated criticism from opponents in the distribution system and advocates of consumer interests. First, they argued that these cooperatives were too powerful to continue in business under the rules of free and fair competition; then, as the organizations cemented their monopolies over supply and marketing, these opponents contended that cooperatives were taking advantage of farmers rather than acting in their interests. How could it be reasonable, they demanded, for a farmers' cooperative to control and fix prices for the entire nation? The president of the California Associated Raisin Company disarmingly turned aside these complaints in 1920: "Call us a trust, if you will, but we're a benevolent one." [20] In his view, the benevolent trust was neither an oxymoron nor a legal fiction but a real answer to the problems the new economy inflicted on farmers.

The very idea of the benevolent trust raised the question of whether a corporation with all the powers of Standard Oil could keep its promise to act on behalf of the farmers it represented, save consumers money, and improve the efficiency of marketing, all without unjustly enriching anyone. For the most part, Americans wanted to believe cooperation's promise. Except for the private firms and distributors that lost business to the cooperatives, no one really wanted to believe that agricultural trusts posed a threat to American consumers. Even when the California cooperatives explicitly made monopoly the hallmark of their organizations, it took an extraordinary conjunction of circumstances to overcome cooperation's exemption from federal antitrust prosecution. Only after World War I, when cooperatives appeared to have an inordinate impact on price, did the federal government decide that litigation was legally justified. The 1920 antitrust prosecution of the CARC—the first ever against a farmers' organization—forced its officers to defend the economic and legal assumptions of the benevolent trust model of cooperation. But the threat to the cooperative's existence and operations was ultimately fleeting. When Congress essentially mooted the raisin antitrust case with new legislation embodying the precepts of the new cooperative style, it was clear that cooperation had become an inseparable aspect of agricultural enterprise and of the intellectual and political foundations of agricultural policy.

This book accordingly has much to say to business and agricultural historians as well as legal scholars. The cooperative movement provides a discrete though interdisciplinary analytical framework through which to understand changes in the organization of agricultural business and its relationship to law and the larger market system. My case study, an extensive history of the raisin growers' organization, follows the method of Alfred Chandler and other business historians who explore the impact of economic change on firms and industries. The case study's contribution to the understanding of agriculture as a business enterprise is particularly relevant for business historians of the "organizational school," who are evincing interest in a more interdisciplinary approach to the history of American business enterprise. Scholars in this vein have redeemed the business person from the traditional role of corporate villain.[21] I attempt to do the same for the agents of "agribusiness" by revealing the dimensions of law and ideology in the cooperative organization of agricultural marketing.

The term "agribusiness" has been a shibboleth, and a pejorative one at that, for decades. Carey McWilliams invented the term in 1939 to describe California agriculture by the 1930s: "factories in the field," dominated by large-scale intensive cultivation, utterly exploitative of human labor and the natural environment.[22] The image of the California that greeted John Steinbeck's Okie migrants has remained with historians of the state for the better part of a century. Walter Goldschmidt's study of the relationship between agribusiness and large farms, though now somewhat discredited, influenced agricultural historians for the better part of two generations. When scholars such as Gerald Nash, Donald Pisani, and Cletus Daniel considered the historical record of California agriculture, they tended to characterize it in implicitly evolutionist terms: from small farms to large, from individual to corporate ownership, from diversified to specialized operations, from subsistence to commercialized agriculture. And they implicated the cooperative movement in this process, folding the organization and market behavior of the major California horticultural cooperatives into the general story of agribusiness. According to this view, cooperatives helped achieve and reinforced the changes in marketing, transportation, and technology that were themselves instrumental in the rise of commercial horticulture. As Sucheng Chan has put it, "California has been a pioneer in organizing marketing cooperatives that control national and worldwide markets through the use of sophisticated ways to gather and relay market information and to adjust supplies instantaneously in response to fluctuations in demand in various places as the fresh produce is being moved across the country." Indeed, there seems to be little scholarly dispute that cooperatives arose in response to marketing inefficiencies over long distances, disparities in bargaining power between the growers and agents of the distribution system, and farmers' firm conviction that they were being cheated out of a just share of the profits from their crops. Historians believe that California was the font of agribusiness and the cooperative movement simply another ingredient in its recipe.[23]

More recently, some scholars have begun to reexamine this long-held understanding. New work on agricultural labor and on farm size sheds new light on the relationship between agriculture as an economic system and the individuals whose labor and capital contributed to that system. Sucheng Chan's work on the Chinese in California between 1850 and 1920, for example, reveals that the corporate consolidation taking place in California agriculture did not eradicate the meaningful economic activities of even the most marginal groups in California society.[24] As I dug deeper into my research, I found myself wondering who belonged to the marketing cooperatives in California, what kind of farmers they were, and what their economic status was. As it turned out, it was not possible to answer all of these questions, but my study uncovers long overlooked information about the growers who toiled between migrant labor and corporate agribusiness.

What I found changes how we understand the history of California agriculture. The cooperative movement in California horticulture was built on the backbone of thousands of small farms, farms that became more, not less, numerous over time. The growers who owned these farms did not fit the profile of the typical California "land baron." Most fruit growers were men and women of modest means, not absentee landlords holding cowed tenants and migrants in prolonged dependency. These small landholders were just as important to the industry's prosperity—and to the cooperative movement—as the social and economic elites whose large holdings made them wealthy and powerful. Since farm size was not always a perfect indicator of class, race, or ethnicity, locating the center of power within the industries is not a matter of finding the largest, richest growers. Rather, it is a matter of understanding the relationship between a small group of large landowners and a very large group of small growers in the context of a sustained drive for industry-wide collective action.[25]

The history of cooperation in the raisin industry directly contradicts the evolutionary story of the rise of agribusiness and, by implication, the role of law and policy in fostering the commercialization of agricultural enterprise. Instead of relentlessly marching toward agribusiness, the raisin industry relied on several different economic subsystems occupying the same space and time, related to each other but generating distinct socioeconomic cultures.[26] Small growers were essential players in the movement to organize cooperatives and to establish monopolies in the marketing of their commodities. Thus the presence of a widespread and growing cadre of small farmers between 1870 and 1940 makes all the difference in the story we can tell about the rise of cooperative marketing.

The story of cooperation in the raisin industry is not just one of farmers against distributors, of the rural world taking on the nonrural world. It is also a story about the social conflict that resulted when small independent growers refused to be dominated by industry leaders. During the Progressive Era, all the California horticultural industries were forming cooperatives, learning from and copying each other in the process. I chose to focus on the raisin growers because they were the most legally innovative and because, in the words of a present-day California cooperative official, "Sun-Maid's the one with the skeletons in its closet."[27] Opening that closet, as it turned out, pointed the way into the world of turn-of-the-century agricultural producers, a world defined in large measure by their relationships to their industry and the market, certainly, but no less fundamentally by their relationships to each other.

The presence of many small farm owners and operators in the raisin industry decisively affected the dynamics of cooperative organizing and the outcome of the CARC's attempt to corner its market. Initially, many growers were shepherded into raisin industry cooperatives by the strong-willed, wealthy growers who made up the industry's elite. Cooperative leaders liberally applied corporation law to the incipient law of cooperative organizations; they then took law into their own hands to hold cooperatives together against fierce opposition and the disparate interests of grower-members. The actions of the night riders who coerced growers into joining the CARC showed the lengths to which individuals were willing to go to ensure the cooperative's success. Unchecked by local and federal authorities, the night riders victimized any grower, regardless of social status, farm size, or ethnicity, who dared to assert economic independence. The prevalence of Armenian Americans among the independent-minded added a racial dimension to the violence, but the mobs' primary aim was forcible inclusion to cement the monopoly.[28] Neither violence nor formal law, or the two together, could preserve the "benevolent trust" in the face of unstable economic conditions whose causes lay beyond the acceptable bounds of regulation.[29]

This book is a history of the cooperative movement that integrates its legal, social, and economic dimensions. It weaves together sources from economics, history, and law. It draws on archival research in the records of the Justice and Agriculture Departments in the National Archives, manuscript collections of people and firms in the raisin industry, and a host of agricultural newspapers and magazines, including the publications of the CARC and Sun-Maid. The research design is necessarily interdisciplinary. The politics of antitrust law, the reform currents of the Progressive Era, the intellectual debate over farming's place in the new economy, the social and economic aspects of the post–World War I "farm crisis," and the comprehensive alterations the Great Depression imposed on state and society alike all shaped the cooperative movement in significant ways. In turn, the cooperative movement exerted no small influence on these developments, helping to reframe the image of American agriculture throughout a period of unsettling and disturbing economic change. The California raisin industry furnished the exception that became the basis of the rule. The California Associated Raisin Company defined modern agricultural cooperation. The national political and legal responses to cooperation's benevolent trust reformed the relationships among agriculture, the law, and the state in twentieth-century society.

This book encompasses four overlapping parts. The broad intellectual, political, and economic settings of the study are described in Part I. Chapter 1 defines the "farm problem" at the turn of the century as the problem of marketing agricultural commodities in a national economy, lays out the breadth and variety of cooperative practices in late nineteenth-century America, and traces the roots of the dominant mode of cooperation that emerged in California in the twentieth century. Chapter 2 describes the California context. It details the social and economic underpinnings of commercial horticulture and explains why California growers turned to the corporation as a way of bringing their distinctive approach to cooperation to fruition. I argue that small farms furnished the essential and lasting foundation of the horticultural economy, which had significant implications for the cooperative movement.

Part II sets out the legal and historical framework for the case study that follows in Part III. Chapter 3 analyzes the legal status of the cooperative in antitrust law at the state level. Until a legislative determination of cooperatives' legal character could be made, the state courts treated them as illegal combinations engaged in restraint of trade. These rulings did not deter experimentation with cooperation's legal form in the raisin industry, as Chapter 4 makes clear, but they did shape growers' attitudes toward the maverick leader who flouted the law in creating the first industry-wide cooperative. The federal antitrust and equal protection issues raised by farmers' cooperatives and the beginnings of a new federal policy are considered in Chapter 5. Together hese three chapters show that growers experimenting with cooperatives in California both relied on and pushed forward emerging developments in state and federal law.

Part III focuses on the CARC's legal and economic innovations. The CARC tested the vague antitrust immunities in federal law by assuming a corporate structure that did not conform to existing legal categories and by engaging in market practices that were of questionable legality for any business corporation. Chapter 6 chronicles the CARC's rise between 1910 and 1918, the beginnings of night riding associated with cooperative membership campaigns, and the U.S. Justice Department's ineffectual supervision before World War I. Chapter 7 details the organized campaign by competing raisin packers to force the Justice Department to prosecute the CARC, the legal and political obstacles to prosecution, and the changed conditions after World War I that led to the lawsuit in 1920. The Capper-Volstead Act of 1922, which essentially mooted the prosecution, marked federal legislative acceptance of the "benevolent trust" model of cooperation. Chapter 8 sets out the practical limits of that acceptance. Despite the sympathetic intervention of the U.S. Department of Agriculture (USDA), which kept the Justice Department out of the cooperative's affairs after Capper-Volstead, the CARC was unable to maintain its control over the raisin industry. Record crops and the corresponding collapse in prices did more to destroy industry-wide collective action than even the most zealous mob could repair. By 1928, the days of the benevolent trust were over.

Part IV demonstrates the impact of the new model of cooperation on state and federal agricultural policy before and during the New Deal. Chapter 9 lays out the close relationship between cooperation and the associationalist state of the 1920s. Thirty-eight states adopted laws that permitted cooperatives to engage in monopoly, to control growers through "iron-clad contracts," and to assert industry-wide control over the marketing of agricultural commodities. The new legal conception of cooperation proved unequal to the task of saving agriculture from its prolonged economic distress. Chapter 10 traces the implications of the cooperative movement for the New Deal. It showcases two state-level programs, one to control prices and the other to control production, and assesses their effectiveness in light of their reliance on cooperative practices and goals. As I argue in the conclusion, the transformation of cooperation went beyond its legal form. It had become subsumed within the administrative state, a party to the socialization of agriculture's enduring economic dysfunction. In the end, cooperation could not assure agricultural self-sufficiency. Yet the image of cooperation—a stable, prosperous agriculture as the basis of modern industrial society—remains a vital aspect of its appeal.

PART ONE

Cooperation and Agriculture

in American Culture

1865–1910

# 1 ∿ The Farm Problem

*Association is the master word of modern days; it is the key to efficiency, power*

*and equality. Corporations are associations of capital for profit; labor unions are*

*associations for the profit of its [sic] members; co-operative associations are for*

*the profit of all the people who choose to associate.* — Sun-Maid Herald, *1918*

At the turn of the twentieth century, American farmers were in trouble. Social unrest and economic stagnation were rife in the rural sector, from dairy sheds in upstate New York to washtubs and summer kitchens on Great Plains wheat farms, from the tobacco-drying barns of the Appalachian piedmont to the irrigated drylands of California. Whatever their ties to and interests in rural life, people fundamentally disagreed not just over the solution to the farm problem but over the nature of the problem itself.

Economists and social scientists thought the farm problem stemmed from inefficiency on the individual farm. Increasing agricultural production through the application of the principles of scientific management and agricultural economics, they believed, was the key to agriculture's survival in the industrial age: "The success of the modern farmer depends . . . upon his capacity to select and produce that crop or combination of crops which, one year with another, will make the farm yield the largest net return." Sociologists blamed the farmer's character for agriculture's economic backwardness: "There must be something wrong in [the farmer's] own make-up, some failing on his part to see what the conditions of life are. . . . The failing is in himself. If you want to reach the root of the farmers' difficulties, you will have to begin with the farmers' minds." In this view, farmers held their fate in their own hands; an agricultural college professor patronizingly lectured that improved farm management would come "by improving the [moral and physical] qualities of farmers."[1]

Politicians framed the issue in ways that exploited cultural perceptions of agriculture. They thought that regardless of how the industrial economy changed, farming had to remain a small-scale, localized activity. Even Herbert Hoover, who championed agriculture's need for better organization and a fairer marketing system, publicly cherished the Jeffersonian agrarian ideal: "Farming is and must continue to be an individualistic business of small units and independent ownership." [2]

Farmers saw the matter differently. They worked hard, under difficult and unpredictable natural and economic conditions, for less money than they felt they deserved. They believed that the root of the farm problem lay not on the individual farm but in the market. Farm leaders pointed to Frank Norris's novel *The Pit* as an accurate description of the market's inherent unfairness: " 'If I go out there in Kansas and raise a crop of wheat, I've got to sell it, whether I want to or not, at the figure named by those fellows in Chicago. And to make themselves rich, they make me sell it at a price that bankrupts me.' " Most farmers believed that they were losing, and losing badly, in the great competitive war imposed on them by the new industrial conditions. [3]

Farmers' critique of the market was grounded in day-to-day interactions with the corporate actors and agents who determined the profits of their labor. The industrial economy permanently altered the relationship between producers and consumers. To farmers, the rise of the "middlemen" to a permanent place in the marketing system gave middle merchants disproportionate influence over agricultural prices and profits. Farmers believed that railroads, grain elevators, stockyards, and distributors unfairly restricted their access to consumers. As the economist Joseph G. Knapp summarized, the farmer "blamed much of his plight on an unfair system of 'interchange,' whereby he was forced to pay excessive toll in marketing his products, and excessive prices for his purchased supplies; and he was vehement in his resentment of all 'monopolists' and 'middlemen.' As the farmer saw his condition, he was 'fleeced both coming and going.' " [4]

The farm protest movements of the late nineteenth century brought cooperation and the marketing problem together in ways that proved of more lasting significance for the cooperative movement than for the farm organizations that championed it. Both the National Grange of the Patrons of Husbandry (better known as the Grange) and the Farmers' Alliance (which gave rise to the Populist Party in certain states) dedicated themselves to solving the marketing problem. Both used cooperation as the foundation of their programs for political and economic reform; in the end, however, neither was able to transform the market. These movements were fundamentally political in nature; they enlisted economic ideas in attacking the interests they claimed were oppressing farmers, but they ultimately invested their organizational capital in electing candidates to public office, particularly at the state level. Their focus on political action diverted these movements from the essential task of adapting the legal status and economic function of cooperation to new market conditions.[5]

Yet the protest movements fed the process of innovation taking place in cooperation around the turn of the twentieth century. This process proved vital in places where, as in California, it was imperative to solve the marketing problem. The sheer physical distance between horticultural producers and eastern consumers made growers determined to control processing and marketing. Between 1880 and 1910 they sifted through the varieties of cooperatives then in existence, tried out a series of legal innovations in form and practice, and emerged with a new kind of cooperation that evoked the movement's traditional values while transforming the organization into a potent economic weapon. That the new kind of cooperative emerging in California hardly resembled the common perception of cooperation had serious implications for its social meaning and legal status. It was that common perception, arising out of a tradition that had already proven itself unworkable in the national industrial economy, that rural and urban interests alike continued to invoke to justify their reliance on cooperation as the solution to the farm problem.

The new model of cooperation raised critical questions about its place in the new economy. For one, cooperation's origins in radical philosophical traditions made it unwelcome in the conservative marketplace of ideas in the nineteenth century. For another, organizing farmers "as perfectly as capital is organized," as William Smythe wrote in 1905, had the potential to transform the relationships between farmers and distributors, retailers, and consumers. As farmers toyed with variations on the traditional form of cooperation between 1865 and 1910, they faced several related challenges. They had to adapt traditional cooperation to the economic needs of their industries without appearing to betray the ideals for which traditional cooperation stood. At the same time, as they created a new legal form for marketing cooperatives, they had to persuade courts and other skeptical observers that their form of collective action did not violate the prevailing norm of economic and constitutional individualism. Either way, for cooperation to gain acceptance as a legitimate mode of enterprise, farmers would have to suppress its potential for shifting the balance of power within the capitalist market.[6]

The cooperative tradition dates back to antiquity. The salient models for American cooperation originated in western Europe and England during the nineteenth century. A pioneering workers' cooperative store at Rochdale, England, especially appealed to Americans because it aspired to reconcile competitive capitalism with the cooperative principle of equality among participants. The U.S. cooperative movement's gains after 1870 owed much to the promise that the Rochdale form of cooperation would solve the marketing problem while preserving free competition.[7]

The Rochdale workers transformed the basic idea of cooperation in the mid-nineteenth century by devising both a broad philosophical foundation and a specific organizational framework for it. After repeated strikes failed to bring improvements in wages and living conditions, twenty-eight weavers established a cooperative store in 1844. What set the Rochdale store apart from earlier cooperative purchasing societies was the way it governed financing, profit sharing, and membership. At once simple and revolutionary, the plan emphasized participation according to contribution of goods or patronage rather than investment of capital. Of the twelve original Rochdale principles, first published in 1860, four emerged repeatedly in subsequent practice. They included business or services performed at cost, on a cash basis, with net returns paid to members in proportion to business transacted, known as patronage dividends; democratic control—one person, one vote; limited dividends on invested capital; and ownership limited to patrons—those who patronized the store or who belonged to the occupation the cooperative was established to serve. Also popular were limits on the amount of stock an individual could own and a social duty to educate people in the tenets of cooperation.[8]

The Rochdale principles converted the private corporation into a public, nonprofit collective. Each of the principles revised an aspect of private market relations. The first principle eliminated the need for extensive management and accounting practices and reduced interest expenses; individuals profited according to their participation rather than their wealth. The second principle ensured that no individual would have a greater say in the affairs of the organization than anyone else, regardless of invested capital, patronage, or shares owned. By restricting interest dividends and limiting membership to like-situated individuals, the cooperative could restrict benefits to participants. In short, unlike regular corporations, in which ownership was divorced from management, Rochdale cooperation maintained a close and direct relationship between the two. Rochdale cooperation had potentially radical implications but took an essentially conservative turn. The English historian of cooperation George Jacob Holyoake stressed cooperation's conservative implications: "[British cooperation] 'began in a desire for equality; but not by pulling down the rich to the pitiful level of the poor, but by teaching the poor how they might raise themselves to the level of the rich. The early cooperators sought equality through equity.' "[9] The idea of equity did not entail systemic redistribution of capital but improvement of the lot of the poor through self-help and fair distribution of the fruits of their collective efforts. The goal was not to attack the capitalist system but to enable more people to share in its benefits.

The Rochdale emphasis on profit sharing and self-sufficiency appealed to Americans. By the time of the Civil War, the Rochdale store had become an international success story. In 1859, *New York Tribune* editor Horace Greeley, a supporter of communitarian alternatives to industrial society, published Holyoake's book on the Rochdale pioneers, *Self-Help by the People*. The book fed the rising interest in cooperation that was developing before the Civil War. In the postbellum era, agrarian leaders believed that cooperation might teach farmers to be better entrepreneurs, raise their standard of living, and flourish in industrial society.[10]

The agrarian protest movements of the late nineteenth century relied on this appeal. They were less certain, however, of how to put its precepts into practice. The problem, as it turned out, was not a failure to implement the Rochdale ideals faithfully; it was that Rochdale cooperatives were designed for collective buying, whereas the primary need of American farmers was collective selling. As the Grangers and the Alliance discovered the legal and economic impediments to cooperative marketing in the late nineteenth century, farmers realized that Rochdale cooperation, however intellectually attractive, could not solve the marketing problem.

## AGRARIAN PROTEST AND COOPERATION

Cooperation was one of many planks in the Grangers' reform platform. During the 1870s, the Granger movement sought to use the political process to protect agriculture from marketing and transportation monopolies. Its primary strategy was to secure legislation regulating railroads at the state level. Rectifying the imbalances in the marketing system, while a chief concern of Grange rank and file, was a secondary priority of the national leadership. This ordering of priorities hindered the development and success of Granger cooperatives.[11]

Beginning in 1871 and 1872, state and regional Granges in the Midwest and South set up cooperative purchasing agencies and consumer stores to deal in farm equipment. As the Grange grew, the locals expanded into marketing, concentrating in grain, cotton, livestock, tobacco, and wool. Most Granger cooperatives were unincorporated; they delegated most transactions to local agents. Many attempted to address all the needs of their members, marketing "nearly everything their members produced, from green onions to dressed beef." The failure to specialize and to adopt a more rigorous legal form took its toll on Granger cooperatives. By 1875, most had failed entirely; others converted to purely commercial operations or were reduced to doing business for nonmembers. Grange membership dropped off considerably, in part because of the disillusionment of farmers impatient for help.[12]

The Grange attempted to head off dissatisfaction over these failures. In November 1875, its executive committee propounded a new set of rules that specified patronage dividends, limited capital, cash sales, and one vote for each member. The move was popular, but it came too late to sustain the Grange as a national political movement. Nor did the new cooperatives fare much better. Stores organized according to the 1875 plan overextended themselves. They incorporated joint stock companies that expanded into the manufacturing of farm equipment, sustained heavy losses, and were out of business by the end of the decade. In effect, as historian Solon J. Buck concluded, the Grangers adopted the winning strategy too late: "It was a distinct misfortune, both for the order of Patrons of Husbandry and for the advancement of cooperation, that the National Grange did not begin to propagate the principles of the Rochdale system until after the ineffectiveness and often disastrous consequences of the early attempts had disgusted many of the Patrons with the idea of cooperative stores."[13]

Contrary to Buck's analysis, the failure of the Granger cooperatives did not redound to the disadvantage of the cooperative movement. Midwestern farmers blamed the Grange for failing to adhere to the Rochdale principles from the start, but they were not disenchanted with cooperation once the Grangers ceased to exert political influence. Indeed, one writer credits the Grangers with "ingrain[ing] the cooperative movement deep into the hearts of American farmers." When the Grange movement collapsed around 1880, it left behind a widespread familiarity with the tenets of Rochdale cooperation.[14]

After the decline of the Grange, farmers continued to put the Rochdale principles to work in a variety of ways. The Farmers' Alliance, which began to form in the late 1870s, was the immediate beneficiary of the Granger legacy. The Alliance originated in northeastern Texas, scored some impressive successes with cooperatives there, and soon strung a network of cooperatives across the South and Midwest. Although these organizations relied far more on the Rochdale principles than their predecessors, they, too, had their weaknesses.[15]

The Alliance initially copied the Grangers, organizing local cooperative stores, grain elevators, and cotton sales yards. By the late 1880s, the crown jewel of Alliance cooperatives was the Texas State Exchange, founded by Alliance leader C. W. Macune in 1887. This cooperative was decidedly non-Rochdalian; it had a capital stock of $500,000, held by twenty-five stockholders who served as trustees for Alliance members. Yet Macune promoted the exchange's adherence to Rochdale egalitarianism. The exchange, he told the Alliance's 1888 convention, was "pure and simple cooperation. . . . It is calculated to benefit *the whole class*, and not simply those who have surplus money to invest in capital stock; it does not aspire to, and is not calculated to be a business for profit in itself." As a business, the exchange was a failure. Problems with credit—banks and merchants began to disable the exchange in 1888—and the inability of impoverished farmers to pay their assessments contributed to its downfall the following year. The Alliance's purchasing and marketing cooperatives in other states recorded a "mixed" record of accomplishment. After 1890 the Alliance turned its attention to the problem of rural credit and focused its political energies on the greenback issue. The fatal flaw of Alliance organizations was their close identification with the Populists' political and currency reform programs. In Texas the cooperatives collapsed when state and private banks withheld credit and in the Midwest when farmers deserted the Populists.[16]

Although leaders such as Macune spoke in Rochdalian terms, and notwithstanding the Alliance's political kinship with "labor radicalism and Greenbacker ideology," the Populists did little more than use cooperation as a rallying cry. Like the Grange before it, the Farmers' Alliance was unable to produce cooperatives that were commercially viable for more than a few years. They failed in part because they lacked sufficient capital and credit and in part because of the political emphases of the protest movements. Instead of focusing on exclusive service to members and devising organizations to address specific marketing problems, which varied among crops and regions, Granger and Alliance cooperatives aimed high, attempting to replace manufacturers in the 1870s and bankers in the 1880s.[17]

Moreover, rank-and-file farmers learned through the failure of these organizations the difference between cooperative ideology and economic function. While both Granger and Populist leaders relied heavily on cooperative ideology, they assumed that the Rochdale model, which had been devised for collective purchasing, could be used for marketing without much modification. That was a critical misjudgment. Designed for collective purchasing, Rochdale cooperatives restricted their clientele and their business to a volume that members' pooled capital could support. The restriction on capital stock was the very feature that supposedly distinguished it from profit-making corporations. Collective marketing, in contrast, required large outlays of capital well in advance of actual returns. This need grew in proportion to the scale of marketing operations. Regional or national operations required larger investments of capital, skill, management, communications, and other inputs. If farmers decided to begin or continue their collective action at the processing stage, they needed money to buy or build these facilities. Limited by statute to nominal amounts of capital stock, Rochdale cooperatives did not possess the resources necessary to undertake these activities. The Granger and Alliance organizations testified to this limitation even as they expanded Rochdale's political and ideological appeal.[18]

In sum, American farmers increasingly relied on the Rochdale model to unite producers after the Civil War. At the same time, they rejected specific features of Rochdale cooperation as unsuited to national marketing. The Granger and Populist movements each reinforced different aspects of the cooperative tradition, helped to popularize specific cooperative practices among rank-and-file farmers, and left behind a legacy of experimentation that proved more lively in the wake of these movements than it had in their heyday. The tension that resulted from farmers' continued invocation of the Rochdale principles as they departed from those principles in practice made recourse to the cooperative ideal increasingly problematic over time. Further, the variation in cooperative practices across the country led observers to question the legitimacy of cooperation generally. If farmers' cooperatives significantly departed from Rochdale precepts, what made them truly cooperative? When did a cooperative cross the line from legitimate collective action to unlawful monopoly and combination? Although the legal and economic form of cooperation was changing constantly during this period, there was one widely shared assumption that did not: cooperatives should never monopolize or restrain trade. It was in California that agricultural producers put that assumption to the test.

THE COOPERATIVE IDEAL TRANSFORMED

As soon as the state entered the Union in 1850, California farmers looked beyond their local markets. The thousands of miles that separated them from their customers made marketing the first and predominant economic problem for California fruit growers. The cooperatives they formed between 1870 and 1912 demonstrated their awareness of the challenges marketing posed and their willingness to overhaul the Rochdale model to address those challenges.[19]

For growers in what soon became the state's largest, most important horticultural industries—citrus, nuts, and grapes—the economics of marketing raised a question that long divided growers and distributors. Was the issue overproduction or simply a lack of markets? Orange growers, for example, were outproducing their home markets as early as the 1870s. Like most American farmers, California growers believed that only good results could come from producing as much as possible each year and more each year than the year before. Until growers secured trade channels through which to sell what they grew, however, they would be constantly subject to the caprices of middle merchants. From the start of commercial agriculture in California, growers defined their central occupational concern as getting crops to consumers with a minimum of loss.

Initially, fruit buyers bought the crop outright from growers and made their profits selling to wholesalers and retailers. Then, in the early 1890s, these buyers shifted to a commission system. They acted as agents for the growers, charged flat fees for packing and selling, and passed all risks and losses to the producers. A University of California economist described the result: "This arrangement relieved the shipper of all risk and insured him a profit, but placed the growers at the mercy of an uncertain market, and finally resulted in such low returns that the entire fruit industry of the state was threatened with ruin."[20]

As horticulture expanded commercially, growers recognized that spoilage, disease, and poor postharvest handling caused appreciable losses. Refrigerated railroad cars and improved communications solved some of the problems associated with marketing perishable fruit across the country, but then another limiting factor in expanding agricultural profits became apparent: consumers' buying habits. Making specialized, expensive fruits and nuts into widely available, affordable foods that Americans could consume year-round was the primary goal of California growers after 1870, and it also became the mission of the organizations growers formed to market their goods.[21]

Horticultural growers experimented not only with the economic functions of cooperatives but with their structure as well. Spreading rapidly across southern California, the citrus industry pioneered the federation of local associations into district- and industry-wide marketing agencies. The locals raised money to construct packinghouses and establish systems to grade and pack fruit. In 1895 the Southern California Fruit Exchange (SCFG) emerged as the first industry-wide marketing cooperative. It was incorporated as a capital-stock cooperative under an 1878 law giving associations the power to regulate voting and shares—an early attempt to codify the Rochdale principles.[22] Local associations also arose in the walnut and almond industries during the late nineteenth century; early in the twentieth, industry-wide associations were

Map 1. California agricultural production regions, 1909. (Source for agricultural data: U.S. Bureau of the Census, *Thirteenth Census of the United States, 1910*, vol. 6: *Agriculture: Reports by States, with Statistics for Counties* [1913; rpt. Washington, D.C.: U.S. Government Printing Office, 1915], 160–65; George Robertson, "The Raisin Industry," in *Report of the State Agricultural Society* [Sacramento: State Printing Office, 1910], 28–31)

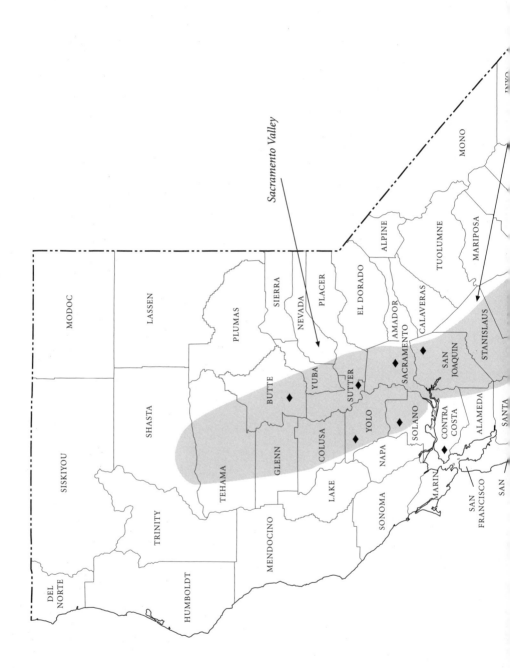

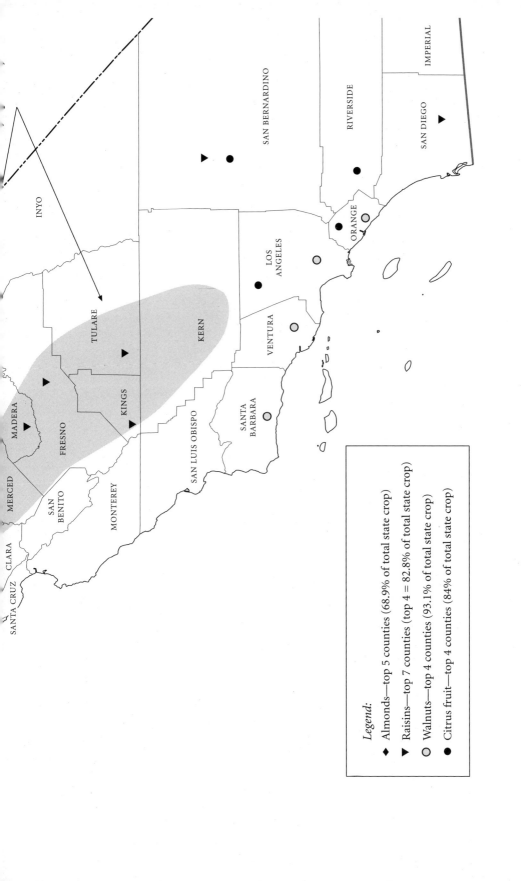

INYO

SAN BERNARDINO

RIVERSIDE

IMPERIAL

SAN DIEGO

ORANGE

LOS ANGELES

TULARE

KERN

VENTURA

KINGS

MADERA

FRESNO

SANTA BARBARA

SAN LUIS OBISPO

MERCED

SAN BENITO

MONTEREY

CLARA

SANTA CRUZ

*Legend:*

♦ Almonds—top 5 counties (68.9% of total state crop)

▶ Raisins—top 7 counties (top 4 = 82.8% of total state crop)

◉ Walnuts—top 4 counties (93.1% of total state crop)

● Citrus fruit—top 4 counties (84% of total state crop)

formed: the California Almond Growers' Exchange in 1910, the California Walnut Growers' Association in 1912, and the California Associated Raisin Company also in 1912. These industry-wide associations did not have to travel far to find growers because of the pronounced localization of the fruit-growing industries. The geographic location of each industry depended largely on the soil and climate requirements of each crop. Unlike the cooperatives formed in California later in the Progressive Era under the administration of Governor Hiram Johnson and State Director of Markets Harris Weinstock, the citrus fruit, almond, walnut, and raisin associations were the result of years of trial-and-error experimentation at the grass-roots level, with no direct assistance from the state aside from the benefits of incorporation under state law (see Map 1).[23]

That body of law was constantly changing, and its fluidity proved a definite asset to horticultural growers. The ability to raise capital and issue capital stock was essential to the marketing, advertising, and selling activities of the horticultural associations, and California had laws on the books permitting the formation of several different kinds of organizations, all termed "cooperative." Growers exploited the cooperative identity of their organizations while adopting overtly entrepreneurial practices. In 1905, the SCFE was reorganized and renamed the California Fruit Growers' Exchange (CFGE). The CFGE quickly became the dominant selling agent of citrus fruit in the United States. Its share of the crop grew from 46 percent in 1905 to 72.5 in 1921. The CFGE established a sophisticated network of marketing brokers and transportation routes that enabled it to track every carload of fruit sent to urban auctions. Economist Ira Cross reported in 1911 that its directors knew "exactly how much of the crop is being put upon the market each day, where it is being sent, and what prices it is bringing. . . . The Exchange has reduced the cost of packing, selling, and collecting by one-half, and as a consequence during the past year netted a clear gain of over $3,000,000 for its members."[24]

Led by general manager G. Harold Powell, a former USDA Bureau of Plant Industry economist, and Los Angeles attorney George Farrand, an expert in corporation law, the CFGE carefully navigated the statutory changes pertaining to cooperative organization. After the Clayton Antitrust Act was passed in 1914, exempting nonstock cooperatives from the federal antitrust laws, the CFGE reorganized as a nonstock organization under California law. Innovative advertising campaigns familiarized the American public with the "Sun-Kist" brand name. By 1920, wrote Herman Steen, the CFGE had become "the world model for cooperative marketing of perishable products."[25]

The CFGE's "striking success" prompted wide imitation. The walnut, almond, peach, table grape, and raisin industries began paralleling the CFGE's

Sunkist advertisement, *Ladies' Home Journal*, February 1919.
(White House Antiques, Addison, Illinois)

tracks in cooperative organization during the 1890s. These organizations all hired experienced business managers, established brand names for their products, and initiated national advertising campaigns to expand demand. "By judicious advertising, for instance, the demand for raisins, largely confined to Thanksgiving and Christmas, has been spread over the year." Organizers learned to build bridges to the banks, a connection both agriculture and capital found beneficial. Indeed, some bank officials were also horticultural growers with direct interests in the prosperity of both.[26] And all preached what was becoming the hallmark of California-style cooperation: monopoly control over the crop. By 1917, the California Walnut Growers' Association controlled 70 percent of the California walnut crop and counted 75 percent of the state's producers as members. The California Almond Growers' Exchange controlled 80 percent of the crop by 1917. The California Associated Raisin Company topped them all, with almost 90 percent of the growers in 1918.[27]

The raisin industry rose to this pinnacle after a rocky start. An initial boom during the 1880s sent land prices soaring from $50 to $200 an acre and annual profits as high as $450 per acre. These successes spurred even more planting. By 1890 growers in Fresno County had planted raisin vines on thirty thousand acres of land—half the state's total raisin acreage. In 1890 and 1891, twenty thousand more acres were planted. These raisins hit the market in 1893 and 1894, producing record crops of 85 million and 103 million pounds, respectively (Table 1.1). Around that time, California's raisin production overtook that of Spain for good. The industry's progress during its first decade backfired in the early 1890s, when larger crops brought prices back down and land values dropped accordingly.[28]

The hardships of the 1890s led growers to view the marketing system as their enemy. The vast distance between producers and consumers meant that growers had little influence over retail prices. Frequently, communication and transportation problems led to gluts in some cities and shortages in others. Fresh and dried fruit was a luxury to most Americans, who consumed it, if at all, only around Christmas. The railroads charged freight rates that growers decried as exorbitant and discriminatory. Before the state enacted grading and quality standardization laws, growers sometimes shipped infested and rotten fruit to market to avoid losses. The switch to commission buying in the early 1890s hit the raisin industry hard and made growers regardless of economic status want greater control over marketing.[29]

The combination of distinctive economic and geographic circumstances and a willingness to experiment with the legal form and economic philosophy of cooperation enabled California's fruit growers to take the lead in reconfiguring cooperation. California cooperatives focused primarily on marketing,

TABLE 1.1. California Raisin Production, in Tons, 1884–1898

| Year | Crop | Year | Crop | Year | Crop |
|------|------|------|------|------|------|
| 1884 | 1,750 | 1889 | 9,870 | 1894 | 51,500 |
| 1885 | 4,700 | 1890 | 19,000 | 1895 | 45,500 |
| 1886 | 7,230 | 1891 | 26,500 | 1896 | 34,000 |
| 1887 | 8,000 | 1892 | 28,500 | 1897 | 46,500 |
| 1888 | 9,430 | 1893 | 42,500 | 1898 | 40,000 |

Source: Annual Report of the California State Board of Agriculture (Sacramento: State Printing Office, 1913), 134.

enlisted the assistance of the financial establishment, and entirely avoided political activism. Unlike the Populists, for whom cooperation was a pillar of their program of protest and economic reform, California fruit growers used cooperation as the legal and economic framework for reorganizing agriculture as a competitive, prosperous enterprise.

The nation certainly wanted to see farmers prosper. "The campaign for improved marketing, rural credit and cooperation is assuming almost the proportions of a political campaign," wrote the Country Gentleman in 1913. At the same time, the country did not unreservedly embrace this new form of cooperation. Cooperation for the benefit of individual farmers who shared equally in the risks of their collective enterprise was one thing; avowed monopolies were quite another. As fluid as cooperation's meaning had become during the last quarter of the nineteenth century, it remained anchored to values Americans automatically associated with rural life: self-help, decentralized economic power, and individualism. The new mode of cooperation awakened concerns that combination and monopoly were spreading like a virus from industry to farming. After 1900 the debate over the nature of "true" cooperation intensified as cooperation became more firmly established in the agricultural economy.[30]

## THE COOPERATIVE IDEAL IN AMERICA, 1890–1910

Just as the American farmer was ideally a special kind of entrepreneur, cooperation was in theory a distinctive way of doing business by dint of its emphasis on equity and participation. Cooperation promised to reconcile capitalism and justice, which social theorists, utopian socialists, and farm leaders depicted as polar opposites. A columnist for the Arena saw cooperation as the antidote to unbridled competition: "Competition means a struggle for victory over fellow-competitors; cooperation means a united struggle for victory over toil and sorrow and ignorance." Cooperation preserved individualism, according to Herbert Croly: "In aiding the accomplishment of the collective purpose by means of increasingly constructive experiments, [individuals] will be increasing the scope and power of their own individual action." The idea that cooperation did not disturb the structure of capitalism got its imprimatur from the British historian of cooperation, George Jacob Holyoake: "'Cooperation may mitigate reckless competition, but it does not destroy competition itself.'"[31]

All this sounded good, but proponents knew that it remained essential for the movement to sidestep its historical association with nineteenth-century utopian socialism. Cooperation had been centrally important to the American Fourierist movement, one branch of socialist thinking. It had figured prominently in the communal experiments of Robert Owen and his followers. And it was closely linked to the work of radical socialists such as Karl Marx and Friedrich Engels. By mid-century, prominent Fourierists such as Horace Greeley had departed considerably from the more extreme philosophical pillars of the movement: "To even the most reform-minded Whig leaders, . . . the cooperative alternative [was] meant to coexist in a competitive world." The triumph of free labor ideology in the Civil War ensured that even the milder form of communitarianism would be unacceptable to a society committed to industrialism and corporate capitalism. Part of the ongoing challenge facing farmers and cooperative promoters, therefore, was to reframe the changes in cooperation as congruent with the aims of late nineteenth-century capitalism.[32]

One example of this reframing arose in the relationship between farmers and labor. Proponents of cooperation in agriculture began using the labor union as an analogy during the 1890s, when the Farmers' Alliance publicly supported the Knights of Labor. A publication of the Farmers' Union urged wheat farmers to federate their local organizations under the umbrella of a large central unit, "with an executive head to represent and defend their common interest as Mr. Gompers does the American Federation of Labor." The idea got a stamp of approval from Harvard economist Thomas Nixon Carver, who observed that cooperation might "do for farmers what the American Federation of Labor has done for wageworkers." A columnist for the *Nation* noted that cooperation's "affiliation with the labor movement is becoming closer every day." In fact, after 1900, the alliance between farmers and labor became increasingly fragmented, as cooperation's legal status became more secure and labor's remained problematic. By invoking the labor union as a metaphor for farmers' organizations, cooperative promoters sought to identify with the progressive impulses of the labor movement and deflect attention away from the emerging corporate character of the new cooperative style.[33]

That new style was threatening to hoist cooperation on its own petard. The California model preached monopoly control in the name of preserving market freedom, but critics did not buy this argument. The public and the legal order were not prepared for a brand of cooperation that applied combination to the problem of suppressed competition and speculation in agricultural marketing. Observers measured the California associations against the Rochdale standard and found them wanting: "Such associations do not in the least differ from the familiar combinations of small manufacturers, 'who attempt to raise profits by joint advertising and marketing and by controlling supply, until sometimes a trust is created.'" A Stanford economist put the point more bluntly: "The California Fruit Growers' Association [sic] exists solely because it can earn greater profits for its members than can be obtained by any other marketing arrangement. It is difficult to see wherein this point of view differs from that which characterizes the stockholders of the ordinary business corporation." Those who believed farming should be the last bastion of self-sufficiency concluded that an agrarian monopoly posed more of a threat to free competition than an industrial one.[34]

Defenders of the new cooperative style turned this argument on its head. Not only were farmers taking recourse to the organizational forms and trade practices of industrial business, they claimed, but it was their right to do so. Edward Adams, the agricultural writer for the *San Francisco Chronicle*, saw no difference between what growers were doing and what the middle merchants had long been up to: "The principle of cooperation in [agriculture] for marketing purposes is identical with that of the cooperation of capitalists in what are called 'Trusts.'" The manager of the Farmers' Grain Company of Omaha folded all forms of corporate collectivism into cooperation: "All combinations among capitalists are but instances of cooperation among those who are interested." Cooperation, these writers urged, was the only economically viable means of preserving the independent producer and the family farm.[35]

This argument put the farmers on dangerous ground, however. It gave rise to accusations that the new form of cooperation defiled the hallowed tradition out of which it emerged. Resorting to the same tactics as industry undercut farmers' long-standing protest against industrial trusts and monopolies and raised troubling questions about the changing nature of cooperation. During a time when courts and legislatures were arguing over the meaning of free competition and corporations were in the midst of their greatest merger movement to date, the new form of cooperation undermined its philosophical claim. How could farmers criticize industrial business for occluding market freedom when their new cooperatives placed a premium on doing just that?

People central to the movement disagreed over how to resolve this issue. CFGE president G. Harold Powell, an agricultural traditionalist, was torn between adherence to the Rochdale ideal and the economic expectations of the citrus growers he led. In 1914, he argued that the only legitimate form of cooperation was a nonstock organization that operated for mutual benefit, distributed voting power equally among members, limited membership to producers, and paid dividends based on patronage. Six years later, he had changed his mind. Cooperatives could hold capital stock as long as members' contributions and interest on dividends were limited and only if the capital was employed for the mutual benefit of all participants.[36]

For Rochdale believers like Powell, capital stock polarized the debate over true cooperation; for others, it mattered little. Powell's counterpart at the California Walnut Growers' Association, Charles C. Teague, defined cooperation in purely functional terms. For Teague, the purpose of cooperation was to eliminate chaotic marketing conditions, to make enterprise more predictable and less risky for producers. The distinction between capital stock and nonstock organizations — so essential to Rochdale cooperation — he noted only in passing. By the end of World War I, some were openly voicing the view that Rochdale had no place in agricultural marketing. In 1922, Herman Steen, editor of the *Prairie Farmer*, lectured Secretary of Agriculture Henry C. Wallace that national policy should not be predicated on Rochdale: "The Rochdale plan has demonstrated its unfitness as a means of cooperative marketing in literally every civilized country in the world. . . . [It is] frequently advocated for cooperative marketing by those who oppose cooperation, with the knowledge that the Rochdale system will not succeed sufficiently to affect their business."[37]

California farmers were not the only ones torn between their need for capital stock and the existing emphasis on nonstock organizations. In midwestern and plains states, farmers initially organized cooperative grain elevators as capital stock corporations, later reorganizing under new cooperative incorporation statutes. These laws did not force farmers to adopt the nonstock form; instead, they required cooperatives to pay patronage dividends. The Rochdale paradigm continued to influence the course of the cooperative movement, if only to set the standard from which cooperatives deviated in practice. Marketing organizations operating as joint stock corporations found it difficult to maintain equality among their members; in 1920 a Nebraska writer observed that "the stock became gradually concentrated in the hands of a few members until they lost such semblance of truly co-operative companies as they might have had."[38]

Yet for cooperation to succeed as a business, cooperatives needed to act like business organizations. Once marketing, not collective purchasing, became farmers' organizational goal, they needed centralized capital — the very feature that supposedly distinguished the cooperative from the corporation — to pay processing, storage, and distribution costs. To conduct marketing operations, they required a legally recognized form of organization and the authority to execute bargains and commercial transactions, including but not limited to corporate charters, contracts, bureaucratic and managerial capacities, and even the power to use trusts. Farmers' experiments with the legal form and economic practices of cooperatives succeeded in making cooperation economically conservative. Having shed the more radical aspects of cooperation's heritage, farmers also unwittingly undermined the notion that the Rochdale principles constituted the only "true" form of cooperation.

Thus, as farmers obtained the passage of new laws to give cooperatives the legal powers of corporations in the late nineteenth century, cooperation's definition became more a matter of law than ideology. A cooperative's legal origins, which mattered hardly at all before 1890, determined the legal status of a cooperative organization thereafter. The legislative movement to change the legal character of cooperatives, described at length in Chapter 3, both antedated and outlasted the nineteenth-century protest movements, but the Grangers and Populists contributed to the emerging legal definition of cooperation. During the 1870s, the Grangers helped to enact several laws to distinguish the incorporation of cooperatives from the statutory regime for business corporations. Two decades later, the Populists' main contribution to the movement for separate incorporation was to support laws for nonstock cooperatives. In California, the Populists secured passage of such a law in 1895, only to see it ignored by every major horticultural association formed thereafter. In the race to find an acceptable formula for cooperation, the Populists bet on the wrong horse. Although the Rochdale principles remained the intellectual touchstone for the cooperative movement, the economic requirements for marketing pushed farmers toward the corporation. When state legislatures embraced this move, the courts were forced to determine whether the new, economically conservative model of cooperation transgressed legal norms about the use of collective action in a competitive market economy.[39]

The widespread support for cooperation that existed during the Progressive Era was for a specific form of cooperation that reflected not only the traditional model's emphasis on equity and democracy but also the cultural image of the farmer. The commercial nature of California agriculture and the aggressively corporate approach of California horticultural cooperatives exemplified the paradox of the larger farm problem: to preserve traditional farming and rural life, farmers believed they had to adopt the forms and practices of industrial business. Yet they saw a vast moral difference between "corporations, or syndicates, . . . organized . . . to limit competition" in transportation, manufacturing, and the sale of agricultural commodities, on the one hand, and the "simple cooperative companies or associations of farmers or producers organized for mutual benefit and protection," on the other.[40] Nonfarmers — eastern consumers as well as middle merchants — agreed that the two kinds of organizations differed. They simply thought the California cooperatives belonged to the first category rather than the second.

Until American society came to terms with the marriage of cooperation and combination, California growers had to defend the union against accusations that the new form of cooperation corrupted its hallowed tradition. The California solution to the farm problem turned on the continuing involvement of thousands of individual farmers, whose beliefs and actions had much to do with the success of marketing cooperatives in the Golden State. Though Jeffersonian agrarianism never gained a foothold in California, the Jeffersonian icon of the family farm did.

# 2 ∾ From Agribusiness to Family Farms

*If you desire a **delightful home** and **prosperity** Get a Fruit Farm in Fresno*

**County**. *Surrounding the town of Fresno, there are 34 colonies, embracing areas*

*of 320 to 9,000 acres each, now occupied by thousands of prosperous families, in*

*fruit farms of 10 to 40 acres each. More than half of these people never farmed before.*

—*M. Theo Kearney settlement circular, 1887*

On January 24, 1848, gold was discovered near Coloma, California. First to bring the news to the infant city of San Francisco was John Bidwell, a prospector who had driven the first American wagon train into Mexican California in 1841. Though anxious to beat the rush for gold, Bidwell was also a visionary who recognized that the land had many sources of wealth. En route to San Francisco, he stopped at San Rafael to collect budwood from the old mission orchards there. Bidwell's errand symbolized the importance of fruit growing not only in California's agricultural past but also in its economic future. Fruit, not cattle or grain, would prove to be the state's most lasting source of agricultural wealth. Horticulture enabled family farms to flourish in California well into the twentieth century, and the independent producers who operated those farms determined whether marketing cooperatives would succeed or fail.[1]

California's fruit industries were surprisingly diverse, both economically and socially. In nineteenth-century California, as a few people greedily accumulated land in massive quantities, others were occupied with subdividing their large holdings into smaller parcels. Subdivision was associated with fruit growing, large tracts with field crops and ranching. For the most part, these areas did not overlap. The proliferation of small farms and the rise of fruit growing, moreover, had the effect of increasing the social diversity of the hor-

ticultural industries. The influx of racial and ethnic minorities into the horticultural districts, coupled with the high value of fruit crops grown on small
farms, created a socioeconomic system in which status was not automatically
correlated with farm size.

When horticulturalists turned to cooperation to solve the problems associated with marketing perishable fruit across the country, they triggered a
struggle for social control in their communities. The congruence of production regions and community boundaries meant that an organization designed
to coordinate the economic lives of the growers profoundly affected their
social status. Large growers and growers of high-value crops had much to lose
from the volatility of an uncoordinated market. Those with a greater stake
in the profitability of horticulture had a greater interest in imposing controls on the conditions under which fruit was sold by growers and bought by
consumers. Thus large growers tended to support industry-wide cooperatives
because such organizations imposed order on the market at both the producers' and consumers' ends. In contrast, small growers had some economic
flexibility. Some had to sell what they grew immediately to pay their debts,
but others—including members of minority groups who maintained cultural
and economic insularity—had sufficient capital to ride out the chaos at harvest and obtain better prices later. The flexibility of these smallholders proved
to be the most serious obstacle facing cooperative organizations and their
leaders. In the fruit-growing districts, large farms did not necessarily confer
corresponding economic advantages on their owners.

In the cooperative movement, small growers held the balance of power. The
ubiquity of small farms created an upwelling of social and economic pressure against the influence and power of the wealthy white men who assumed
leadership roles in the cooperatives. Cooperation was a top-down movement;
larger growers sought to control the small fry who undermined the market with their independence, their unpredictability, and their willingness to
undercut prices.[2] Whether the small growers stayed or left ultimately determined the fate of the organizations. Some smallholders were willing to ally
with industry elites, motivated at times by a belief in cooperative ideology and
at other times by economic interest. Others did not automatically fall in line.
The presence of small farmers explains why large growers supported cooperation as the best way not only to control the market beyond California but
also to control the market before the market—the point at which all growers
jockeyed to sell their fruit at the same time of the year. The interests of all
growers were not the same, as cooperative leaders and organizers discovered,
sometimes to their consternation. Thus, although to the rest of the nation the

California cooperatives seemed like powerful monopolies, beneath the edifice of monopoly was an unstable association of fractious growers.

## THE HISTORICAL UNDERSTANDING OF
## THE FAMILY FARM IN CALIFORNIA

Historians have suggested that family farms in California died a quick death, supplanted by agribusiness, the economic system characterized by land monopoly, corporate ownership of farms, the exploitation of migrant laborers, the use of industrial cultivation practices damaging to the environment, and, perhaps most damning, farms that were deviantly large.[3] The national perception that California agriculture was aberrant in both its economic structure and the social costs of the wealth it produced began to form early in the state's history; the unique circumstances surrounding its settlement and economic development only reinforced the perception. The gold rush drew thousands of people to the state overnight, not to farm but to mine the Sierras. The rush lasted only a few years. By the 1860s, enterprising landowners had begun planting thousands of acres of wheat, but this rush, too, was short-lived. In the 1870s, the settlement of Mexican land claims that had tied up ownership of millions of acres made possible yet a third stage in California's economic development, one that has proved to be the most lasting: the commercial production of fruit (tree and vine) crops.

With the beginnings of commercial horticulture in the last quarter of the nineteenth century, the historical claim runs, California turned its back on the Jeffersonian tradition of family farms and began its inexorable march toward agribusiness. "The idea of small-scale family farming lost out to agribusiness in California," Cletus Daniel wrote, in a scathing attack on the "intensive nature of . . . fruit and vegetable cultivation." Conventional wisdom ascribes to growers a "generally unalloyed commercial mentality" and depicts them as either corporate or absentee. They are the "land barons" who operate the "factories in the field," in Carey McWilliams's memorable phrase. The assumption that small, family-owned farms died at the hand of corporate farming is reflected in the title of Donald Pisani's *From the Family Farm to Agribusiness*.[4]

Even those scholars who recognized, in the words of Paul Gates, the "survival" of small farms in California as late as 1969 were, like Gates himself, far more attracted to large farms as an object of study and analysis. This fascination stems from a variety of cultural and intellectual phenomena. Large farms are the bête noire of American agriculture. Western farms have always been larger than those in the Midwest and South. The exploitation of migrant

TABLE 2.1. Average and Median Farm Size in California, 1860–1940

| Decennial | Total Number of Farms | Average Farm Size (in Acres) | Median Farm Size (in Acres) |
|---|---|---|---|
| 1860 | 14,046 | 622 | 234.9 |
| 1870 | 23,724 | 482 | 283.5 |
| 1880 | 35,934 | 462 | 421.7 |
| 1890 | 52,894 | 405 | 388.9 |
| 1900 | 72,542 | 397 | 317.0 |
| 1910 | 88,197 | 317 | 260.3 |
| 1920 | 117,670 | 250 | 210.2 |
| 1930 | 135,676 | 224 | 180.3 |
| 1940 | 155,216 | 230 | 173.2 |

*Source*: ICPSR Data Sets, 1860–1940. Median farm size was calculated by hand from ICPSR data.

labor reached industrial proportions in the West. And only in the West did agriculture depend on costly and complex irrigation projects. Economists and historians have advanced several explanations for the presence of large farms. The costs of irrigation required large economies of scale; the efficiency of large farms and the presence of migrant laborers have been favorite themes. As Sucheng Chan has said, "The need to find economies of scale in marketing . . . propelled California agriculture towards bigness."[5]

The focus on bigness and large farms has generated a picture of California agriculture as economically and socially undifferentiated. Its interpretive weakness stems in part from an evidentiary fallacy: the statistics of average statewide farm size. As Table 2.1 shows, averages are a misleading way of measuring farm size. Average farm size in California was indeed large (397 acres in 1900, 317 in 1910), but averages magnify the importance of a few very large wheat farms and livestock ranches. Medians are a better measure of group characteristics. They mark the point at which half of the farms in the sample are larger and half are smaller; thus a few very large farms do not skew the result. Median farm size was much smaller than statewide average farm size, dramatically in 1860, less so by 1940, but by around 70 acres each decennial. Medians reveal the presence of more small farms than historians have previously realized.[6] Beneath the monolith of industrial farming lies unexpected complexity. In the fruit-growing sectors, growers exhibited a wide diversity of interests and farm ownership was surprisingly widespread. The presence of many small farmers in the fruit industries significantly affected the dynamics of organizing cooperatives and maintaining the marketing monopolies that were central to the California approach to cooperation.

## AGRICULTURAL DEVELOPMENT IN
## NINETEENTH-CENTURY CALIFORNIA

The origins of small farms in horticulture were rooted in both the eighteenth-century Spanish missions and the system of agricultural colonies established in the Central Valley during the 1870s. When Europeans began arriving on the North American continent in the sixteenth century, they brought a multitude of cultivated plants with them — 147 distinct species and varieties of field crops, fruits and nuts, vegetables, herbs, medicinal plants, and ornamentals, by one historian's reckoning.[7] The Franciscan missions were bountiful. According to University of California pomologist Edward J. Wickson, "The California missions had fruit growing establishments greater in variety and commercial value of fruit products than any similar enterprises on the Atlantic side."[8] When the Mexicans took California, they secularized the missions, ending the careful cultivation of the mission orchards and vineyards, but the hardier plants endured.[9] The Spanish and Mexicans limited their land interests to coastal areas; the unfriendly desert in the interior valley deterred settlement until after California joined the Union.

As long as mining occupied the entrepreneurial energies of most of the state's residents, farming was a local concern, conducted on small farms clustered near the major towns, usually by Chinese who were excluded by law from mining. The small farms near Los Angeles and Sacramento produced vegetables, hogs, and poultry for miners and urban residents. As the state looked to diversify its national and international trade relations after the gold rush ended, the process of transforming the land and the agricultural economy began in earnest.[10]

This transformation was not a matter of simply putting plow to ground. The legal and social construction of land ownership had as much impact on the direction of agricultural development as soils, climate, and geography. The tangled web of Spanish and Mexican land claims, whose settlement Congress bungled during the 1850s, created a system of land ownership that pitted a few hundred absentee claimants against thousands of settlers, developers, and speculators eager to obtain clear titles.[11] The legal uncertainty surrounding the Mexican land claims prompted Americans to scramble to obtain large parcels, which they immediately turned to the most profitable uses available:

cattle ranching and wheat farming, two lucrative commodities that did not tax scarce water supplies.[12] Wheat, the state's first cash crop, linked California to international markets and brought in needed capital during the 1860s, but bonanza farming — planting as many as three crops per year — exhausted the soil. By the 1880s, increased competition from Canada and Russia lowered wheat prices worldwide and cemented the industry's decline.[13]

Wheat farming instigated the first public debates about the future of democratic landholding in California. Newspaper editors rained a torrent of criticism on bonanza farmers for violating the ideals of Jeffersonian agrarianism. As early as 1853, Governor John Bigler argued that only family farms could curb "the power of unscrupulous capitalists and speculators to monopolize the very necessaries of life."[14] Californians expected the wheat industry to conform to the American ideal of the family farm, not because they unblinkingly believed in Jeffersonian ideology but because the accretion of so much wealth to so few individuals offended ideas of fair opportunity for white male citizens. Yet contemporary critics overlooked two important points. First, Americans hardly approached the newly opened California frontier with anything resembling a Jeffersonian ethic. The intensive cultivation of initial settlement was largely a continuation of Spanish and Mexican agricultural policies, which were designed to extract wealth from the land as cheaply and profitably as possible. Second, the wheat kings, after all, were both good capitalists and good at capitalism. It was easier to demonize them in the name of social reform and democratic landholding than it was to implicate capitalism's role in thwarting an orderly resolution of the underlying cause of land monopoly — the messy, corrupt transition from Mexican land grants to American homesteading policy.

The wheat bonanza laid the economic and structural foundation for commercial horticulture. Once the wheat boom peaked, the more astute of the wheat kings realized that a fine way to make money was to subdivide and sell off their vast spreads. They used some of their profits to make the vast arid interior ready for fruit and nut cultivation. Without irrigation, most California farmland could support only dryland field crops. Without the capital that wheat farming infused into the state, the extensive irrigation works essential for a horticultural economy could not have been built. And without horticulture, no family could make a decent living off a family-sized farm in California. Thus the shift from wheat to horticulture that was already under way by the 1870s was not merely a reaction to the decline of wheat; it funneled capital accumulated through the wheat trade back into land development, colonization plans, and irrigation companies. The progression from wheat to fruit marked the path from agribusiness to family farms.[15]

The rise of fruit growing in California was no accident. Land developers recognized that fruit farms did not have to compete with large ranches in order to prosper because small fruit farms provided families with a comfortable living. Varieties of locally adapted fruit crops were already close at hand. The mission orchards and vineyards, though not maintained, still contained living, high-quality varieties of tree fruits, nuts, and grapevines. The hunger of the state's growing population for fresh fruit was all the impetus enterprising farmers needed to revive the mission orchards or, like John Bidwell, to take breeding stock and start orchards of their own. Many immigrants, particularly those from eastern Europe, brought budwood and vines with them when they came to the United States, along with their accumulated knowledge of and experience with cultivating, harvesting, and marketing fruits and nuts. The real challenge was experimenting to see what crops could grow in the hot inland valley. Before statehood, horticultural pioneers William Wolfskill, William M. Stockton, and Thomas Davis planted experimental orchards in citrus and deciduous fruits in the Sacramento Valley. The state's first commercial raisin vineyards also were planted north of the San Francisco Delta, in Davisville and Woodland, where G. G. Briggs and R. B. Blowers first marketed raisins in 1873. Once the land barons realized that subtropical and deciduous fruits could grow in the valley, the major hurdle to colonization was convincing people that fruit growing was both affordable and profitable.[16]

The campaign to attract fruit farmers had no more dramatic example of horticulture's economic feasibility than T. S. Price's 1893 report to readers of the *Pacific Rural Press*. He purchased twenty acres for a vineyard near Fresno in 1888 for $1,900. Carefully recording his expenses for irrigation, vines, planting, cultivation, and harvesting, Price spent a total of $1,809.63 over four years. When the vines came into bearing after two years, his receipts totaled $2,140, for a net of $330.37. After six years, he asserted, "a conservative estimate will place the owner entirely free from debt on the property and a bank account of not less than $1000. The property should be worth at least $300 per acre." The upshot was clear: even men of "modest means" could make a go of it under these conditions. Other writers made the same claim for other horticultural crops. Robert Glass Cleland estimated the net profits from a ten-acre orange grove in the 1870s at $11,700. Horticulture, he said, "was recognized, almost from the beginning, as having almost unlimited possibilities."[17]

Fruit growing did not prove as easy as the promotional literature promised. In addition to coping with aridity (a challenge sufficient to overwhelm even the most experienced farmers), settlers and immigrants had to deal with uncertain land titles, water rights that evaporated in the heat of intense and controversial litigation, and the vagaries of nationwide marketing.[18] The California landscape was uninviting, the climate harsh, the hills treeless and brown, according to lemon grower Charles C. Teague, and the initial disappointment at not encountering a tropical paradise was amplified by the difficulties of growing fruit for profit. The recurrence of diseases and pests meant that no grower was safe from devastating crop losses. Vineyardists had to cope with phylloxera, a particularly insidious infestation that attacked the vine at its roots. Phylloxera came into California with grapevines imported from France; growers, scientists, and legislatures spent decades trying to control it.[19] Growers organized trade associations during the 1880s and 1890s as much to combat natural threats as to do battle with the railroads and middle merchants.[20] Fruit growing posed great risks and promised high rewards, but public rhetoric focused entirely on the latter because so many people believed that the state's progress depended on rapid settlement and aggressive economic development and that fruit growing was the key to both.

The wheat kings subdivided great expanses of the Sacramento and San Joaquin Valleys and southern California in the 1870s. San Francisco land speculators William S. Chapman, Isaac Friedlander, and John Bidwell and the cattle barons Henry Miller and Charles Lux, among others, undertook ambitious schemes to establish agricultural colonies. They did so to benefit themselves, of course, but their actions also transformed land ownership in rural California. Chapman, perhaps the state's largest individual landowner at that time, was a shrewd entrepreneur who foresaw that wheat would give way to more diversified farming. Friedlander, the "Grain King," was a major investor in Chapman's San Joaquin and Kings River Irrigation Company, formed to supply water to new agricultural settlements in and around Fresno. Bidwell, in many respects a progressive thinker, belonged to the Grange, encouraged the state's infant fruit and nut industries, and "spoke out on behalf of small farmers." All of them placed fruit growing at the center of their plans for immigration, settlement, land subdivision, and family farming.[21]

These men quickly prepared small parcels for fruit farming. They persuaded friends at the Central Pacific and Southern Pacific Railroads—when they did not own the lines outright—to build railroads through the valley. They incorporated irrigation companies to claim water from the San Joaquin and Kings Rivers.[22] They established an extensive system of farming colonies in and around Fresno, consisting of clusters of small farms with water rights appurtenant to the land.[23] They even supplied some tracts with canals and irrigation works that would carry river water to the peach orchards and grape vineyards that settlers planted after the mid-1870s.[24] In Fresno County alone, developers connected more than three hundred thousand acres to private canals during the 1870s. Moses J. Church, a member of the Chapman/Friedlander syndicate, formed the Fresno Canal and Irrigation Company to provide "a comprehensive system of irrigation for the district," which laid the foundation for the cooperative irrigation district the county established after the turn of the century.[25] Other private irrigation works sprang up in the Central Valley as subdivision proceeded.[26]

The agricultural colonization schemes touched off an immigration boom and created a diversified landholding pattern, both of which were crucial to the rise of commercial horticulture. Between 1875 and 1890, former wheat barons laid out more than twenty-five colonies in Fresno and Tulare Counties. Towns such as Selma, Kingsburg, Reedley, and Sanger began to dot the landscape along the Southern Pacific's main route and branch lines. Promotional literature extolled in effusive terms the benevolent climate and the profitability of fruit farming: "Fresno County, A Wonderfully Prosperous District in California. The Land of Sunshine, Fruits, and Flowers. No Ice. No Snow. No Blizzards. No Cyclones." The town of Fresno, a burgeoning railroad terminus advantageously sited between the San Joaquin and Kings Rivers, quickly became the social and trade center of the San Joaquin Valley. Local boosters agreed that the colony system was essential to the town's rapid growth: "The colony idea supplied the first real impetus to Fresno's development and laid the foundation of the county's future prosperity. This farming on a community plan, which became quite a craze, would not have been possible without the crude but still effective irrigation system" supplied by the developers (Map 2).[27]

Promotional literature emphasized the entrepreneurial, as opposed to ideological, mission of these settlements. The *Fresno Republican* boasted that the colonies produced "a system of ideal rural homes and communities, where the science of tillage is carried to its greatest perfection." The first of these settlements, the Alabama Colony, was built on land Chapman and Friedlander sold at twice the government's price of $1.25 per acre to southern expatriates who wanted a fresh start after the Civil War. In 1875, Chapman and Bernard Marks organized the Central California Colony, populated primarily by Scandinavian immigrants. A second Scandinavian colony was established in 1878; it soon accepted settlers from all over. The Washington Irrigated Colony was marketed in Germany, Sweden, and Australia. The Nevada Colony was not restricted to citizens of the Silver State, but Moses J. Church's Temperance Colony was open only to those who could abide prohibition by land deed.

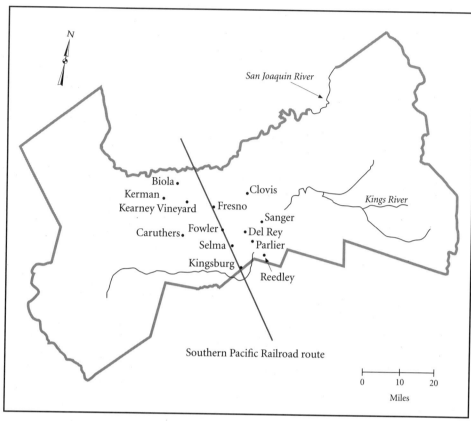

Map 2. Fresno County, California, ca. 1920

South of Fresno, on the Southern Pacific line, the town of Fowler spawned four colonies of its own. Selma, Reedley, and Sanger became trade centers for the surrounding settlements. The town of Kingsburg had its own colony to the northeast; both colony and town were peopled primarily by Swedes. The colonies planted in the San Joaquin Valley "a population so varied in respect to nationalities that no country on the globe appeared to lack a representative," according to the historian Virginia Thickens. In addition to transplants from the eastern United States, discrete communities of Germans, Italians, British, Scandinavians (especially Danes), Armenians, Portuguese, Russians, Mexicans, and Chinese together made up the infant raisin industry. Armenian, Italian, and Portuguese settlers brought their skills at grape growing, wine making, and raisin drying (Map 3).[28]

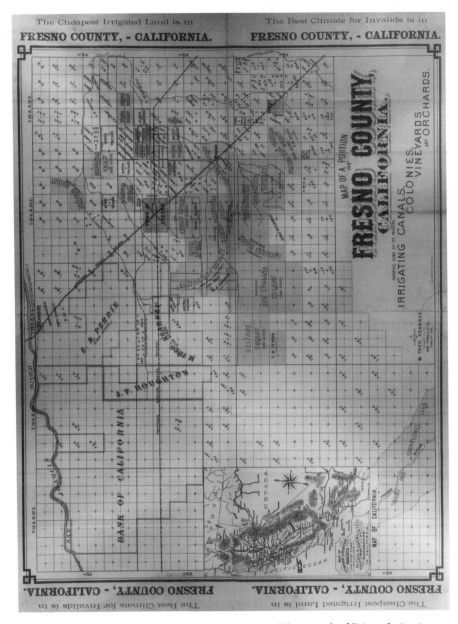

Map 3. Agricultural colonies in Fresno County, ca. 1875. (Photograph of "Map of a Portion of Fresno County, California . . . Published by M. Theo Kearney"; Fresno Historical Society Archives)

The overlap between industry and community, which would prove essential to cooperative organizing, was particularly apparent in the raisin industry. In the nineteenth century, the raisin industry was scattered across California; growers planted grapevines in the upper Sacramento Valley, the Napa Valley north of San Francisco, the El Cajon Valley near San Diego, and Orange, Los Angeles, and San Bernardino Counties. Fresno County quickly emerged as the industry's nucleus. By 1886, one historian wrote, "the raisin was recognized as the principal industry of the county." To serve the growing population of growers, processing and marketing firms set up shop in the city of Fresno. It was no accident that the raisin industry developed where and when it did, and it was even less coincidental that the raisin industry, more so than citrus, which spread across southern California, was highly localized. Society and economy, very nearly congruent from the outset, remained so into the twentieth century.[29]

The main reason for the location of the raisin industry was the favorable natural environment. Immigrants chose the area because of their familiarity with the climate and soil needs of grapevines in their countries of origin. They chose well. The area of the Central Valley in and around Fresno was and remains "the one district in all California [that] is most suited to [raisin] culture." The three main varieties of raisin grapes, Muscat, Sultana, and Sultantinia (known as Thompson Seedless), thrive in the sandy loam of the piedmont alluvial plain. Raisin vines require a dormant season with moderate frosts; winter temperatures in Fresno rarely drop beneath 25°F. Fresno's dry summers are ideal for fruit maturation, although high winds and excessive heat pose occasional dangers. The Central Valley's aridity becomes an asset at harvest in August, when for two to three weeks the grapes lie in the sun to dry. Although rain in September is rare, the risk of such a disaster late in the season is real. Sun-drying the grapes not only saved growers money but also meant that the Central Valley, rather than California's wet and cool coastal valleys, would be home to the commercial raisin industry.[30]

In addition to the favorable natural environment, starting a vineyard during the 1870s and 1880s required the "gradual capitalization of several assets of both the social and the natural environment of that district." Capitalizing the "social environment" meant securing a reliable and constant supply of labor for both year-round vineyard maintenance and the big job of harvesting. The grapes were and still are picked by hand, usually when sugar content is about 22 percent, and then laid in wooden trays to dry. After about ten days, workers invert the trays to expose the undersides, then they stack the trays to "sweat"

An early Fresno raisin-packing plant, ca. 1890.
(Kearney Collection, Fresno Historical Society Archives)

the raisins, equalizing the moisture throughout the fruit. Large and small growers alike depended on transient workers, but wages varied. Archibald Grant, owner of a forty-five-acre vineyard, hired Japanese workers to pick his crop and paid them two and a half cents per tray, while clearing at least that much per pound. "Owners of smaller tracts do not always succeed in securing [labor] contracts at figures so favorable," he told the *Fresno Republican* in 1898. In the packinghouses, women did almost all the work of cleaning, processing, and packing.[31]

The colonies began to disband in the late 1880s and early 1890s, but land-holding patterns endured. Contemporary observers boasted about the importance and extent of small farms in the Central Valley. George West, the state viticultural commissioner for the San Joaquin district, remarked in 1891, "Large tracts of land in all the counties of the valley are being subdivided and sold in tracts of from ten to fifty acres, many of which are bought by people who intend settling upon the land purchased and embarking their all in the raisin business." By decade's end, M. Theo Kearney had extended this pattern to lands outside the original colonies: "Thousands of families," he boasted, "are now settled on twenty and forty acre fruit farms about Fresno." The same could also be said for the fertile valleys in and around Los Angeles. The economies of scale in fruit growing made small farms both manageable and profit-

Raisins drying in the sun, ca. 1900.
(Kearney Collection, Fresno Historical Society Archives)

able: "The small raisin grower has advantages which are beyond the reach of the big grower. Where the owner does practically all the work, the profit per acre is higher, and one man can handle as many as forty acres without assistance except in the harvest period. At that time, even on a ten-acre vineyard, help must be hired." The plans of the wheat barons and colony organizers had a lasting effect on landholding in California.[32]

Data from the published U.S. census show conclusively that small farms not only existed in California but also flourished over time. Fruit farms required a relatively high initial capital investment but labor costs were comparably low, except at harvest. The value of the commodity per acre was much higher for fruit than for field crops, meaning that fruit farms could yield sufficient income to support families.[33] In the fruit-growing regions of the state, there were more farms, and more smaller farms, than in areas of the state where fruit was of little or no economic importance. A statistical analysis of U.S. census data, detailed in the Appendix, indicates that fruit farming was highly correlated with farms of fewer than one hundred acres. Median farm size in California's most commercially important fruit-growing counties was less than fifty acres, and most of these farms were operated by their owners. This remained the case throughout the first half of the twentieth century (Table 2.2).

The prevalence of small farms in central and southern California contra-

Raisins drying in sweat boxes lying between rows of grapevines, 1940, with the Sierra Nevadas at the horizon. (Historical photographs from the C. "Pop" Laval Collection, Fresno, California)

dicts conventional wisdom about farm size in the state. Small farms constituted the dominant mode of agricultural production in the fruit-growing counties. They were not economically obscured by the big field crop spreads or livestock ranches because as fruit growing arose in the San Joaquin Valley, field crops and livestock moved into counties less geographically suitable for horticulture. Large orchards and vineyards were not uncommon in Fresno County and other counties where horticulture was the primary agricultural industry, but fruit growing was dominated mostly by smallholders. In short, horticulture encompassed some growers with very large holdings and many growers with much smaller investments. Moreover, economic differentiation and social diversity went hand in hand.

The openness of colony settlements to persons from diverse racial and ethnic backgrounds contrasts sharply with the reception tendered to those whose labor supplied the backbone of the emerging horticultural economy: migrant workers. All fruit growers had to hire labor several times per year, depending on conditions set by the crop, the weather, and the market. Often faceless and nameless, the Chinese, Japanese, and Mexicans who harvested fruit were interchangeably exploited as a supply of labor and excluded from immigration, naturalization, and eventually landholding.[34] The antagonistic relationship between the commercial imperatives of horticulture and the agricultural labor force made rural California an intolerant place. White Califor-

TABLE 2.2. Farm Size in Five Fruit-Growing Counties, 1900 and 1920

| | Percentage and Number of Farms, 1900 | | | |
| --- | --- | --- | --- | --- |
| | % 1–19 Acres (N) | % 20–99 Acres (N) | % 100–499 Acres (N) | % 500+ Acres (N) |
| Fresno | 11 (354) | 51 (1,690) | 26 (866) | 12 (380) |
| Los Angeles | 44 (2,909) | 36 (2,359) | 16 (1,082) | 3 (227) |
| Orange | 33 (793) | 53 (1,254) | 10 (250) | 4 (91) |
| San Bernardino | 50 (1,165) | 35 (827) | 13 (298) | 3 (60) |
| Santa Clara | 39 (1,562) | 37 (1,495) | 19 (759) | 4 (179) |

| | Percentage and Number of Farms, 1920 | | | |
| --- | --- | --- | --- | --- |
| | % 1–19 Acres (N) | % 20–99 Acres (N) | % 100–499 Acres (N) | % 500+ Acres (N) |
| Fresno | 14 (1,229) | 72 (6,439) | 11 (967) | 3 (282) |
| Los Angeles | 60 (7,505) | 29 (3,401) | 11 (1,309) | 2 (229) |
| Orange | 57 (2,385) | 33 (1,388) | 3 (322) | 2 (93) |
| San Bernardino | 53 (2,123) | 31 (1,247) | 14 (569) | 2 (84) |
| Santa Clara | 48 (2,408) | 40 (1,982) | 10 (508) | 2 (118) |

*Source*: ICPSR Data Sets, 1900 and 1920.

nians wanted it both ways: they desired cheap labor, but they did not want to live next to their workers. The concerns of workers played no role in the cooperative movement. Marketing cooperatives were the domain of growers and producers, not the hired hands, a fact that marks an ironic limit to the embrace of the cooperative gospel.[35] This book does not address labor's relationship to cooperative organizations. Rather, my focus is on the importance of small growers to horticulture and the cooperative movement. In the raisin industry in particular, the relationship of small growers to the cooperative was not unlike that of workers to their employer. And as the cooperative movement grew, those belonging to ethnic and racial minorities found themselves in the peculiar position of being economically essential but socially undesirable.

That most farmers and nearly all landowners were racially white was to contemporary Californians a notable achievement, not an index of racism. The colony system that brought many of these white farmers to the Central Valley was no doubt intended to preserve white social and economic prerogatives. Yet, despite increasing nativism and legal exclusion of Asians in California, the colony settlements did not erect permanent barriers to upward mobility of people not barred by statute from owning land.[36] Between 1880 and 1920, the population of Central Valley farmers grew not only in number but also in racial and ethnic diversity.

Among the small landholders who held the balance of power in the California raisin industry, one ethnic group turned out to be singularly important. The Armenians in the Central Valley were noted, even reviled, for their cohesiveness, economic resilience, and independence. Fresnans looked on them as a "racial unit, separate and distinct from other inhabitants of the county." Ethnically white, they were treated by Euro-Americans as racially nonwhite. Long described as a "hidden minority," Armenians played a highly visible role in the history of cooperation in the raisin industry.[37]

Armenians exemplified the strengths and weaknesses of the position of small growers. Because they acted as a more or less cohesive group within the industry, their refusal to participate limited a cooperative's ability to monopolize the growers. Armenians' economic independence reinforced contemporary whites' views of them as a separate racial group. When night riders attacked growers for refusing to join cooperatives, they used the cooperative ideal to impose economic conformity. The attacks on Armenians layered contemporary racial prejudice on economic self-interest.

Armenians were no different from many immigrant groups who came to the United States in search of economic, social, and political freedom. Armenians began to enter the country in significant numbers after the Civil War, sometimes following Protestant missionaries returning home. Immigration escalated in the late 1880s and 1890s after the Turks murdered over one hundred thousand Armenians and soared again after 1900 in response to continuing massacres, especially the genocide of 1914–15, in which more than one million Armenians died. By World War I, about eighty thousand Armenians had entered the country.[38]

U.S. immigration records categorized Armenians variously as Turkish (usually understood as a European classification), Oriental, or Asiatic (originating from eastern Turkey). In view of the increasing nativism of U.S. immigration law generally and the anti-Asian practices employed by immigration agencies

Armenian immigrants harvesting wine grapes, ca. 1885.
(Kearney Collection, Fresno Historical Society Archives)

in particular after 1900, the ambiguity about Armenian racial identity proved troublesome to exclusionists and government agencies charged with enforcing immigration laws. Ultimately, the federal courts ruled that Armenians were European and thus legally white for purposes of immigration and natural-ization.[39] These judicial victories set Armenians above Chinese, Hindus, and other East Asians in the immigration pecking order, but in less formal venues Armenians continued to face racial discrimination.

In California, which led the nation in hostility to Asian immigrants, Arme-nians' economic success was met with animosity, resentment, and discrimina-tion. Though many had not been farmers in the old country, those who settled in California came to make a living in agriculture. The first Armenian arrived in the Central Valley in 1882. By 1900, Armenians began to rival the popula-tions of groups who settled in the agricultural colonies. By 1908, Armenians outnumbered all other ethnic and racial groups in Fresno County except the Danes. At the outbreak of World War I, Armenians made up 25 percent of the foreign-born population in the Central Valley.[40]

Armenians quickly became an economic force in the area. A surprisingly large number brought significant amounts of money with them. Attracted by the agricultural colonies, nearly all of these people invested in land and grew figs and raisins. Those without ready capital to purchase land began as laborers in packinghouses and vineyards, gradually buying land with their savings or on installment. Armenians soon branched out into the packing business as well, shipping figs and raisins across the country by rail. Successful and independent Armenians willingly supported other Armenians by hiring them as field or packinghouse laborers. Packinghouse work was a ticket to economic mobility because it was relatively high paying, and many Armenians used it to make the transition from immigrant to landowner.[41]

How many of these Armenian landowners became raisin growers? How many tenants or lessees growing raisins were Armenian? It is not possible to determine exact numbers. The manuscript population census schedules do not consistently list occupations, and the manuscript agricultural schedules are no longer extant. Contemporary studies of the raisin industry contain anecdotal evidence of the number of Armenian raisin growers, but these numbers are little more than time-slice estimates. A more systematic assessment of their numbers is necessary.

Samples drawn from the manuscript population schedules for the four top raisin-producing counties in California for 1900, 1910, and 1920 indicate that Armenians made up a relatively small proportion of the populations there. Between 1900 and 1920, the proportion of Armenians in the population of Fresno County rose from 1 percent of the total population to just over 6 percent. Over half of all Armenians in the state lived there, and most of the Armenians in the Central Valley were concentrated in Fresno County.[42] Tulare County had the second highest proportion of Armenians in its population, just over 1 percent for 1910 and 1920. These estimates indicate that while Armenians may have been one of the larger of the foreign-born groups in the area, their absolute numbers were small. Data for Armenian populations in specific towns reinforce this impression (Tables 2.3 and 2.4).

Because Armenians achieved economic success relatively rapidly, growers and cooperative leaders treated them as a bigger force than their numbers might warrant. Armenians controlled a disproportionately large share of the industry. In 1908, the U.S. Immigration Commission found that Armenians owned twenty-five thousand acres of land, of which between sixteen thousand and twenty thousand acres were devoted to vineyards. If accurate, this estimate means that Armenians owned about one-sixth of the industry's productive capacity; the estimate does not include Armenian tenants or lessees. A student of the industry in the 1930s noted that between four and five thousand Armenians remained on their farms in the San Joaquin Valley. Armenians' economic success was sometimes compared to that of the Japanese, who were vilified for their quick transition from laborers to landowners. If anything,

TABLE 2.3. Proportion of Armenians in the Populations of Four Raisin-Growing Counties, 1900–1920

| County | 1900 | | 1910 | | 1920 | |
|---|---|---|---|---|---|---|
| | Population | % Armenian | Population | % Armenian | Population | % Armenian |
| Fresno | 27,350 | 1.0[a] (n = 276) | 65,600 | 3.9 (n = 2,548) | 133,200 | 6.1 (n = 8,078) |
| Kings | 10,000 | <1[b] | 17,700 | 0.2 (n = 38) | 23,100 | 0.4 (n = 102) |
| Madera | 6,400 | <1[b] | 9,500 | 0.3 (n = 32) | 12,850 | 0.6 (n = 72) |
| Tulare | 18,900 | <1[b] | 37,300 | 1.2 (n = 447) | 53,500 | 1.3 (n = 695) |

Source: Proportions calculated from samples taken from Twelfth, Thirteenth, and Fourteenth Censuses of the United States, Census Microfilm Publication Numbers T623 (1900), T624 (1910), and T625 (1920).

[a] Sampling error ± <1 percent for each county and decennial for which it could be calculated

[b] Numbers too small to calculate proportions

TABLE 2.4. Armenian and Total Populations in Raisin Industry Towns, 1918–1920

| County | Number of Armenians, 1918 | Total Population, 1920 | % Armenian |
|---|---|---|---|
| Fowler[a] | 1,000 | 1,528 | 65 |
| Fresno[a] | 4,000 | 45,086 | 9 |
| Kingsburg[a] | 600 | 1,316 | 46 |
| Reedley/Dinuba[a] | 700 | 5,847 | 12 |
| Sanger/Centerville[a,b] | 200 | 2,578 | 8 |
| Selma | 500 | 3,158 | 16 |
| Turlock | 400 | 3,394 | 12 |
| Visalia/Tulare | 400 | 9,292 | 4 |

Sources: 1918 Armenian populations: Wilson Wallis, Fresno Armenians (to 1919), ed. Nectar Davidian (Lawrence, Kans.: Coronado Press, 1965), 37; 1920 town populations: Fourteenth Census of the United States, 1920, vol. 1, Population: Number and Description of Inhabitants (Washington, D.C.: U.S. Government Printing Office, 1921), 183–85.
[a] Town located within Fresno County
[b] 1920 population figure for Sanger only

Armenians frustrated white landowners more than the Japanese. Despite the efforts of Fresno's representatives, the legislature did not include Armenians in the Alien Land Laws, probably because they had no significant presence outside of Fresno County. In addition to owning land, they were eligible for citizenship. Indeed, Armenian leaders urged the community to naturalize as quickly as possible and vote in every election.[43]

At the same time, Armenians maintained social habits and traditions that defiantly asserted their culture in the face of overwhelming pressure to assimilate. One of the more obvious markers of social separation was Armenian religious practices. The Armenian Apostolic Church quickly became a fixture in Fresno; Armenian converts to Protestantism built their own churches in Fresno when they were forbidden to worship with whites. Since Armenians were excluded from buying homes in certain areas of Fresno, they had all the more reason to settle together in de facto residentially segregated areas.[44] Armenians published Armenian-language newspapers, established their own social clubs, patronized Armenian businesses, and for all intents and purposes created what William Saroyan called a "New Armenia." As the Fresno Asbarez noted in 1911, "Any Fresno Armenian who has a good 40 acres or a brisk trade can dress in his Sunday shoes and clothes more healthily, comfortably, and attractively than King Solomon."[45]

Armenians' assertion of their culture drew active discrimination from Americans of western European descent. In addition to housing and religious discrimination, "[Armenians] were excluded from social clubs and professional associations; . . . they were ill-treated by clerks, salesmen, and other townspeople in their daily transactions. Armenians were called 'Turks,' a label they resented bitterly, and other equally demeaning slurs." The epithet "Turk" was often embellished as "black Turk," making its racial connotation explicit. One California Armenian remembers being called a "Fresno Indian."[46]

The Armenian presence in the raisin industry and in Fresno society is significant in helping to explain what happened when the cooperative movement intersected with American racial prejudice. One reason why a racist majority discriminates against a minority is to impose separation in meaningful facets of life, to claim a limited resource exclusively for itself, or both.[47] In contrast, cooperation inverts the usual relationship of majority to minority; its logic requires the participation of all its constituents or else it will fail. In Fresno between 1900 and 1930, the Armenian minority was strong enough to maintain its social identity, and cooperative loyalists saw that separation as a danger to the cooperative's monopoly. In the context of the cooperative movement, forcible inclusion weakened rather than empowered the minority because it denied individuals the ability to act autonomously. To resist inclusion was to claim economic independence. The Armenian experience is all the more ironic because by the time it occurred, the federal courts had defined Armenians' racial identity as white, but in Fresno they were treated as not white. As the history of cooperative organization in the raisin industry reveals, neither the idealism of the cooperative gospel nor the imperative of cooperative monopoly could transcend contemporary conceptions of race.

In the raisin industry, cooperation's economic purpose, which was to impose order on chaotic marketing conditions, took precedence over its social mission, which was to unite individuals. The cooperatives' success in raising prices created ample incentive for small and minority growers to free ride, an economic term meaning that they enjoyed the benefits of the cooperation's existence without sharing the attendant costs or risks. By fixing prices and establishing monopolistic marketing channels for the growers, the cooperative held an umbrella over the whole industry, including those who did not belong. As long as the cooperative held the umbrella, growers wishing to cheat could do so profitably.[48] And because of the innovations they made in corporation law, the raisin industry's cooperatives empowered themselves to hold a very large umbrella over most of the industry for a long time. Taking advantage of frequent changes in the law of cooperative organization between 1870 and 1910, the raisin industry—and other agricultural entrepreneurs in California and elsewhere in the country—redefined the legal status of agricultural cooperation. Anxious to placate their agricultural constituents and entranced by the cultural ideal of farming, the states proved willing partners in this legal transformation.

PART TWO

---

The Legal Status of Cooperatives

1865–1914

# 3 ∾ Voluntary Associations or Corporate Combinations?

*We are familiar with the duties of the farmer and the cares and trials of his business*

*life, and appreciate highly the customary compliments paid by mankind to the rural*

*yeomanry of the land. . . . Yet what is there about it all to entitle him to the privilege*

*of combining in restraint of trade?* — In re Grice, *1897*

As in so many other areas of regulation, the states provided the laboratories for experiments in the legal status of agricultural cooperation that took place between 1865 and 1920. As farmers began to organize on a wider scale after the Civil War, they discovered that law gave corporations distinct advantages over unincorporated associations. The more that farmers' cooperatives appropriated the forms and practices of the regular business corporation, however, the more their opponents tried to use the image of traditional cooperation and the authority of antitrust law to hold them back. Legislators, judges, lawyers, and farmers disagreed sharply over the legal status of cooperation. In particular, they clashed over whether cooperatives could be truly cooperative when they looked and acted like regular corporations.

Determining what constituted true cooperation became increasingly important during the Progressive Era because many states wanted to treat farmers differently under the antitrust laws. Fearing that agriculture would be unable to compete with the well-organized industrial corporations that dominated transportation, communication, and marketing in the new economy, state legislators believed that granting farmers immunity from the antitrust laws would enable them to market their goods collectively. These lawmakers assumed that farmers would perform collective marketing through cooperatives, that these organizations would be readily identifiable as cooperative, and that cooperatives would pose no threat to the free market because, unlike regular corporations, cooperatives did not centralize capital. This logic stood on an unstable foundation: the notion that cooperatives would always adhere to a distinct form of organization. Equating Rochdale cooperation with "true" cooperation, many in government and agriculture assumed that cooperatives adhering to the Rochdale principles would automatically come within statutory antitrust exemptions. By the time these exemptions were enacted, however, farmers had developed new legal forms of cooperation that did not strictly apply the Rochdale principles. Interpreting the federal equal protection clause narrowly, federal courts treated cooperatives with capital stock or other non-Rochdale features as cartels rather than individual firms and refused to honor state antitrust exemptions for farmers. During the formative era of cooperation's legal status, state legislative attempts to broaden farmers' access to the legal tools of industrial business were thwarted by the federal courts.

## CORPORATION LAW AND COOPERATIVES

When farmers began organizing cooperatives on a large scale after the Civil War, the law offered little guidance. The agricultural economist Edwin G. Nourse observed in 1927, "Many . . . of the earlier Grange undertakings were not incorporated at all, but operated simply as voluntary associations. Where they did incorporate, it had in most instances to be done under the general incorporation laws." General incorporation laws facilitated the organization of for-profit commercial corporations. Some state laws and constitutions required all corporations to adhere to such orthodox practices as voting according to number of shares and distributing profits according to investment. The farm movement spurred the process already under way to pass statutes providing for separate incorporation of cooperatives. Early examples of these statutes gave farmers access to basic corporate features.[1] An 1865 Michigan statute gave associations the power to regulate shareholding, voting, and dividend payments and authorized $500,000 in capital stock. In 1866, Massachusetts capped capital stock at $50,000, required members to purchase all shares in cash, and limited each member to one vote. By 1895, nine other states — California, Pennsylvania, Minnesota, Connecticut, Ohio, Kansas, Wisconsin, Montana, and Tennessee — enacted variations on the Massachusetts "one-member, one-vote" law.[2]

These statutes were essentially hybrids, offspring of rudimentary elements of the corporation and various Rochdale principles. All of them retained one of the corporation's most fundamental attributes: capital stock, the standard measure of ownership, investment, and assets. Seen in an extreme light, the cooperatives organized under these statutes were no more than "ordinary stock corporations serving members and nonmembers alike, being 'cooperative' in no other respect than that their stock was owned by farmers." At the very least, as the economist Edwin G. Nourse astutely noted, "neither profits as such nor capital as such were eliminated." The roles of profits and capital, however, were greatly altered. When cooperatives distributed profits according to patronage, they placed a premium on participation and rewarded members in proportion to the risk they bore. One-member, one-vote statutes ensured that no individual had greater influence over the organization than anyone else, regardless of the number of shares held. Further, most Rochdale laws limited capital stock to nominal amounts, usually $5,000, and restricted the transferability of stock certificates to preserve the permanent relationship between members and organizations. Despite statutory declarations that capital stock played a different role in cooperatives than in regular corporations, Nourse argued that its presence in cooperatives "perpetuate[d] the ordinary corporation idea that capital stock is the primary claimant to benefits from the operations of the organization."[3]

To counteract this impression, states slowly began eliminating capital stock from the separate incorporation laws. California led the way, passing the first nonstock law in 1895 with the support of the Farmers' Alliance.[4] The law construed membership as personal, nontransferable, and, above all, equitable: "In such association[s] the rights and interest of all members shall be equal, and no member can have or acquire a greater interest therein than any other member has." Fourteen years later, a new law limited participation in nonstock cooperatives to persons engaged in "the production, preserving, drying, packing, shipping, or marketing" of agricultural products. Together, the two laws purported to set agricultural organizations apart from other corporations and to restrict cooperatives' distinct status to farmers.[5]

The problem with nonstock laws was immediately obvious to farmers in California and elsewhere. Without capital stock, a cooperative could not pool the capital needed for storage, grading, packing, marketing, and shipping. The need was particularly acute for commodities such as milk, tobacco, and fruit, all of which incurred high costs for handling and processing, but it also arose in staple crops such as wheat and corn. While seven states copied California's nonstock laws, nineteen passed versions of a law with more generous provisions for capital stock. The capital stock laws of Wisconsin and Nebraska served as models, with the result that these laws achieved greater uniformity from state to state than the first Rochdale laws.[6] Most of the states copying the California nonstock laws also passed capital stock laws, demonstrating farmers' preference for the capital stock form.[7] The capital stock laws contained many of the same elements of Rochdale practice as the nonstock laws: restricting membership to farmers, requiring one person, one vote, and using patronage dividends to divide up profits. The popularity of the capital stock laws suggests that lawmakers believed that these attributes of Rochdale cooperation were compatible with the presence of capital stock.

As it charted a path for the organization of rural enterprise, the law evolved from regular corporation law with capital stock, to separate statutes without capital stock, and then back to capital stock but under separate cover from regular corporations. The decentralized, noncapitalized organizations suited to cooperative purchasing were essentially useless in the national marketing of agricultural commodities. The presence of capital stock, however, signified a corporate rather than a cooperative purpose. Farmers had to demonstrate that their use of corporate forms and practices did not produce the anticompetitive effects commonly attributed to "evil" or "bad" trusts. The more they availed themselves of the organizational and economic advantages of capital stock organizations, the more difficult this task became.

## ANTITRUST LAW AND COOPERATIVES

The common law's apparently unadulterated hostility toward monopoly and restraint of trade has played a contrapuntal theme in most accounts of the rise of the trusts.[8] Embodying key tenets of classical liberalism, the common law's posture on competition widely influenced the development of state and federal antitrust statutes.[9] At common law, agreements in restraint of trade were illegal if they tended to create a monopoly; to increase prices; to suppress new entry or continued enterprise of others in the industry; or to constitute a concerted effort by a few to control the market participation of others.[10] Such prohibitions did not mean, however, that all combinations or cartels were inherently illegal. Rather, the conception of economic freedom reigning in the late nineteenth century assumed that free exchange should produce greater efficiencies for the parties to the bargain. What the common law prohibited was, in economic terms, agreements with negative externalities that had the effect of suppressing competition in the larger industrial world.[11] As Martin Sklar has argued: "The overriding principle at common law in the United States was not unrestricted competition, but the natural liberty principle of freedom of contract: that is to say, the right to compete, not the compulsion to compete."[12]

State legislatures conducted an antitrust movement of their own both before and after the federal Sherman Antitrust Act of 1890. In addition to codifying common law prohibitions on contracts and combinations in restraint of trade, the state laws adopted a more precise definition of monopoly than that embodied in either the common law or the Sherman Act. Many states specifically outlawed price cutting and exclusive dealing contracts, which trade associations typically employed to decrease competition.[13]

In contrast to the statutes' broad charter of competitive freedom, some state legislatures insulated farmers from the laws' penalties. Between 1889 and 1897, eleven states—all but one in the Midwest and South—exempted farmers from their antitrust statutes.[14] Farmers' success in swaying legislatures in the Midwest apparently owed as much to a long-standing tradition of agrarian political participation as to Populist influence. Jeffrey Ostler has showed that north of the Mason-Dixon line, farmers' political destinies did not rest solely with the Populists; rather, midwestern farmers tended to work through the two-party system, particularly in states where antimonopoly sentiment was strong.[15] This may explain the inclusion of exemptions for farmers in statutes that strictly criminalized anticompetitive behavior. Farmers in the Midwest, moreover, had specific economic needs. They wanted relief from exorbitant railroad rates, monopolistic grain elevator companies, and dairy processors that played producers off against each other to drive down prices. Licensing cooperatives to represent combinations of farmers offered a solution to these problems without jeopardizing established political norms and institutions.

Still recovering from the physical and economic ravages of the Civil War, southern farmers turned to agricultural collectivism to compete with better-established and financed northern producers and processors. Securing legal relief from an oppressive marketing and currency system was the central aim of the Southern Alliance. Yet only five of the eleven states that historian Lawrence Goodwyn cited as major centers of Populist activity passed antitrust exemptions for farmers. In the South, too, the legislative aspirations of the cooperative movement did not walk in lockstep with the region's largest agricultural protest movement.[16]

Regardless of their political origins, the state exemptions shared a common legal dilemma: they granted the general privilege of antitrust exemption but failed to specify how or by whom that privilege was to be invoked. Most exemptions stayed the operation of the statutory penalties against "agricultural products while in the hands of the producer or raiser" or, in a few states, against "combinations of farmers." In some cases the exemptions applied to producers and in other cases to farm products. The laws permitted farmers to fix prices and enter into agreements to secure control over marketing but did not specify whether farmers should engage in these practices through cooperatives, corporations, unions, or other organizations. The economic consequences of the exemptions were similarly ambiguous. The statutes, John Hanna concluded, were "particularly severe as to all conduct designed to raise the price or diminish the production of food-stuffs."[17] Designed to enable agricultural producers to avail themselves of organizational efficiencies, the exemption apparently would protect farmers only if prices did not rise as a result.

The ambiguity of the state antitrust exemptions rendered the position of cooperatives in the law and in prevailing conceptions of the market more problematic. Most judges, lawyers, and legislators learned what they knew about cooperatives from newspaper accounts of the Granger and Populist movements. The lack of a recognizable separate legal form for cooperatives rendered them invisible even to legislators anxious to do farmers a favor. As a result, when cooperatives sought to enforce arrangements with members, or when nonfarmers raised constitutional objections to the exemptions, the legality of the exemptions came under attack.

By the mid-1890s, it became clear that state legislatures and courts talked about cooperation in entirely different languages. Legislatures saw cooperatives as a valid means of achieving collective efficiencies; courts saw them as combinations that restrained trade. The Illinois Supreme Court's 1895 decision in *Ford v. Chicago Milk Shippers' Association* was the first sign of trouble. In *Ford*, a dairy cooperative sought to enforce an exclusive dealing contract with a member who sold milk to a commercial dealer. The plaintiff argued that he was not bound by his agreement with the cooperative because the association fixed prices in violation of the state's prohibition against combinations that affected price or production. Although the Illinois antitrust act exempted "agricultural products or live stock while in the hands of the producer or raiser," the court held that the association's price fixing impeded the free flow of commerce: "The association [and] its members . . . carry out a scheme [by which they] limit the amount to be sold within the corporate limits of the city of Chicago."[18] To determine whether the milk dealers' association restrained trade, the court examined its market behavior rather than its intent or philosophical purpose. The cooperative was liable under the antitrust law because its effect on the market was to prevent one independent entrepreneur — the milk producer — from selling to another — the milk dealer. As antitrust decisions go, this one was unremarkable.

From the perspective of farmers, however, the court's refusal to consider the cooperative's purpose as relevant to its effect on competition was alarming. The association's self-professed object was to secure for producers "a just return for the sale of [their milk]; to rid the field of city distribution of irresponsible and dishonest dealers; to establish a central bureau of information for the shipper's benefit; and to secure to the dealer of milk a pure, wholesome, honest quality of that product." That the cooperative's intent was to equalize the bargaining positions of dealers and farmers held as much weight as the arguments of union organizers seeking better wages and working conditions for industrial laborers. A defense based on the argument that the association's purpose was to secure higher returns for its members garnered little judicial sympathy in light of the antiunion sentiment that settled on the state after the Pullman strike and the Haymarket riots.[19]

More significantly, the court refused to apply the statutory exemption for agriculture to the milk association. The opinion is silent on its reasoning; apparently the court read the statute literally. The law applied to "the producer or raiser" rather than organized associations of individual producers. Organized as a regular corporation, the cooperative did not generate any intellectual connection to the exemption, and the court treated it as a combination whose market practices unreasonably interfered with competition. Like many contemporary courts that were concerned with maintaining property rights in a free market, the *Ford* court focused on applying the proper statutory rules to anticompetitive behavior rather than determining whether farmers deserved the exemption or, for that matter, whether farmers could claim the benefits of the exemption when acting through cooperatives. Other state jurisdictions with antitrust exemptions for farmers came to the same conclusion, and states without them drew analogies from industry and manufacturing that entirely obscured the cooperative nature of the association.[20]

The similarity between labor unions and cooperatives in the context of the "liberty of contract" ideal was evident during the 1890s. The courts' animosity toward cooperative arrangements to fix minimum prices for their members' produce tracked judicial treatment of laws setting maximum hours for workers. Both interfered with a natural liberty guaranteed to individuals by the Fourteenth Amendment. Both also constituted unnatural concentrations of entrepreneurial activity whose very existence augured ill for the contemporary idea that competitive equilibrium ultimately governed industrial development. The court in *Ford* focused on individual market relations because that was the prevailing norm in restraint of trade cases, and the cooperative's obvious anticompetitive effect signaled the court to treat it like any other illegal combination. The cooperative lost the case because it could not persuade the court that the antitrust statute treated farmers differently.[21] *Ford* set the tone for the courts' consideration of this broader constitutional issue.

## COOPERATIVES AND EQUAL PROTECTION

In a series of cases decided between 1897 and 1902, nonfarmer litigants asserted that the exemptions for farmers in the Texas, Illinois, and Georgia antitrust statutes violated the equal protection clause. Since the states did not specify how cooperatives were to be categorized under the antitrust laws, courts applied the prevailing standards of restraint of trade and equal protection law. The result was a resounding declaration that farming was no different from other trades or occupations.

In 1897, a federal circuit court sustained an equal protection challenge to Texas's agricultural exemption in *In re Grice*. The petitioner, who was not a farmer, argued that it was unconstitutional for the state to convict entrepreneurs of acts that were not crimes for farmers. The weight of precedent lay with the petitioner. Essentially, any statute creating a separate classification was treated as suspect unless a rational basis existed for distinguishing among citizens.[22] Here, the Texas legislature provided no reasonable justification for permitting agricultural producers to do what members of all other trades were prohibited from doing: "What ground can there be for setting aside, as this act has done, four-fifths of the citizens of Texas as an exempt class from the punishment for felony, because they are producing farmers?" The court could see no constitutionally valid distinction between agriculture and any other trade. "[The act] is aimed to favor the agricultural class, and is against the merchant and mechanic, and all the others, without either reason or justice."[23]

In 1902, the U.S. Supreme Court followed the *Grice* court's reasoning in *Connolly v. Union Sewer Pipe Co.* Speaking for a seven-to-1 Court, Justice John Marshall Harlan ruled that the Illinois legislature could not grant antitrust immunity to farmers without violating the Fourteenth Amendment. Quoting with approval the Court's decisions in *Barbier v. Connolly* (1885) and *Yick Wo v. Hopkins* (1886),[24] Harlan held that the Illinois statute violated the right of non-farmers — including corporations — to equal protection of the laws because it accorded farmers privileges that the state denied to everyone else. Unlike the situation in *Yick Wo*, where a statute was applied selectively against Chinese laundry operators, the state was not singling out farmers for unfair discrimination. Unlike *Barbier*, where the statute was neutral on its face, the exemption clearly set farmers apart as a class without providing a rational basis for the distinction. The cultural importance of farmers did not justify treating them as a distinct class that deserved special legal privileges.[25]

Accordingly, Harlan saw no difference between agriculture and other forms of economic enterprise for purposes of antitrust regulation. Rather, he said, farmers, merchants, and traders "are all in the same general class, that is, they are all alike engaged in domestic trade, which is, of right, open to all, subject to such regulations, applicable alike to all in like conditions, as the state may legally prescribe." Thus, when a state regulated combinations in trade, it could not constitutionally prohibit them in one business belonging to "the same general class" and permit them to exist in another: "'Arbitrary selection can never be justified by calling it classification.'"[26] The Illinois law erected one such "arbitrary selection": "If combinations of capital, skill, or acts, in respect of the sale or purchase of goods, merchandise, or commodities, whereby such combinations may, for their benefit exclusively, control or establish prices, are hurtful to the public interests and should be suppressed, it is impossible to perceive why like combinations in respect of agricultural products and live stock are not also hurtful."[27] Farming was like any other industrial activity; therefore, farmers could not escape the antitrust laws through statutory exemptions. Other state supreme courts immediately followed the rule.[28]

Only Justice Joseph McKenna argued for treating farmers differently from other entrepreneurs. Classifying agriculture separately, he said, could be justified on the basis of the distinction between agricultural producers and commercial traders: "Might not the legislature see difference in opportunities and powers between the classes in regard to the prohibited acts?" McKenna was willing to recognize that farmers' combinations, dealing as they did with the product of the farmers' own labor, could conduct trade of a character that was both distinguishable from other monopolies and beneficial in ways that public policy might legitimately recognize. The Court had so ruled in an earlier case, and McKenna saw no reason not to follow that precedent in *Connolly*.[29]

The decisions holding that statutory exemptions for agriculture violated the equal protection clause were framed less by judicial antagonism toward cooperation per se than by an insistence that legislatures follow established constitutional norms for distinguishing agriculture from other forms of commerce when they dispensed legal privileges. Indeed, almost no antitrust exemptions survived judicial scrutiny during this time. The imperatives of antitrust law and the Fourteenth Amendment dictated that all trades be treated alike unless legislation explicitly provided a "rational basis" for doing otherwise. The courts' conclusion that agriculture did not warrant the legislative privilege of antitrust immunity followed directly from the legislatures' failure to provide the constitutional basis for the favors they desired to bestow.[30]

In light of judicial decisions that it was unconstitutional to exempt farmers explicitly, a few states tried more subtle ways to accomplish their purpose. California, one of the few states not to pass an antitrust law before 1900, finally did so in 1907; amendments in 1909 exempted agreements to form combinations or associations, "the purpose and effect of which shall be to promote, encourage or increase competition in any trade or industry." In industries such as citrus, raisins, wine, peaches, and apricots, growers had resorted to combination and monopoly. They assumed that this vague instruction sanctioned their method, and agricultural lawyers in the state publicly agreed with that interpretation. The amendment went untested until the early 1920s, when the courts upheld the law.[31] Colorado adopted an identical provision in 1913.[32]

Kentucky tried a different and ultimately less successful tack. There, the legislature determined that tobacco buying was concentrated "in the hands of a small number of buyers practically dominated by a single 'trust.'" In 1906, the legislature declared the existence of a state of emergency and authorized field crop producers to "combine, unite, or pool . . . for the purpose of obtaining a greater or higher price therefor than they might or could obtain or receive by selling said crops separately or individually."[33] In effect, the statute declared that the solution to the buyers' monopoly was to permit the producers to form one of their own. The state supreme court agreed that the emergency justified the legislation, held that the act was "classification but not exclusion," and ruled that granting farmers the privilege to engage in pools and combinations was a reasonable exercise of the police power. The state court held its ground even when nonfarmers charged that the statutes were invalid under *Connolly*.[34] The U.S. Supreme Court struck the law on grounds of vagueness and thus did not reach the constitutional issue.[35]

In sum, courts sanctioned neither the explicit exemptions contained in the antitrust laws nor the separate statutory authorization for farmers' pools. Instead, judges examined the market behavior of cooperatives under the same precedents used for other private entrepreneurs accused of monopolizing supply or restraining trade. Even when a state avoided the *Connolly* equal protection problem and provided a constitutional basis for the classification, the Supreme Court insisted that the classification fit into the larger web of antitrust law. Except in Kentucky, where judges were sympathetic, and California, where judges had no opportunity to rule on the statute, state courts were even more inflexible. To them, neither the nature of the restraint on competition nor the identity of the entrepreneurs conducting the restraint seemed to matter. Many judges believed that farmers posed just as much a threat to free markets as the robber barons.

These rulings did not stop farmers' ongoing experimentation with the legal form of cooperation. Although the courts saw the presence of capital stock as a per se indicator of corporate character, farmers continued to employ capital stock and to use other corporate and legal devices to protect their investments in their marketing businesses. For the courts, this practice raised the question of whether cooperatives could possess capital stock and still operate in a cooperative manner. The capital stock incorporation statutes did not explain how cooperatives could use operating capital and the nature of cooperative profits without threatening free competition. Accordingly, nothing in the statutes dislodged the assumptions about market relations guiding the courts. As cooperatives resorted to contracts and fines against members to protect their market control, opponents again used antitrust laws to stop them.

## PROTECTING COOPERATIVES' PATRONAGE

By 1910, the major wave of separate cooperative incorporation statute writing had fairly well run its course, and cooperatives were using these new laws to become more powerful players in the market. In the Midwest, competing elevator companies, grain dealers, and meatpackers, "abetted by the railroads," sought to crush cooperatives and reduce their influence over the sales of agricultural commodities. They boycotted cooperatives and any buyers who did business with them. Cooperatives fought back by restricting their members' selling practices through "maintenance clauses"—so named because they maintained members' loyalty to the cooperative by imposing fines on members who sold outside the cooperative. The clause effectively worked a boycott against competing buyers. Between 1890 and 1920, the maintenance clause came into "considerable vogue" as cooperatives retaliated against boycotts and other practices designed to undermine their market control.[36] In turn, elevator owners and dealers attacked the maintenance clauses as impeding their freedom to buy from all farmers. This allegation resonated with judges who were distrustful of cooperatives' impact on the market.

In Iowa, for example, the conflict between the legislature's prerogative in promoting cooperation and judicial antitrust rationales torpedoed cooperatives' attempts to protect their patronage. The case of *Reeves v. Decorah Farmers' Cooperative Society* (1913) tested the legality of the maintenance clause, raising the larger question of whether cooperatives could restrict their members' selling and thus interfere with individuals' economic freedom for the sake of agricultural economic stability. The litigation was staged by a large Chicago meat dealer who set up one of its local buyers as the plaintiff. The cooperative's legal status was typically murky. Ignoring Iowa's capital stock cooperative incorporation law, the Decorah cooperators organized as a for-profit corporation under the state's general incorporation statute. They set their capital stock at $20,000, but at the time of litigation, 350 members had subscribed just $4,000. Moreover, the Decorah Society's purpose echoed the cooperative canon: to "establish a market where the farmers would receive for their hogs what they were worth here in Decorah" by acting "primarily as a selling agency for the members of the society." All members had to sell all their livestock to the organization, and any member selling to a competitor had to pay the cooperative five cents for each one hundred pounds sold.[37]

The *Reeves* case squarely presented the question of whether a maintenance clause constituted anticompetitive behavior. The hog buyer argued that the maintenance clause belied the Decorah Society's benign purpose and was void under both common and statutory law because it imposed a penalty for breach of contract. He accused the society of attempting to drive him and all hog dealers in Decorah out of business. To remain competitive, he claimed, he had to offer prices that were at least five cents higher than the society's—the cost of the penalty imposed by the maintenance clause.[38]

The Iowa court examined the cooperative's structure and found that the Decorah Society was an ordinary for-profit corporation.[39] This finding foreclosed any consideration of a cooperative's special status under Iowa law, and the court proceeded to treat the suit as an ordinary antitrust case involving two corporate entrepreneurs. The controlling antitrust issue was the maintenance clause's impact on the market freedom of other hog buyers: "To our minds, this was undue restraint of competition, or, as the term is now understood, 'restraint of trade.'"[40] The court permanently enjoined the "enterprising Decorah hog raisers" from enforcing the exclusive bargaining clauses in their membership agreements. To contemporary economists, the court's action reinforced the idea that "free competition between individual sellers and buyers must be upheld, even if the latter possessed the advantage of quasi-monopolistic bargaining power."[41] In this case, a corporate packing company threatened competitive prices less than a producers' cooperative organization.

The underlying problem was that the legal status of cooperatives was highly ambiguous under the new laws. The lack of legislative guidance left courts to rely on prevailing standards of review in resolving the issues of cooperatives' impact on the market. The court's conception of farmers' relationship to their cooperative paralleled the Progressive Era juridical understanding of the relation between employees and their labor unions. In this view, no common interest united the individual and the organization; rather, the organization encumbered the market freedom of the individual. In no respect did the organization represent the collective interests of its members; in fact, the organization coerced them to join and held them there by dint of an expensive fine. In this sense, *Reeves* was the agricultural equivalent of the Supreme Court's decisions in *Adair v. U.S.* (1908) and *Coppage v. Kansas* (1915).[42] In those cases, the Court struck legislation prohibiting employers from discriminating against workers who joined labor unions on the theory that the government could not infringe on the rights of both laborer and employer to deal freely with one another in the market. While the labor cases involved constitutional due process rights and the cooperative case statutory and judge-made antitrust law, the result was the same and the underlying rationale consistent. The *Reeves* court subscribed to the prevailing concern of the *Lochner* age: ensuring the right of farmers and laborers to participate in a market unhampered by unnatural and anticompetitive restraints imposed by corporate agents.[43] That cooperatives' use of a maintenance clause offended this sensibility was made clear in *Reeves* and other state court decisions that followed the Iowa ruling.[44]

As of 1914, state courts refused to permit cooperatives with capital stock to hamper free trade even if they sought to accomplish cooperative goals. Despite variations in cooperatives' capital stock and dividend policies, the courts treated the presence of capital stock as indicating automatically that the organization was corporate and not cooperative. As long as they funded their operations with capital stock, cooperatives would find it difficult to free themselves of their corporation law origins. The judiciary insisted that a cooperative disavow any intention of accumulating commercial profits or capital in any form and forbear from any market practices — as necessary as they were to the cooperative's existence — that interfered with the freedom of competitors. In other words, farmers feeling oppressed by transportation and marketing monopolies could find no safe corner in conducting monopolies of their own.

Litigation challenging the use of maintenance clauses, liquidated damages, and capital stock focused the judiciary's attention on cooperatives' corporate structure. Accordingly, the legal status of agricultural cooperatives was defined almost entirely by the separate incorporation statutes. These statutes, however, embodied the broader uncertainty about what made cooperation distinctive as a way of organizing economic enterprise. Did capital in any form put cooperatives in the same boat as corporations? Did the use of maintenance clauses cancel out a cooperative's goals of self-help and voluntarism? If cooperatives became influential actors in the market, did they compromise the farmer's special place in society and culture?

Farmers themselves were divided on these questions, and contemporary observers tended to paint the issue using mutually exclusive categories. "One great difficulty with cooperation is that some members want to spell it corporation," wrote the *Country Gentleman* in 1913. The next year, an observer of the new developments in state law carefully listed those "powers [that] are essential to the success of a cooperative business organization." Nowhere did the writer mention maintenance clauses or other practices that affected competition; instead, he argued that the most common attributes of the new laws drew directly from Rochdale. Farmers had begun to pay a price for appropriating the corporation and its legal powers for their own use.[45]

Progressive Era courts concluded that there was no legal or constitutional basis for the antitrust exemptions that discriminated in favor of farmers and against other entrepreneurs. The courts lagged behind the legislatures in ratifying what farmers were doing to change market relations and cooperative practices, hardly a surprising story in the social and regulatory history of the period.[46] What the story reveals in this context is that when farmers wanted to change the legal construction of their place in the market, they turned to legislatures to adapt the corporation, a creature of state law, to their purposes. Legislatures were more susceptible than courts to the political pressure farmers could bring to bear, and farmers were adept at working within the legal system to obtain extraordinary official declarations of sympathy with their economic plight. The strength of prevailing equal protection and antitrust rules, however, circumscribed farmers' ability to use cooperatives to alter the competitive balance in the market. Most courts rejected the notion that the farmer's unique status offered sufficient protection to society against monopolization and price fixing, probably because the legislatures never coherently developed a legal conception of this status that could overcome the judiciary's preference for individual market freedom.

As ensuing events would disclose, there was some reason for judicial skepticism. The cooperatives that California horticulturalists formed beginning in the 1890s seemed to fuel the perception that farmers' monopolies not only were possible but in fact posed a substantial threat to the interests of urban consumers. One in particular exemplified the gains that growers could make by imaginatively applying corporation law to the unique situation of horticultural marketing. The organizations that surfaced in the raisin industry beginning in 1898 promised to give farmers unprecedented power — and lent credence to courts' doubts about agricultural monopolies.

# 4 ❧ A Growers' Trust

*This is an era of trusts and combinations. . . . Our sole aim from now on should*

*be to apply this principle of combination in all our affairs so as to secure the greatest*

*economy in production and in placing our products on the markets of the world.*

—M. Theo Kearney, 1899

In California toward the end of the nineteenth century, new developments in cooperative marketing flew directly in the face of the courts' adverse rulings and explicitly appropriated the corporate trust as a legal and economic tool. The composer of this rogue variation on the cooperative theme was a wealthy grower and landowner named M. Theo Kearney, who approached the task of organizing an industry-wide monopoly by deliberately flouting the rules that the courts were strictly applying to farmers elsewhere. Kearney brought the industry to the brink of corporate combination only to see his organization collapse for lack of grower support. The lesson of the Kearney era was that while growers wanted to assume greater control over marketing through cooperation, they had reservations about the antitrust problems that Kearney's overtly monopolistic, trustlike model created.

The history of raisin industry organizations between 1898 and 1906 was almost entirely a reflection of Kearney's own personality and style. He was an individual of extraordinary imagination and foresight with just as great a capacity for alienating people. Kearney personified the connection between cooperative ideals and the legal and organizational innovations he believed were necessary for effective collective action. His legal creativity produced the California model of cooperation adopted by other horticultural industries during the Progressive Era. This model featured industry-wide monopoly, regular incorporation with hundreds of thousands of dollars in capital stock, and the use of trust devices to centralize economic power. The model promised big dividends for the industry, but it posed substantial risks as well. Kearney's attempt to create a growers' trust was thwarted not only by packers, bankers, and merchants but by growers who deserted Kearney when he treated them more like hired hands than equal partners. The California Associated Raisin Company's organizers drew directly on Kearney's legal innovations when incorporating their organization in 1912.

### A MOSES FOR RAISIN GROWERS

The 1890s were difficult years for the raisin industry. The commission system, the depression of 1893, and the abolition of tariffs on imported raisins took the luster off raisin growing. Prices declined sharply as crops became larger. These developments had predictable effects on land prices and values. In 1890, growers received an average of eight cents per pound; the next year, they were lucky to get two.[1] According to assessors' records, growers pulled up fourteen thousand acres of vines in 1894 and 1895, apparently convinced that planting something else made better sense. In 1897, according to the *Pacific Rural Press*, "The most fertile land in California can be bought to-day for thirty percent of what was asked for it in 1890." The *San Francisco Chronicle*'s agricultural editor observed, "Not only were small raisin-growers in real distress, but the largest financial interests of the raisin districts had become thoroughly alarmed." By 1894 the industry had lapsed into a steep depression. Growers concluded that their only hope of economic survival lay in cooperation, which would enable them to fix the price that packers and merchants would pay for raisins: "[If] the California growers and the California handlers unite against the system of consigning to the East, there is no serious obstacle to [the] destruction [of the consignment system]."[2]

Cooperation was easy to invoke in mass meetings and community gatherings but far more difficult to put into practice. "On all theoretical grounds," the *Pacific Rural Press* editorialized, "co-operation would meet the evils, but this cannot be done practically because producers either cannot or will not combine." As prices continued to fall, the question became not whether to cooperate but who would join in such an effort and what legal form it would take. Early in the decade, growers established informal pools to sell to packers and merchants. Organized along traditional cooperative principles, these pools lacked enforcement powers and were easily broken by the commercial raisin packers at Fresno.[3]

Legislative help to abate this annoyance was unavailing. Governor Henry S. Markham vetoed a bill to enforce the pooling arrangement on the grounds that a combination of growers, commission brokers, and packers would lead to monopoly. Growers then turned to larger, more formal organizations, setting up cooperative warehouses under the state's general incorporation statute. These organizations could raise capital through the sale of stock and enter into formal contracts with growers for their raisins. Often, however, processors, packers, brokers, and other nonproducers purchased shares of stock for speculative purposes or to subvert the cooperative's ability to act in the growers' interests.[4]

By 1894, it had become clear that local cooperative associations could not raise adequate capital or obtain credit for packing and marketing operations. Further, they provided an unstable foundation for an effective industry-wide association. Unlike the citrus industry, in which local cooperatives were already beginning to form the nucleus of an industry-wide alliance between growers and packers, in the raisin industry the local organizations and the packers never overcame their mutual distrust. The short-lived associations of the 1890s could not reconcile the economic and ethnic divisions among the industry's three thousand producers, some twenty commercial packing firms, and Fresno's banking and mercantile establishment.[5]

The problem, essentially, was that most growers regarded packers, bankers, and merchants as interlopers in an industry that rightfully belonged to producers. The vast majority of raisin growers were full-time farmers holding small plots. At the top of the industry was a cadre of wealthy men — local bankers, merchants, and lawyers — who saw agriculture as a side venture or an investment. These men were socially and financially allied with the heads of private packinghouses in and around Fresno. Naturally, they resented any inference that their stake in the industry was somehow illegitimate. As the pools collapsed, year after year, these individuals and the packers pointed out that no pool could succeed without them, because "*they* were as legitimately related to the raisin business as are the growers themselves." These men believed that cooperation had to embrace the entire industry — not only growers but also packers, merchants, and bankers. Indeed, bankers in Fresno, Los Angeles, and San Francisco supported the idea of an industry-wide cooperative because stable prices meant current mortgage payments, and the banks could make more money lending to producers in prosperous times. But the wealthy gentleman planters who might have bridged the social and economic gap between growers and packers were too closely associated with the region's financial establishment to earn the trust of smallholders.[6]

The industry's failure to agree on a formula of cooperation led its leaders to reconsider their approach to cooperation. They wanted to retain the Rochdale idea of excluding nonproducers from membership. Yet they knew any cooperative had to contend with the commercial packers and brokers who con-

Formal portrait of M. Theo Kearney, 1903.
(Kearney Collection, Fresno Historical Society Archives)

trolled processing and marketing. To dislodge the packers' control over selling and distribution, a cooperative required sufficient capital to fix prices and finance advance payments to growers. For the organization to adhere to the Rochdale form, capital stock had to be distributed widely to avoid the concentration of shares in a few hands, and no shareholder could have more than one vote, regardless of number of shares held. These features would ideally prevent speculators and nonproducer investors from taking control of the growers' organization. To do all of these things, the growers required, in the words of a Fresno reporter in 1917, "a Moses to lead them out of the wilderness."[7]

The raisin growers' unlikely Moses was M. Theo Kearney, a shrewd entrepreneur to whom the economic potential of cooperation meant far more than its tradition. Born in Ireland in 1841, he arrived in California in 1870 and promptly made his way into San Francisco society. He made a fortune investing in the Fresno County colony settlements. During the 1880s he sold off hundreds of acres in small tracts, only to foreclose on many of these farms after the depression of 1893. The reversions left him with more than a thousand acres of raisin vineyards, "which at prevailing prices promised to ruin him." His accumulation of so much land at the expense of the less fortunate earned him a reputation for being unsympathetic to the plight of ordinary growers. He apparently had few friends and made enemies easily. A social elitist, he did not hide his disdain for what he believed was the growers' unsophisticated approach to cooperation. He was not significantly involved in the cooperative movement before 1898 because he considered the attempts to organize unsound. Only when the foreclosures put hundreds of tons of raisins in his hands did he decide to assume a leadership role in the industry.[8]

## THE CALIFORNIA RAISIN GROWERS' ASSOCIATION

When Kearney began to assert a more visible role in the industry, he saw himself as the "J. P. Morgan of the Valley" rather than a new Moses. In May 1898, Kearney told several hundred leading growers that they should adopt the form and practices of an industrial trust: "The only practical remedy," he said, "is to pool our crops, control the quality and quantity when offered for sale, and [fix] the price at which the output shall be sold." With two thousand growers assenting, Kearney organized the California Raisin Growers' Association (CRGA) under California's regular incorporation law with half a million dollars in capital stock. Shareholding was limited to growers, who purchased shares for $5 in proportion to their share of the crop. The CRGA represented its members in dealings with commercial packers. Packers who wanted to purchase CRGA raisins signed contracts that bound them to buy and sell raisins at CRGA prices.[9]

All these features were innovative. The radical feature of Kearney's cooperative was that its operation was contingent on the contractual participation of 75 percent of the growers. This was the first time in California or in the nation that anyone expressly linked cooperative organization to market control. Kearney believed that cooperative monopoly served the public as well as the growers, and he wrapped his monopolistic cooperative in the banner of American political ideals: " 'I maintain,' " he asserted, " 'that the public good takes precedence over all the private interests, and the liberty loving people of our country have pretty thoroughly established the principle that the will of the majority shall be the law of the land.' " The rhetoric proved effective. By the 1898 harvest 90 percent of the raisin acreage was committed to the pool through contracts with growers. Prices doubled from the previous year, from one and a half to three cents per pound. No doubt it helped that the 1898 crop was the smallest in six years.[10] Kearney created an organization with enough capital to keep the packers at bay but which maintained the growers' special relationship to the cooperative. By stressing participation, he also united a fractious group of growers. The CRGA's successful first season was, according to the *Pacific Rural Press*, "one of the most significant things in the recent history of our agriculture" (Table 4.1).[11]

The CRGA got off the ground not only because of the growers' support but also because Fresno banking interests did not undermine Kearney's efforts. Indeed, bolstered by the fact that the CRGA's leader was a person of social prominence and economic stature, the bankers believed that cooperation would strengthen the market and enable growers to pay their mortgages. As president, Kearney appointed some friendly bankers to the CRGA's board of directors. Three of the seven trustees—T. C. White, John McMullin, and Louis Einstein—were growers as well as bankers. As for the unfriendly bankers, in the words of a Fresno historian, "Kearney's main policy was . . . to force better loan conditions for the farmers by threatening to establish his own banking system." Kearney himself put the matter bluntly: "We loosened [the packers'] grip on the needy growers by arranging with the banks to let the growers have money on orders [from the CRGA]."[12]

Initially, the packers were not terribly disturbed by Kearney's activities. They were willing to relax their control over selling and distribution as long as their own business investments were protected: "With [this] essential granted, they were willing that an organization of growers should control pretty much everything else, including the fixing of prices, and were ready to enter into such contracts with the growers' organization as should assure to the latter

TABLE 4.1. California Raisin Production, in Tons, 1899–1913

| Year | Crop | Year | Crop | Year | Crop |
|------|------|------|------|------|------|
| 1899 | 35,500 | 1904 | 37,500 | 1909 | 70,000 |
| 1900 | 45,000 | 1905 | 43,500 | 1910 | 57,500 |
| 1901 | 37,000 | 1906 | 47,500 | 1911 | 60,000 |
| 1902 | 53,000 | 1907 | 70,000 | 1912 | 92,500 |
| 1903 | 60,000 | 1908 | 65,000 | 1913 | 65,000 |

Source: *Annual Report of the California State Board of Agriculture* (Sacramento: State Printing Office, 1913), 134.

complete control of the business." The tacit agreement was that the packers would forbear from obstructing the cooperative if the growers stayed out of the packing business.[13]

Kearney's ambitions soon upset this uneasy compromise. He did not trust the packers to act in the growers' interests. "In my judgment," he told the state fruit growers' convention in December 1898, "to make our organization a complete success, it is absolutely necessary to eliminate all conflicting interests. . . . We cannot blend the interests of the commercial packer with the interests of the grower any more than we can blend oil and water." Kearney envisioned a growers' organization that would control not only selling and distribution but also packing and brand-name advertising—in short, complete vertical integration. At the 1899 California Fruit Growers' Convention, Kearney urged the growers to embrace the corporation without reservation: "This is an era of trusts and combinations. . . . Our sole aim from now on should be to apply this principle of combination in all our affairs so as to secure the greatest economy in production and in placing our products on the markets of the world."[14]

Early that year, Kearney proposed that the association build its own packing plants, funded by $2 assessments on each acre under contract to the CRGA. The commercial packers concluded that this proposal reneged on the CRGA's agreement not to interfere with the packers' business. After the CRGA's initial success, Kearney raised his sights even higher. He funneled the growers' frustration with the national market into an attack on the local market—the processors, packers, and bankers who controlled relations between the growers and the trade. The *San Francisco Chronicle*, half admiringly, half warningly, called the CRGA "a growers' Trust."[15]

Vertical integration of this scope was possible because the CRGA had only nebulous connections to traditional cooperation. The fact that the CRGA was organized under the regular incorporation laws, not the state's 1895 non-stock law, became a double-edged sword. The CRGA could raise the capital it needed as a regular corporation, but its exercise of corporate powers and privileges could be regulated under the antitrust laws. This possibility worried many growers. At a CRGA meeting in early 1899, members rejected Kearney's initiative and voted to change the CRGA from an outright pool to a marketing agent.[16] Growers' fears of antitrust prosecution played into the hands of the directors. They obtained the growers' consent to reorganize the CRGA under the 1895 nonstock law, bringing it more clearly under the banner of traditional cooperation and limiting its ability to raise capital. This move gave the bankers greater leverage against Kearney in directing the cooperative's affairs.[17] It also signaled that a fight was on—not between the growers and the national trade but between the cooperative's president, the growers, and the local financial establishment.

Kearney took the reorganization as a personal rebuff, but falling prices enabled him to resume the offensive. In January 1900, he announced a plan to expand the CRGA's commercial operations. The plan featured a new growers' contract known as the "yellow slip." By its terms the growers "agreed to convey to the directors of the association as trustees half a cent a pound on the next season's crop for investment in packing houses."[18] The CRGA was going into the packing business. By April 1, eleven hundred of the CRGA's twenty-five hundred members had signed yellow slips. The CRGA directors then incorporated the California Raisin and Fruit Packing Company and sold half its $1 million capital stock to the commercial packers. The CRGA purchased the other half with loans using the yellow slip proceeds as security. These developments were not universally welcomed. The new company was not cooperatively organized; further, the directors were taking full salaries from both the CRGA and the new packing company. At a growers' meeting Kearney defended the decision to incorporate the packing company under the joint stock law by attacking the 1895 nonstock law as "a very unsatisfactory business law." Kearney also told the growers that the double salaries were necessary to keep qualified officers in the association's employ.[19]

Meanwhile the packers were cultivating opposition to the yellow slip contracts. Fiercely opposed to any attempt by the growers' organization to build its own packing plants, they "had no intention of going out of business, at least [not] without a struggle, and they had many supporters among the growers and businessmen, many of whom believed it inadvisable to attempt to eliminate the packers." The packers saw to it that fewer than 60 percent of the growers signed the new contracts by May 1900. To stabilize the coopera-

Mass meeting of the California Raisin Growers Association, Fresno, California, 1900. Kearney is seated at the table at center. (Kearney Collection, Fresno Historical Society Archives)

tive's operations, the banks mediated a new contract between the CRGA and the packers to cover the 1900, 1901, and 1902 seasons. In a circular letter to the growers, Kearney angrily denounced his opponents, whom he identified as bankers and speculators "animated by an intense desire to tear down the man who has stood as the rock of Gibraltar, guarding and protecting the interests of the growers during their progress from poverty to prosperity." When the growers refused his request for an increase in salary, he resigned as president.[20]

In Kearney's absence, the packers and bankers proceeded on the basis of their mediated agreement. The contract required the CRGA to control 90 percent of the raisin acreage before the packers would be obligated to pack exclusively for the cooperative.[21] Meeting this condition proved difficult: "Many of the growers held out to the last, and only signed after all the influence possible was brought to bear upon them. . . . Many of these held out on account of their sympathies for Mr. Kearney, others because they believed they had certain advantages in being outside of the association." Only 75 percent of the producers signed with the CRGA, but the packers went ahead with the agreement. Their generosity did the CRGA little good. Its price of six and a half cents collapsed because growers made heavy sales outside the association, imports

Kearney riding in chauffeur-driven Mercedes-Benz limousine, Paris, 1904. His annual vacations in Europe did not endear him to growers. (Kearney Collection, Fresno Historical Society Archives)

increased, and there was a fourteen-thousand-ton carryover from 1899. At a mass meeting in 1901, Kearney returned from a self-imposed exile in Germany and proposed that the weaknesses of the CRGA be remedied by increasing both its borrowing power and the president's salary. The subsequent election demonstrated Kearney's enduring popularity. The growers again elected him president but rejected his proposals. The *Fresno Morning Republican* smugly reported, "The outcome of yesterday's raisin election is a personal victory for Mr. Kearney and a defeat for his policies." [22]

## KEARNEY'S COOPERATIVE TRUST

Developments in the courts, however, gave Kearney another chance to seize the initiative. In a 1901 case involving the California Cured Fruit Association, a prune cooperative, the California Supreme Court limited the damages the association could recover when a grower refused to deliver. Contemporary reports indicate that California trial courts circumscribed the ability of marketing cooperatives to combine growers by refusing fully to enforce crop contracts. The prune case alarmed the CRGA directors; they used a similar contract. But as Kearney informed eastern newspapers, the CRGA had not lost any lawsuits "because we understood the contract better than the prune people and have, therefore, conducted our legal affairs in a more successful manner." The decision emboldened Kearney to try to remedy the legal weaknesses of the crop contract and to restructure the CRGA's corporate identity.[23]

To evade the limitations of the California nonstock cooperative law, Kearney came up with his wildest idea yet: reincorporating the CRGA under the laws of New Jersey. To evade the restraint of trade problem with the crop contracts, Kearney offered a substitute: a lease agreement under which growers would rent their vineyards to the CRGA, which would pay them 95 percent of the net proceeds as wages for services performed in producing the crop. Legally, the CRGA would be the owner of the crop and thus could not be accused of restraining trade because it would no longer constitute a combination of independent producers. Indeed, this approach would have entirely wiped out the restraint of trade problem. Instead of a cartel, Kearney would run a corporation whose workers — the growers — were paid wages that were remarkably like closed prices. J. P. Morgan would have appreciated this piece of legal craftsmanship.[24]

It was in fact J. P. Morgan whom Kearney sought to emulate. Kearney told the growers that Morgan's trust provided the perfect example for the CRGA to follow: "The Steel trust, and many other very wealthy corporations, have chosen to be incorporated in New Jersey, and we have closely followed the forms and papers of the Steel trust on file . . . for the obvious reason that these papers were prepared by some of the most able corporation lawyers in the country." Kearney was aware of the then current trend in incorporation law, that New Jersey's incorporation statutes had the fewest restrictions on corporate structure and the lightest corporate tax burdens of any state in the Union. And he saw no legal barrier to applying New Jersey corporation law to the problem of cooperation in the raisin industry: "The most able business men in American have selected New Jersey to incorporate in, and the capital represented by them is simply enormous. . . . What is good enough for them must surely be good enough for us." [25]

New Jersey incorporation carried still another advantage. The CRGA would have the power to compel growers to abide by their crop contracts without exposing itself to liability under the antitrust laws. As Kearney argued, "New Jersey corporation law gives us all the advantages that we may desire of the cooperative law of this State . . . and it provides us with corporation law which has already been thoroughly established in the courts." Slowly but steadily, Kearney was moving the CRGA from a voluntary association to a mandatory combination. Like leaders of other post-Populist cooperatives, he took a dim view of the growers' willingness to sacrifice for the collective good: " 'The growers . . . as a whole will not keep within [cooperatives] unless they are forced to. They cannot be held together on honor and sentiment alone.' " Thus, for the growers' own good, a cooperative leader had to resort to the law to preserve the organization's existence: "A few—twenty or thirty or fifty or one hundred—manufacturers can get together and fix their shares and sell whatever they have to sell at whatever prices they desire, but it is different with us in putting our farms into one company. . . . [We] should have a law that will bind and hold the growers together, whether they like it or not, after they once sign." Other growers had come to similar conclusions about the impossibility of full voluntary participation: "There is no plan," CRGA director Alexander Gordon told a gathering of fruit growers, "there is no man, there is no power outside of the intervention of the Almighty that can make all of the raisin-growers get together into an association." [26]

Kearney's daring ideas seemed to many growers to be the most viable plan for saving the cooperative. Others, however, did not share Kearney's enthusiasm for the steel trust as a model for an agricultural cooperative. The departure from the Rochdale ideal of participatory industrial capitalism was both clear and extreme. CRGA directors supported the New Jersey lease primarily to preserve the CRGA, not out of loyalty to Kearney or admiration for his ideas. At a mass meeting in August 1901, grower D. D. Allison professed that he "had no use for Kearney" but was willing to support the CRGA rather than "go back to the disasters of years gone by." Alexander Gordon, president of the Sacramento Bank, grew raisins as a side venture. He had set aside his personal dislike for Kearney and helped organize the CRGA. Yet Kearney's turbulent leadership had not shaken his faith in the body they created: "The CRGA has blazed the pathway of success for the raisin grower, and without its aid there is no hope." And Robert Boot promised that the new contract would not interfere with the growers' "personal conduct" of their vineyards. [27]

Outside observers were less sanguine. Edward J. Wickson, University of California pomologist and editor of the *Pacific Rural Press*, wryly noted, "It does not look very wholesome for co-operation when the force of organization is largely occupied in cracking the skulls of recalcitrant members." The packers and banks staunchly opposed any reorganization that gave the CRGA independent packing facilities and financing. Another Kearney salary demand only fueled grower discontent. The rank and file feared that Kearney's plan would take control over their crops, their land, and their organization and give it to the directors under a scheme that hardly seemed cooperative. [28]

Instead of trying to allay the growers' fears, Kearney gave them an ultimatum. He set 75 percent as the minimum acceptable level of grower commitment to the New Jersey lease. The growers failed to meet this standard in time for the 1901 harvest; a month later, a trial court invalidated the growers' old membership contract. On December 10, the directors voted to abandon the lease. A month later, a growers' committee recommended that the CRGA continue operations under its California charter, ending all consideration of the New Jersey plan. In a final ignominy, Kearney himself was voted out of office in March 1902. Accused by Chester Rowell of betraying the association, he bitterly responded in a rival newspaper: "Another instance of Satan rebuking sin." [29]

Ethnic and racial tensions surfaced as the CRGA went down. To complete the monopoly, Kearney welcomed all growers, whatever their racial identity, into the cooperative. His popularity and forceful leadership brought Asians, Armenians, and other minorities into the CRGA fold, at least initially. At the 1901 mass meeting Kearney "told the audience he had the Seropian Brothers to thank for bringing its large acreage and making a success of [the CRGA]." Armenian newspapers urged Armenians to sign cooperative contracts and proudly announced that two hundred Armenian growers — "surely a majority of the valley's Armenian ranchers" — had signed up with Kearney's cooperative. Discontent within the Armenian community surfaced the next year when, citing discrimination, many Armenian growers resigned their memberships.[30] Their departure gave Kearney and future industry leaders fodder for the claim that foreigners did not understand cooperative philosophy and could not be trusted to be loyal. Amplifying the racial tension, Kearney blamed "small growers, renters, and Chinamen" for the defeat of the New Jersey lease. But his refusal to give these marginal members of the industry any security for the risks he asked them to take was responsible for hardening their position.[31]

Kearney's opponents claimed that his defeat meant that the sound notion of cooperation had prevailed over the dangerous method of corporate trust. Robert Boot, the new CRGA president, proclaimed, " 'We have not a trust, but we are an association of men engaged in producing from the soil. We have a right to organize, without seeking to enhance the product at the expense of someone else, but to get what [we] are entitled to.' " Bankers on the CRGA board threw their weight behind Boot's downscaled association. Alexander Gordon, the bank president, tied the question of rural credit directly to the CRGA's fate: " 'If the association should fail it would make a difference of 50% of the amount of money the Sacramento Bank will loan on Fresno vineyard property.' " Despite the backing of these directors, the CRGA found the going rough. The 1902 crop yielded over thirty million pounds more than the previous year, leaving the association with five thousand tons of unsold raisins and substantial losses from paying growers more than the raisins brought. The CRGA sued over six hundred growers who refused to repay the cooperative, and in 1911 the California Supreme Court upheld trial court judgments in the CRGA's favor. Nevertheless, observers who favored the older model of voluntary, noncapitalized cooperation proclaimed victory. Edward J. Wickson lauded the CRGA's handling of the 1902 crop as "the most brilliant and successful marketing operation yet accomplished by growers in California."[32]

This was hyperbole. The older model did not work in the raisin industry. The CRGA marketed its last crop in 1903. Many growers, including many Armenians, distrusted the new management and refused to renew their contracts. Some delivered their raisins to outside packers under cover of darkness. Unable to sell all the raisins produced by its remaining members, the CRGA announced in 1904 that it would suspend business for that season. A packers' combine took over the unsold stock and financial obligations of the growers' association. For two more years, Kearney tried to establish corporations like the one he had proposed in 1901 but to no avail.[33]

Lonely and embittered, Kearney died at sea on May 27, 1906, "remov[ing] a man of unique prominence and force from the agricultural circles of California" and leaving behind a legacy of contradictory ideas. On the one hand, a "powerful packing-marketing corporation" carried undeniable appeal; on the other, the Rochdale ideal was deeply rooted in American cooperation. Kearney was ultimately unable to persuade growers that his organizations did not betray the ideal. The reason for this failure is more complex than contemporary and scholarly accounts have recognized.[34]

THE KEARNEY LEGACY

At the time of his death, popular opinion was thoroughly settled on the reasons for Kearney's ouster from the industry. "All believed him incompetent, many believed him dishonest and in league with the enemies of the producers for the deliberate purpose of coining wealth for himself through the betrayal of his neighbors," wrote a local scribe. According to this account, Kearney lost control because he was extravagant, he deserted the industry each spring to take lavish vacations in Europe, and he did not bother to conceal his disdain for the growers.[35]

Certainly Kearney was undone by his own limitations as a leader. By all accounts, he lacked charisma. His personal abrasiveness kept friends at a distance; his imperious confidence in his own ideas alienated the bankers and packers on whose support cooperative organizations relied. His salary demands and his desire to exercise autocratic control over the industry's affairs undermined his standing with the growers and hastened his exit from the scene. A leading packer, recalling Kearney's contributions in 1919, argued that the industry was doomed as long as its fortunes rested on the man's character: "[He was] a very arbitrary man who had a peculiar faculty of antagonizing nearly everybody with whom he came in contact, businessmen, the community, the newspapers and everybody else and the result was, it was only a year or so before he had the whole raisin business in a turmoil."[36]

The standard line of what little scholarly reflection there is on the issue takes a different tack. Kearney's biographers, Schyler Rehart and William K. Patterson, argue that the growers were simply not ready to embrace Kearney's advanced ideas. Growers "were particularly opposed to an organization that would demand funds from them to create a grower-operated packing concern. . . . They had little interest in any ambitious and expensive enterprise." This view was also prevalent in James Bragg's early economic study of the raisin industry: "[The raisin] community was not ripe yet for understanding . . . advanced ideas." Although more sympathetic to Kearney, this explanation also faults Kearney's overbearing style for diverting growers from seriously considering his innovations.[37]

Both these interpretations miss the point. The Kearney cooperatives did not fail because growers distrusted Kearney; his enduring popularity makes that thesis improbable. Neither did they fail because growers were incapable of grasping the intricacies of Kearney's program; the debates at CRGA mass meetings and the California State Horticultural Society evince a widespread understanding of cooperation and antitrust law among raisin producers. Rather, the CRGA ran aground because the complex array of interests embedded in the industry could not unite behind a cooperative that appeared not to work in their interests. Growers, packers, bankers, merchants, and other affiliated entrepreneurs all generally supported the idea of an industry-wide monopoly. But each group had its own idea of such a cooperative, how it should be structured, how it should work, and who would hold its purse. To many growers, Kearney's vision of a raisin "trust" merely substituted one unaccountable packing company for another. His models—the steel trust and out-of-state corporation law—were unfamiliar and legally suspect. And he made no attempt to allay fears about attendant antitrust problems, even when trial courts ruled against cooperatives on the legality of crop contracts. Kearney's proposals for assessments on member vineyards and million-dollar corporations required the growers to put their own earnings in danger. As much as the growers liked Kearney and wanted him as their leader, in the end they concluded that his organizations imposed greater financial and legal risks than they were willing to assume.

Ironically, Kearney's deepest desire—to ensure that only growers profited from the growing and selling of raisins—reflected an essential assumption of traditional cooperation, but the means he used to achieve this end had little in common with prevailing understandings of cooperative practices. Indeed, he was only too successful in emulating J. P. Morgan. While Kearney believed that farmers were justified in their use of trusts and monopoly, neither the legal system nor society generally saw much difference between a monopolistic agricultural cooperative and an industrial trust. And he never could persuade the growers, whose support was essential to a cooperative trust.

Lost in the wreckage of the CRGA's messy demise was the larger significance of Kearney's accomplishment. By putting corporation law, monopoly, and trust into the hands of agricultural producers, Kearney demonstrated that a new conception of cooperation could facilitate collective marketing, enable growers to compete with packers and processors, and furnish the organizational foundation for the state's agricultural economy. While the raisin industry was unwilling to follow Kearney during his lifetime, he started the evolution of cooperation into a new legal form and laid the foundation for the transformation of the legal structure of agricultural enterprise. When others exported this model of cooperation to other agricultural industries in California and the rest of the country in the decades to follow, nationwide legal battles ensued over its legitimacy. Nowhere would this model be put to more spectacular use and tested more thoroughly in the courts, however, than in the raisin industry where it originated.

# 5 ∾ Cooperatives and Federal Law

*If ever we are to solve the rural problem it must be done by a recognition of the fact*

*that the state lines, while offering many constitutional obstacles, have no economic*

*significance whatever. — John H. Gray, 1912*

The ongoing legal experimentation with cooperation in California and else-where enabled growers to relate to the market as a more or less concerted group rather than as scattered individuals. This development unsettled na-tional perceptions of farmers and agricultural enterprise by revealing the ways in which farmers could deploy the corporation, the trust, and monopolistic trade practices to serve their own economic goals. Judicial reaction to the new form of cooperative fit logically within the larger constitutional and regulatory context of the late nineteenth century. When courts cut back on the powers of these new organizations, farmers turned to the legislatures for separate laws to govern cooperative incorporation. In litigation contesting the mean-ing of these statutes, courts hammered out the legal status of cooperatives as a species of corporation.

This body of law raised significant questions for state economic regulation between 1890 and 1920. The state cooperative incorporation laws, on the one hand, acknowledged that farmers required organizations with the power to conduct large-scale business transactions and, on the other, sought to limit how those powers would be exercised. Permitting farmers to fix prices by im-munizing them from state antitrust laws seemed like a good way to facilitate agricultural efforts at self-help. Courts thwarted this policy by declining to de-clare that farmers constituted a special class entitled to legal privileges denied to other entrepreneurs. The state cooperative incorporation laws neglected to reconcile farmers' access to corporate powers and business practices with the assumptions of antitrust policy.

Frustrated at the turn of events at the state level, farm groups and cooperative promoters began advocating for a national policy on farmers' use of the corporate form. The federal government remained uninvolved in the movement to change the legal status of cooperatives until well into the Progressive Era. When Congress responded to pressure from labor and farmers for relief from antitrust prosecution, federal legislators had to decide how to frame the legal form of such relief. In the Clayton Act of 1914, Congress provided cooperatives with a limited exemption from the Sherman Act of 1890 in language that reflected its preference for a more benign model of cooperation: the nonstock association, which most people saw as more faithful to the Rochdale tradition than capital stock organizations.

Before the passage of the Clayton Act, Congress opened a back door to exempting all kinds of cooperatives with the first of a series of annual riders suspending the Justice Department's authority to prosecute farmers under the antitrust laws. This rider did not specify any particular legal form to which cooperatives had to adhere in order to evade prosecution; in fact, it said nothing at all about requiring farmers to engage in cooperation to earn immunity. Congress's equivocation on the definition of cooperatives in federal antitrust law sent a message to federal agencies charged with promoting cooperation among farmers.

Mirroring both the legal uncertainty in federal law and the larger confusion over the changing nature of cooperation, the U.S. Department of Agriculture took a passive role in regulating agricultural marketing during the Progressive Era. The USDA's Office of Markets, though charged with encouraging the formation of cooperatives, did little to assist farmers in the actual process of organizing and incorporating, in part because of restraints grounded in dual federalism and in part because USDA agricultural economists interpreted their mandate in accord with their training in classical economics. Both Congress and the executive branch were aware of the legal and economic innovations taking place at the state level, but federal policy on cooperation through 1920 relied on an older, Rochdalian notion of agricultural associations, in order to ward off the anticompetitive tendencies of the newer model.

FEDERAL ANTITRUST LAW AND COOPERATIVES

Federal lawmakers responded to the rise of the trusts according to well-recognized principles of dual federalism and common law standards of reasonableness. The Sherman Act of 1890 did little more than codify Anglo-American antipathy toward monopoly and combinations in restraint of trade.[1] Congress was also careful not to intrude on the well-recognized authority of the states to regulate state-chartered corporations and economic enterprise under the commerce clause. The statute applied only to combinations operating in interstate commerce and did not permit firms to incorporate under federal law.[2]

The common law's reliance on subjective judgments of reasonability and the Sherman Act's deference to contemporary dual federalism scruples together gave corporations "de facto legitimacy," in the words of historian Charles McCurdy. The Court made the awkward distinction between production and commerce the linchpin of its commerce clause analysis in *U.S. v. E. C. Knight Co.* (1895). In that case it held that federal authority to regulate combinations did not extend to operations involving production and manufacturing, even when a production monopoly controlled over 95 percent of the domestic supply of a staple commodity such as sugar.[3] Lawyers and reformers immediately attacked the sharp distinction between production and commerce as unworkable and impractical; the Court itself seemed to agree, holding in *U.S. v. Trans-Missouri Freight Association* (1897) that the Sherman Act applied to tight combinations of railroads and industrial manufacturers. After *Trans-Missouri*, the Court gradually chipped away at the *Knight* rule, using such convoluted constructions as the "stream of commerce" idea to broaden the scope of federal regulatory jurisdiction without explicitly saying so.[4] In 1911, the "rule of reason" decisions in *Standard Oil v. U.S.* and *American Tobacco v. U.S.* seemed to abandon the production/commerce distinction entirely in favor of evaluating combinations according to a standard of reasonableness.[5] The Court's central concern was not jurisdiction but the effect of the consolidation. If a trust device diminished competition by impeding the ability of new firms to enter the market, then it was banned by the Sherman Act.[6]

But few people were satisfied with this state of affairs. The rigor of antitrust policy between 1890 and 1911 depended on both a clearly defined law of free competition, which the Sherman Act as interpreted by the Court did not provide, and scrupulous execution of enforcement responsibilities, which the federal executive branch and the states failed to perform.[7] Although the great merger wave ebbed after 1904, the popular belief remained that corporations flouted antitrust law with impunity. The rule of reason decisions signaled that the Court had accepted large-scale corporate organizations as a permanent fixture in the industrial order.[8]

Farmers could not understand why the large corporations that oppressed them seemed to evade reform while their own smaller, less powerful organizations tripped over the antitrust laws. One possible explanation is that the urban public and the legal order were not ready for a brand of cooperation that fought fire with fire, one that applied combination to the problem of suppressed competition and speculation in food marketing. The model of cooperation emerging in California drew farmers into the central legal issues of industrialization: how much combination constituted a monopoly, when did control over competition became unreasonable restraint of trade, and how much influence over price constituted price fixing. More specifically, the new model pointedly raised the problem of agriculture's relationship to industrial society. How would urban consumers protect their interest in a cheap, abundant food supply if large farmers' combinations controlled supply and price? At the same time, agriculture's growing political and cultural influence reinforced the notion that farmers needed special help to survive economically. Agricultural economists, government officials, and many judges wanted to give farmers economic assistance but without endorsing the predatory corporate style of the California associations. As these corporate cooperatives grew in size and influence, more conservative voices called for a return to Rochdale cooperation — which they oversimplified as the nonstock cooperative — as the legitimate basis for national policy. This was the model Congress ultimately accepted as pressure mounted to amend the federal antitrust laws.

## THE FEDERAL ANTITRUST EXEMPTION FOR AGRICULTURE

Federal judicial decisions in two Sherman Act cases forced Congress to reconsider its original decision not to specify whether the Sherman Act applied to farmers and laborers. The first case, *Steers v. U.S.* (1911), involved night riders in the Kentucky tobacco industry. A mob took possession of a shipment of unpooled tobacco as it lay on the Dry Ridge, Kentucky, train station platform bound for Cincinnati. A federal grand jury returned the first criminal indictments ever issued under the Sherman Act against twelve members of the Burley Tobacco Society. Eight were convicted of conspiracy and fined. On appeal, the Sixth Circuit upheld the convictions, noting that the night riders had "wholly stopped" interstate commerce and thus federal jurisdiction was properly invoked. The decision's commerce clause analysis was consistent with precedent; however, the political backlash from the federal government's prosecution of farmers sparked such public indignation that President Wilson later commuted all fines to payment of the costs of the suit.[9]

The outcry over *Steers* was but a whimper compared to the fury aroused by the infamous Danbury Hatters' case, *Loewe v. Lawlor* (1908). Speaking unanimously through Chief Justice Melville Fuller, the Supreme Court ruled that since Congress had not specifically exempted laborers and farmers from the Sherman Act, they were included within its prohibitions: "[Congress] made the interdiction include combinations of labor, as well as of capital, in fact, all combinations in restraint of commerce, without reference to the character of the persons who entered into them." The vulnerability of unions to Sherman Act injunctions spurred organized labor's efforts to amend the law.[10]

The threat to farmers and cooperatives may have been more potential than actual, and in any case it was decidedly less acute than what labor was facing. Aside from *Connolly* and *Steers*, no federal court had yet been called upon to decide antitrust controversies involving farm pools or cooperative organizations. And these two cases danced around the issues central to cooperatives. *Steers* dealt with the extralegal actions of some members of a farmers' organization; *Connolly* involved farmers' cooperatives only indirectly, through its ruling on the constitutionality of the state antitrust exemption for farmers. Nevertheless, the reach of the federal commerce power in the labor cases, particularly the *Danbury Hatters'* case, alarmed agricultural leaders and politicians friendly to farmers. To generate support for amending federal law, they argued that if consumers got angry about high food prices, the federal government might prosecute farmers' associations instead of industrial trusts. It was not clear whether farmers' cooperatives would be considered reasonable combinations under the rule of reason. Although the rule of reason decisions resurrected the common law principle that restriction of competition as such did not necessarily constitute an illegal restraint of trade, the state courts continued to strike down such tools as the maintenance clause on the grounds that they restrained trade.[11]

Early in the Wilson administration, labor and agricultural leaders began agitating for reform of the antitrust laws. The labor movement played a far more visible role in lobbying for reform than the cooperative movement. Labor leaders such as Samuel Gompers, president of the American Federation of Labor, were well known in the halls of Congress, while the agricultural sector's primary lobbyist was the Farmers' Educational and Cooperative Union of America, at best only a regionally oriented association with little national political clout. The Farmers' Union "maintained cordial relations with organized labor," according to Grant McConnell, but politically farmers and labor were already going their separate ways.[12]

Whatever model of cooperation was influencing Congress at this time, it was certainly not that of M. Theo Kearney or the monopolistic California organizations. Section 6 of the Clayton Act defined a federal policy on cooperation for the first time: "Nothing contained in the antitrust laws shall be construed to forbid the existence and operation of labor, agricultural, or horticultural organizations, *instituted for the purposes of mutual help, and not having capital stock or conducted for profit,* or to forbid or restrain individual members of such organizations from lawfully carrying out the legitimate objects thereof; nor shall such organizations, or the members thereof, be held or construed to be illegal combinations or conspiracies in restraint of trade, under the antitrust laws."[13] The section reflected two assumptions about Congress's understanding of cooperation. First, agricultural combinations differed from regular industrial combinations in legally significant ways, particularly their not-for-profit status and their philosophy of mutual benefit. Second, these differences were germane to antitrust policy and freed agricultural organizations from per se liability.[14] This was precisely the argument offered by counsel for the Farmers' Union during congressional debate on the bill: "The exclusive object of . . . capitalistic combinations is invariably predatory profit and not the mutual help, education, cooperative and fraternal production designed to increase principally the all-around individual and collective efficiency of its many members that characterize farmers' cooperative and educational unions and organizations."[15]

In limiting immunity to nonstock associations that met the ideals of mutual help, education, and fraternal membership, Congress revealed its preference for the older model of cooperation. Politically, federal lawmakers wanted to do agriculture a favor, but not at the expense of alienating consumers and urban distribution interests. Legally, Congress was not prepared to embrace the organizations arising under the new state capital stock laws. Few state legislatures had settled the question of whether nonstock associations were inherently cooperative and capital stock organizations inherently corporate. In view of the weight of prevailing judicial opinion that cooperative purpose was incompatible with corporate powers, Congress played it safe, narrowed the scope of the immunity, and endorsed the perception that only nonstock associations conformed to true Rochdale cooperation.[16]

This view was by no means unanimous within Congress. A small minority of the House Judiciary Committee spelled out the problems that Section 6 raised: "Every organization of farmers which aims to cooperatively bargain as to the products of its members is 'conducted for profit,' and many of them have 'capital stock.' . . . As soon as farmers combine to get better prices for their products, or to sell directly to consumers, this paragraph affords them no relief from the antitrust laws."[17] A Columbia University economist echoed this criticism: "[The nonstock] organization is not easily applied to all types of farmers' cooperative associations. Many states do not have laws which provide for this form of organization. Furthermore . . . the universal adoption of the provision of Section 6 would necessitate a reorganization of a large majority of the existing farmers' cooperative marketing companies—an almost impossible task."[18] From the perspective of farmers seeking to compete with middle merchants and processors, the Clayton Act did not provide much protection. The law said that the only acceptable form of agricultural combination was an organization without capital and without the power to enhance retail food prices.

The Clayton Act, however, was not the only legislation Congress enacted on the legal form of cooperatives. The other statute that Congress wrote got almost no attention at the time. Yet an appropriations rider that Congress attached to the Justice Department's annual budget had as much impact on substantive policy as the Clayton Act because it directly restrained the enforcement authority of the Justice Department. Under Wilson, federal antitrust policy began to track the Supreme Court's turn toward acceptance of reasonable combinations and restraints on competition. Although the trust question as such was no longer a burning issue in party politics after the 1912 election, federal legislators were under pressure to give farmers and labor some measure of protection from antitrust prosecution.[19]

In 1913, House Democrats from the Midwest and South attached a rider to the sundry civil appropriations bill containing the $250,000 earmarked for antitrust enforcement. The rider stipulated: "No part of this appropriation shall be expended for the prosecution of producers of farm products and associations of farmers who cooperate and organize in an effort to and for the purpose to obtain and maintain a fair and reasonable price for their products." Similar language extended the same privilege to "any organization or individual for entering into any combination or agreement having in view the increasing of wages, shortening of hours or bettering the conditions of labor, or for any act done in furtherance thereof, not in itself unlawful."[20] Although the rider did not specify the legal structure that farmers had to use to "cooperate and organize" within the meaning of the statute, it did declare that the Department of Justice could not enforce the antitrust laws against farmers unless they obtained prices in excess of "fair and reasonable."

This bill sailed through the House, aided by a rule that prevented amendments unless the entire bill was recommitted to committee. Opponents in the Senate, however, put the friends of labor and agriculture to the test. Businesses flooded the chamber with letters of protest. Some senators contended that the proviso did not endow labor unions and farmers with privileges they did not already have; if they violated the Sherman Act by using, for example, a secondary boycott, the government was free to proceed against them. Others questioned the practice of—and the political motives behind—enacting substantive policy in an appropriations law.[21] In both houses, the debate focused almost exclusively on labor. Even the rider's supporters did not address cooperatives except to aver that the logic used to proceed against labor unions could conceivably be used against farmers as well. After six weary weeks of debate the Senate passed the bill in June 1913, and President Wilson immediately signed it. An identical proviso was included in annual appropriations bills almost every year thereafter through 1927.[22]

The Clayton Act and the rider set out the political sentiment that farmers deserved special treatment under the antitrust laws. Neither piece of legislation, however, substantively advanced the legal status of agricultural cooperatives, according to a USDA official: "The [Clayton] amendment has tended to make the status of farmers' organizations more indefinite and more uncertain rather than clarifying their legal position."[23] The states developed a legal form for cooperation, but courts struck the state antitrust exemptions; federal law provided an exemption but relied on existing state law to define cooperatives' legal form. In setting out the vague goal that farmers should be able to obtain fair prices, Congress expressed the widespread popular belief that agriculture needed the state's help to remain economically viable; but in failing to reconcile the vague goal with the specific issues bound up in cooperation's legal status, Congress reflected the nation's enduring if unrealistic preference for a decentralized, self-sufficient agricultural sector.

Crucial differences between the Clayton Act and the rider illustrate this problem. The Clayton Act restricted agriculture's antitrust immunity to non-stock cooperatives, but the rider contained no such limitation. The rider granted cooperatives practical immunity for all market activities used to "maintain fair and reasonable price[s]." As long as cooperatives were perceived as pursuing that goal, a lawyer for the Wholesale Grocers' Association noted, the rider supplied them with greater protection from the Justice Department and the federal courts than the Clayton Act: "Here is an indirect but very substantial exemption of farmer organizations."[24]

The rider's opponents, however, had less to fear than they anticipated. If the purpose of the rider was to protect farmers from prosecution in most cases, the "fair and reasonable" standard did not do the job. The Grocers' Association lawyer reassured his clients that the Justice Department could easily render the rider a dead letter: "The determination of reasonableness rests solely within the discretion of the attorney-general and the means of enforcement is solely the threat of legal prosecution." And nothing in either the Clayton Act or the rider prevented the Justice Department from prosecuting the individual members rather than the cooperative. When the U.S. attorney secured indictments against members of the Aroostook Potato Shippers' Association in 1915 for staging a boycott against nonmembers who dealt with blacklisted members, the federal courts twice ruled in the government's favor.[25]

Taken together, the Clayton Act and the appropriations rider provided farmers with meaningful safeguards from antitrust prosecution only when neither cooperatives' structure nor their impact on price triggered the residual enforcement powers that federal law preserved. Capital stock denoted the power to manipulate the market, to constrain others' competitive freedom, and to centralize economic activity. To maintain access to capital stock organizations, farmers had to disassociate capital stock from the presumption of illegal market behavior. To do this, farmers turned to the U.S. Department of Agriculture and the extension system for help.

## THE USDA'S MARKETING WORK

The Clayton Act spurred federal administrative promotion of efficient agricultural marketing. Leading officials in the U.S. Department of Agriculture maintained a purposeful detachment from actual organizing, allying themselves with proponents of nonstock cooperatives and individual farm efficiency. Divided between advocates of individual farm management and scholars who stressed farmers' interdependence with the marketing system, agricultural economists on the staffs of the USDA's marketing agencies did not immediately know what to make of the revolutionary legal innovations taking place at the state level. As a result, federal administrative policy grew increasingly distant from farmers' actions in the market.

The cooperative movement's difficulty in mobilizing administrative assistance lay in the USDA's mission to maximize production — a mission that the historian James Malin defined as "making two blades of grass grow where only one had grown before." Up to 1913, the USDA's marketing activities were confined to the study of specific, local markets, defined by commodity. The USDA's philosophy was better marketing through the availability of market information. By market information, the USDA meant the volume of shipments in and out of terminal points and the price paid for each commodity. The USDA used these data to show farmers how to prevent gluts and spot shortages.[26] The USDA's understanding of what made markets operate efficiently reflected contemporary views in and out of government. Its studies were segregated by commodity because its market specialists tended to be experts in single commodities such as cotton, grain crops, or fruit. The USDA's purpose was primarily "the furtherance of agriculture and the products of the soil." Under prevailing constitutional norms, commercial or trade matters belonged in another bailiwick.[27]

Under the Wilson administration the USDA was "unprepared to assume leadership" in the field of agricultural marketing, according to James Malin. Wilson's secretary of agriculture, David F. Houston, was a political economist by training; he followed the British moral philosophers and their belief in minimal regulation rather than the contextualist school in which some of the rising rural economists had trained.[28] Congress was not inclined to widen the USDA's involvement in the public administration of markets except on rare occasions such as World War I and times of extensive crop failures.[29]

Congress did, however, give the secretary authority to create the Office of Markets in 1913. Under Houston, the Office of Markets gathered information in five areas: market surveys, transportation and storage, urban distribution, grades and standards, and cooperative production and marketing. Houston was just not particularly interested in cooperative marketing.[30] The pride and joy of his eight years as secretary was the establishment of a USDA market news service to collect and publish price and shipping information. With branches in ten major cities, it epitomized the Progressive Era application of scientific management and expertise to economic problems. Houston believed it was the government's job to gather and distribute information to all trade participants from producers to distributors to consumers; he left it to farmers to determine how that information could be used to obtain equitable prices. Cooperation among farmers in marketing, he noted, was "essential," but, reflecting federal suspicion about agricultural combinations, he declared that "cooperation can not result in an organization which shall attempt to establish a closed market and to fix prices."[31]

Under Houston, who served until 1920, the Office of Markets did not depart from its express mandate: the study of the marketing problems of American farmers.[32] When it rose to bureau status in 1917, nothing changed: "The regulatory function of the Bureau of Markets consists of enforcing certain rules and regulations which are the result of specific acts."[33] Aside from running the market news service, until 1923 the Bureau of Markets promoted cooperation primarily through the Federal-State Extension Service established by the Smith-Lever Act of 1914.[34] The bureau's technical work on cooperation consisted of a survey of American cooperatives and "a study of the systems of accounting and auditing for such organizations."[35] The Bureau of Markets could do little about farmers' paramount concern: "So far as marketing work is concerned, the activities of the Office, therefor, are limited to the collection and distribution of information." The USDA, after all, had no power to enforce the antitrust laws, to sue individuals or firms in federal court, or to take any legal action against anyone committing market fraud.[36]

In fact, the federal government had its greatest presence at the local level through the county agents, who provided the main point of contact between farmers and experts working in government or the agricultural colleges. A historian of cooperation noted the impact of the extension program on state marketing agencies: "Field agents of the Office of Markets worked in cooperation with state market bureaus, extension divisions of agricultural colleges, commissions of agriculture, and other state agents under cooperative agreements developed by the Office and cooperating agencies." By 1920, much of the bureau's work had been delegated to the states.[37]

The USDA's policy on cooperative marketing under Houston was to affirm it as a solution properly implemented by farmers themselves. This position echoed what cooperation's own proponents said about it. The USDA promoted cooperation as an ideal accommodation of farmers' need for marketing assistance and the political preference for a restrained regulatory state. Before and after the creation of the Bureau of Markets, the USDA's view of what constituted "true" cooperative practice closely tracked the Rochdale principles.

## THE USDA AND "TRUE" COOPERATION

Federal legal change during the Progressive Era stoked the debate over the nature of true cooperation. The federal government's symbolic endorsement of cooperation as a plank of federal farm policy meant that the legal form of cooperatives remained a paramount political and legal issue. In this respect, the USDA faced a regulatory problem typical of the period. When extremely low or excessively high prices signaled a systemic problem, what degree of public intervention could be sustained under contemporary legal and constitutional scruples? The dilemma followed from the belief that prices, even when indicating economic distress, were not to be directly affected by anything that government did.[38] As much as they agreed on the need for farmers to adapt to the new economy, academic and public sector economists believed government should not do much to foster that adaptation. For them, cooperation's emphasis on voluntarism and self-help was the best thing about it, and this view was reinforced by the prevailing ethos of limited federal regulation of production.

USDA officials were captivated by the congruence between cooperation and traditional political and economic values: "Co-operation is much more than real business. Its underlying principles are the substratum of our entire social system. . . . The actual study of Co-operation is the best course which any student may take as a preparation for citizenship in a democratic country." When the USDA conducted a survey of American agricultural cooperatives in 1912, it limited its search to "those that are organized and operate for trading purposes on the Rochdale plan." The USDA asked all respondents to its survey to indicate the state law under which they were organized, how dividends were distributed, how voting was conducted, how much capital stock was subscribed, and whether they conducted business for nonmembers. The survey assumed that cooperatives could never be properly interested in profit as such but should undertake only to achieve efficiencies in production and marketing. What the savings from such operations were called and how they were distributed determined the organization's legitimacy.[39]

USDA economists believed that the Rochdale model was the only proper form of cooperative because Rochdale distinguished cooperatives from ordinary corporate enterprise and reinforced the goal of efficient management of individual farms. Cooperatives that did not conform strictly to the Rochdale principles were by definition suspicious. At a time when state legislatures were busily passing capital stock laws that significantly altered cooperatives' appearance and departed from Rochdale, the USDA sought to maintain the integrity of the "true" cooperative form by "prevent[ing] people [from] 'wrongly' labeling their organization."[40] Since the USDA had no power to police market activities, this was as far as it could go.

The Clayton Act only reinforced the USDA's preoccupation with the proper form of cooperatives. The USDA wanted to help farmers understand what kinds of cooperatives would still be liable under the federal antitrust laws, and so it had a stake in helping to shape the definition of cooperation. After the act became law, marketing experts in the Bureau of Markets clashed with the USDA solicitor over the legal definition of cooperation. The Bureau of Markets interpreted the Clayton Act to mean that cooperatives paying patronage dividends to members came within the exemption because the organizations retained no proceeds after paying their costs. But the solicitor, Francis J. Caffey, thought that the payment of patronage dividends indicated for-profit operations. Caffey did not understand that when a cooperative paid patronage dividends, it retained no profits as such.[41]

Internal conflicts within the USDA also surfaced in discussions of the state court developments. When the Iowa Supreme Court handed down the *Reeves* decision in 1914, holding that a cooperative could not penalize members for breaching their contracts, the chief of the Bureau of Markets immediately recognized the case's implications. "The very life of a selling organization depends on its ability to carry out its contracts," Charles Brand told the solicitor. "Can you suggest a legal way of protecting an association against loss, when they . . . find that the contracting grower has jumped his agreement and sold elsewhere?" Caffey offered no help. Echoing the *Reeves* opinion, he said that imposing fines on members "denies to competitors that right of competition in the market which they would otherwise have."[42]

On the whole, Houston believed that the Clayton Act settled the questions of cooperative structure and antitrust liability. He saw no reason for cooperatives not to conform to the law because, if they did, they gained the privilege of immunity. Indeed, the California Fruit Growers' Exchange, reorganized as a nonstock association after the Clayton Act went into effect. Few cooperatives in California or elsewhere, however, followed suit. Interpreting the Clayton Act even more conservatively than the law's opponents, the solicitor's office "based its instructions to the field organizers of the Bureau of Markets upon the opinion that the acts of the cooperative associations are not immune from federal anti-trust laws even though these associations be organized on a non-stock, non-profit basis."[43] Since the law governing cooperative form was in a state of flux and litigation over cooperative market practices had produced rulings adverse to farmers, the USDA consistently urged caution, advising farmers that cooperation was no panacea, that it could succeed only under certain circumstances, and that experimenting with the established formula would invite prosecution.

Besieged by requests from farmers for assistance on complying with the Clayton Act, the Bureau of Markets began work on a draft of a model state cooperative incorporation law, in consultation with the CFGE's president and lawyer.[44] The bureau's reliance on CFGE officials resulted in a draft that largely resembled the California nonstock law. The bill permitted nonstock coopera-tives to raise capital by imposing "fees, dues, assessments, or charges for the services" performed by the organization.[45] Before he would approve the draft, Caffey, the USDA solicitor, asked the bureau to separate the treatment of capi-tal stock and nonstock organizations "in order to avoid possible confusion or misunderstanding as to the distinction between these two classes."[46]

That "confusion or misunderstanding" already lay at the heart of the law of cooperative incorporation. The Clayton Act assumed that capital stock was incompatible with the cooperative purpose, and the USDA's draft bill did nothing to dislodge that assumption. In fact, CFGE president G. Harold Powell praised the bill for clarifying the difference between a cooperative and a "profit making organization": "In the former, it is operated for the mutual help of the members; in the latter, it is for the profit or advantage of the corpo-ration itself." Powell's view was out of the mainstream, and so was the USDA model bill. Only five states, four of them in the Northeast, passed versions of this bill between 1918 and 1921. By then, many state legislatures had accepted the idea that capital stock was essential to effective cooperative marketing.[47]

Moreover, the USDA bill had no effect on cooperatives' antitrust liability. After the Clayton Act, most courts looked to the intent and purpose of the cooperative to determine what constituted an unreasonable restraint of trade. What constituted intent and purpose under federal law was corporate struc-ture, rather than the aim of securing fair prices. This rule placed cooperatives in a perilous position. If farmers adopted corporate features necessary for national marketing operations, they could not claim federal antitrust immunity. Even though the Justice Department was theoretically barred by the appropriations rider from taking cooperatives to court, state prosecutors were under no such interdiction, and state courts persisted in construing cooperative practice and corporate structure as mutually exclusive.[48]

Yet, while the USDA may not have done as much to promote cooperation as farmers and farm groups demanded, neither did it obstruct the cooperative movement's development. When businesses complained about the effects of the cooperative movement, Houston finessed: "Legitimate and economical collective bargaining on the part of farmers is desirable and promotive, rather than destructive, of the interests of those middlemen who render efficient service on reasonable terms." When farmers criticized the department for failing to do more to help them, Houston repeatedly declared that it was the states' responsibility to regulate marketing. The federal government's role was properly one of investigation and the collection of market information.[49] The Bureau of Markets was often far ahead of the secretary. It actively instructed farmers in the methods and practices of cooperation, but it could not step outside its legally demarcated regulatory authority. The USDA could do nothing about the growing interstate nature of agricultural marketing, despite one scholar's observation that "the state lines, while offering many constitutional obstacles, have no economic significance whatever."[50]

In short, while Congress and the USDA both endorsed cooperation, they agreed that cooperatives should not change the market itself. A leading agricultural economist and proponent of cooperation was perfectly willing to draw the bright line: "We are loath to hand over the countryside" to corporate forms of organization, Edwin G. Nourse wrote in 1917, "unless we are quite certain that this is the only effective means of escape from the inevitable inefficiency of unorganized agriculture."[51] Throughout the Progressive Era, policy makers never believed that time had come. They continued to believe that capital stock cooperatives restrained trade illegally and that Rochdale cooperatives most closely conformed to the Jeffersonian model of small-scale production and virtuous rural life.

The mandate to increase the efficiency and productivity of individual farms, rather than change the dynamics of exchange in the marketing system, underlay Progressive Era federal agricultural policy. That the USDA did little to protect cooperatives from their adversaries in the market or in the courts resulted from the limits of its legal authority, not from active opposition to the cooperative movement. The USDA left farmers to decide for themselves how to interpret federal legal developments. Some, like the CFGE, took their cue from the Clayton Act and reincorporated as nonstock associations in order to meet federal requirements for antitrust immunity. Others saw the appropriations rider as the more relevant statute and assumed that whether or not they retained their capital stock, the federal government could not prosecute them. The fact that authority to regulate the incorporation of cooperatives remained with the states gave farmers freedom to experiment at that level.

Nowhere did farmers take greater advantage of that freedom than in the raisin industry. The rise of the California Associated Raisin Company was more than an economic triumph for its leaders and growers. It blazed a new legal trail. Its monopoly, corporate structure, and market control broke the mold for agricultural cooperatives in a state disinclined to trim the sails of commercial agriculture. It would eventually test the prosecutorial constraints of the appropriations rider. The CARC's role as the leading example of the new agricultural cooperative would force Congress to codify new legal rules to govern cooperatives' market behavior, in turn setting off new rounds of debate over the place of cooperation in the industrial economy.

PART THREE

The Benevolent Trust in Law and Policy

1912–1928

# 6 ᴧ A Ruthless Trust Monopoly

*The fact that the organization is practically a public one, or at the very least a semi-public*

*proposition, has allowed of the use of an appeal to the people of the community as a whole*

*on the grounds of public necessity and has allowed of the use of methods which, in very*

*many instances, would have been denied by law to an individual or group of individuals*

*less complete than a large proportion of the producers in a particular community.*

—California Fruit News, *1916*

Legal developments at the state and federal levels between 1890 and 1910 figured prominently in the continuing movement to organize growers in the California raisin industry. The California state antitrust law of 1907, the federal appropriations rider in 1913, and the Clayton Act of 1914 embraced a specific version of cooperation — the nonstock form — and declared it compatible with the antitrust laws. Despite hostile judicial decisions in Illinois, Iowa, and the U.S. Supreme Court, this form of cooperation remained at the forefront of official agricultural policy. At the same time, many farmers viewed the other model of cooperation — the capital stock organization with corporate powers — as the most promising solution to their marketing problems. The tensions resulting from the conflicting priorities of courts and legislatures at both levels of government within the federal-state system set the stage for what happened in California during the Progressive Era.

The legal setbacks farmers took during the 1890s and 1900s were not, in fact, the most significant obstacle standing in the way of effective cooperatives. The experience of the raisin industry during the Kearney era suggests that growers differed among themselves over what business form and practices their cooperatives should adopt. The firms and entrepreneurs with interests in the industry also had something to say about what a growers' organization should do. In the six years after Kearney's death in 1906, industry leaders annually experimented with industry-wide marketing organizations, but packers, local bankers, elite growers, and the commercial establishments of Fresno, San Francisco, and Los Angeles could not agree on a financial and organizational plan. When, in 1912, a new set of grower-leaders bridged the social and economic divisions in the industry with a new organization that revived many of Kearney's ideas, they launched what would become the most monopolistic agricultural cooperative in American history.

This cooperative, the California Associated Raisin Company, succeeded in bringing almost the entire industry under the umbrella of a single cooperative. The CARC used law astutely and cleverly manipulated the image of traditional cooperation. Behind the rags-to-riches story was a fractious group of growers, some of whom became increasingly disenchanted with the putatively cooperative organization they joined. The CARC was supposed to act in their interests but was in fact consolidating its power at the expense of its relationship with members. The legal immunities for cooperatives in federal law and the complete indulgence for them in California law, however, effectively insulated the CARC from its opponents, the commercial packers and distribution firms that sought to scale back its monopoly.

## THE "MILLION DOLLAR CORPORATION"

After Kearney's death, the raisin industry swung between depression and mild recession. Prices returned to their low pre-CRGA levels; unable to cover the costs of production, some growers deserted their vineyards. Without Kearney, the cooperative movement in the industry rested with the initiative of social and economic elites in the Central Valley. A succession of more traditional cooperatives failed to last more than one season each. William R. Nutting and Wylie Giffen, prominent growers and veterans of the Kearney organizations, became convinced that "no plan to organize the raisin growers could succeed without a ready paid-in cash capital." That meant devising a form of organization cooperative enough to persuade growers it would serve them and corporate enough to persuade the banks that it could handle the financial risks of marketing. The industry's sagging fortunes made everyone increasingly desperate.[1]

In 1912, Nutting and Giffen asked a coalition of prominent growers, lawyers, and bankers to resurrect Kearney's vision of the cooperative as corporate trust. The practical and legal problem that Kearney had encountered still faced them: who would supply the capital? William A. Sutherland, a Fresno attorney and state assembly member, proposed a joint stock company with the growers as owners and shareholders. Attorney H. H. Welsh and bankers I. G. Maxson and H. W. Wrightson argued that the Kearney experience showed that growers would not be able to subscribe all the required capital. This belief apparently won out in the debate over what kind of cooperative to form. In the fall of 1912, a committee consisting of Sutherland, Welsh, Maxson, Wrightson, and growers Nutting, Robert Madsen, Charles Parlier, and William Glass was appointed "to form a company which would be in the highest degree cooperative, and at the same time have power to enforce its crop contracts."[2]

On November 11, more than one thousand raisin growers gathered at a mass meeting in Fresno's main public hall to hear the details of the "Million Dollar Corporation." The committee's design for the California Associated Raisin Company wove the essence of Kearney's ideas into a unique hybrid of cooperative and corporation. The organizers deliberately chose a name and a form of incorporation that repudiated the legal forms available to cooperatives. The CARC was organized under the state's general incorporation statute with a capital stock of $1 million. Membership was not limited to growers. As M. Theo Kearney had done with the CRGA fourteen years before, the plan let anyone purchase shares, including bankers, merchants, and packers. Shareholders would earn 6 to 8 percent dividends on their investments.[3]

To ensure that the stock would not fall into the hands of "interests hostile to raisin producers," however, a voting trust arrangement was established under which twenty-five trustees held and voted the stock on behalf of all shareholders.[4] Elected by the growers, the trustees reflected the CARC's distinctive constituency. Most, but not all, owned vineyards, but only half were full-time raisin growers; the rest worked as bankers, lawyers, merchants, utility managers, and in other professions. One, Karl Emirzian, was Armenian. This slate of trustees proved remarkably cohesive; all but two served throughout the decade. By April 1913, growers representing 76 percent of the raisin acreage signed contracts to deliver their crops to the CARC, and the company raised $800,000 in capital stock subscriptions. The new organization relinquished the intimate democracy of Rochdale membership and voting in favor of a more impersonal structure that mimicked corporate shareholding. The seed that Kearney had planted grew into a distinct species of cooperative organization, a hybrid of cooperative principles and corporate trust devices (Table 6.1).[5]

Why did growers and bankers alike embrace the CARC after so resoundingly rejecting similar corporate creatures? In both structure and purpose, the CARC resembled more an ordinary commercial business than an agricultural mutual help society. The CARC flouted statutory limitations on both the total amount of capital stock that a cooperative could sell and the amount that

TABLE 6.1. Trustees of the California Associated Raisin Company and Their Occupations, 1912–1918

| Trustee | Occupation(s) |
|---|---|
| Hector Burness [a] | Grower, banker, merchant |
| J. J. Eymann | Grower, banker, lumberman |
| Lucien Franscini | Grower |
| Levi Garrett | Banker |
| Wylie M. Giffen [a] | Grower, land investor |
| Hans Graff [a] | Banker, merchant |
| James Madison [a,c] | Grower, raisin packer [a] |
| R. K. Madsen | Grower, utility company official |
| W. B. Nichols [b] | Grower, banker, packing company director |
| J. L. Norman | Grower, banker |
| William R. Nutting [a,c] | Grower, raisin packer |
| E. F. Pickerill | Grower, banker |
| William Parlier | Grower, utility company official |
| A. Grant Robinson | Grower, grain farmer |
| A. Sorensen [b] | Grower, teacher, insurance company owner |
| H. H. Welsh [a] | Grower, attorney, oil investor |
| A. G. Wishon [a,d] | Grower, utility company manager |

*Sources*: W. Y. Spence, "Success after Twenty Years," *SMH* 3 (Mar. 1918): 9; 3 (Apr. 1918): 6; 3 (May 1918): 6; 3 (June 1918): 6–7; 3 (July 1918): 6–7; 4 (Sept. 1918): 6; 4 (Oct. 1918): 4; 4 (Nov. 1918): 6; 4 (Dec. 1918): 6; 4 (Jan. 1919): 4; 4 (Feb. 1919): 4; 4 (Mar. 1919): 4; 4 (Apr. 1919): 4; James M. Bragg, "History of Co-operative Marketing in the Raisin Industry to 1923" (M.A. thesis, University of California at Berkeley, 1930), 64.

*Note*: Occupation unknown: Mark Bassett, S. J. Carroll, Edwin Dudley, Karl Emirzian, George Feaver Jr., J. E. Hall, E. A. McCord, Thomas Martin, Richard Norris, George B. Posson.

[a] Chosen by trustees in 1912 to serve as director

[b] Elected to replace Madison and Nutting in 1917–18

[c] Both Madison and Nutting resigned their partnerships in commercial raisin packing firms before accepting their directorships in 1912.

[d] Wishon did not purchase a vineyard until 1918.

any individual could purchase.[6] Although capital stock and regular incorporation were not unusual among American cooperatives, the CARC's voting trust arrangement was a definite innovation. In most other respects, however, the CARC essentially copied the organizational and financial structure of the CRGA. What made the new organization acceptable when its parent had been repudiated?

The CARC garnered industry support for two reasons. First, in marked contrast to Kearney's alienating style, the CARC's leaders openly invited area bankers to participate not just in financing the company but also in running it. In effect, the CARC co-opted the individuals and institutions most likely to destroy it. It did so, however, at no small cost to cooperative principles. The voting trust agreement, although meant to protect the growers' interests by giving them indirect representation before the CARC's board of directors, introduced nongrower interests into the CARC's management. The CARC's organizers took this step not because they intended to topple the pillars of cooperative ideas but because their experiences with cooperatives did not conform to the ideal. Fruit required processing and packing—two capital-intensive operations—before it could be sold; the Kearney organizations had demonstrated the folly of relying solely on cash-poor growers for raising capital. The economic interdependence of industry subgroups induced the CARC's organizers to create a formula for cooperation that satisfied the financial community: "Business and professional men for the first time saw [something] that looked good from a business point of view."[7]

A second reason for the CARC's acceptance also came straight from the Kearney play book. Giffen announced early in the subscription campaign that the CARC would not begin marketing activities until 75 percent of the industry was under contract to deliver to the cooperative. Again, though in substance no different from earlier years, the CARC used this policy to create an imperative for the entire community. In explaining the plan to growers, CARC leaders took pains to stress that despite its corporate form, the CARC was in spirit a true growers' cooperative organization. Describing the legal character of the association, Sutherland contended that "the new company [is] not purely cooperative but a stock company, while at the same time . . . it [is] to be conducted as the associated effort of the raisin growers." General manager James Madison emphasized the growers' role in the founding of the CARC: "The Association has been created by the largest percentage of growers ever joined together in the history of the state." What mattered most, said CARC president Giffen, was "getting good prices for raisins [rather] than for the dividends." Promoting the growers' role in the CARC was both good politics and good business. Growers had already shown that they would not give much allegiance to an organization that did not rigorously pursue their interests. The CARC leaders knew from past experience that the key to successful marketing was monopoly. They also knew that not every grower would voluntarily participate in a monopoly, cooperative or not, but a cooperative had a much better chance of attracting growers than a commercial raisin-packing firm. So they hid the CARC's commercial structure behind an extensive campaign that proclaimed the cooperative gospel without acknowledging how much the CARC departed from that gospel. Although they did not draw the connection themselves, Sutherland, Madison, and Giffen had all adopted Kearney's vision of the cooperative as trust, boldly expanded its economic reach, and dressed it in the sheep's clothing of cooperation.[8]

One difference between the CARC and previous cooperatives was that the CARC did not profit farmers alone. Its own officers linked the CARC to the commercial establishment and social elite of the Central Valley. Wylie Giffen personified this connection. A man of little formal education and the son of a Presbyterian minister, Giffen brought his family to California from Pennsylvania in 1889. He purchased land on easy terms during the depression of the 1890s, and by 1912 his forty-two hundred acres ranked among the industry's largest individual holdings. After weathering the stormy Kearney organizations, he came to the conclusion that "individualism was a myth." After he became CARC president in 1914, he never took a salary — a fact that endeared him to the growers. James Madison immigrated to California in 1892 from Denmark, where cooperative marketing was the centerpiece of agricultural policy. During the 1890s he acquired several hundred acres of vineyards, purchased an interest in an Alaska salmon canning plant, and was a partner in a Fresno commercial raisin-packing firm. Though he was a packer, "he had the mental attitude common to many Danes which permitted him to accept the idea of cooperation as a perfectly normal practical means of accomplishing an economic purpose." But he was no idealist. He "believed in a free, or competitive market, and held that [the CARC], to be a success, must be able to produce and market raisins under fighting conditions. He believed in neither social nor governmental protection." William A. Sutherland, the lawyer and state legislator, wrote the CARC's articles of incorporation and bylaws and served as the CARC's legal counsel. In 1920, he and Giffen bought a Fresno bank. Throughout, he maintained a lively law practice. His closest connection to farming was through his father's cattle ranch, twenty-five miles south of Fresno.[9]

The CARC's new form raised pressing legal issues. Its articles of incorporation neatly merged traditional cooperative functions and corporate financial powers. The CARC was authorized to handle the crop as the growers' agent and to buy warehouses, packinghouses, and other real property as required.

And like any corporation, the CARC could buy, sell, and hold the capital stock of private corporations; further, it could issue, buy, sell, and borrow against warehouse receipts, bills of lading, and other forms of security, which agricultural processors commonly employed as collateral. It was this corporate power that the state nonstock statute withheld from cooperatives, ostensibly to prevent them from accumulating substantial amounts of capital. In fact, the statute effectively separated the privilege of exemption from the antitrust laws from the power (and the abuse thereof) of owning capital. By organizing under the regular incorporation statute, the CARC's leaders made a deliberate choice, and they recognized the risks that accompanied that choice. The question was whether the growers, in supporting the choice, also accepted the risks.[10]

Chief among these risks was the CARC's antitrust liability. The new generation of raisin industry leaders knew that for their monopoly to be considered within the bounds of the law, everyone — growers, packers, and the state — had to perceive the CARC as operating like a cooperative. Publicly, CARC officials downplayed any suggestion that the CARC was not, fundamentally, a cooperative. Stressing the CARC's uniqueness as an agricultural organization did not dispel concerns about the CARC's legal status, however. At mass meetings, community gatherings, and anywhere that people met to talk about the industry, growers raised questions about the CARC's possible antitrust problems. The first appropriations rider in 1913 seemed to dispose of the threat of federal prosecution. After the Progressive Hiram Johnson was elected governor in 1910, farmers enjoyed an unprecedented degree of protection and promotion from state government. As a result, Giffen, Madison, and the trustees assumed that their grafting of cooperative and corporation was not only just but legal. At a mass meeting in April 1913, a raisin grower inquired about the legality of the CARC's contracts and methods. CARC trustee H. H. Welsh invoked Theodore Roosevelt's famous aphorism in his reply, later recounted by a rival raisin packer: " 'I think we will have to admit that we are a trust, but we claim to be a good trust,' and he smiled as he said it."[11]

VERTICAL INTEGRATION IN THE "GOOD TRUST"

The CARC did its best to behave as a good trust, or at least look like one, even as it departed from traditional cooperative practices. Since stockholding served the purpose of raising capital rather than tying members to the organization, the CARC secured control over the crop through the device of the contract. The CARC's 1913 crop contract kept growers on a short leash, binding them to sell all they produced to the CARC at the guaranteed price of three and one-quarter cents per pound for Muscat raisins. Growers shared in the profits in proportion to their contribution, as was the custom in agricultural cooperatives. But before paying the growers, the CARC retained one-quarter cent per pound to cover "expenses and interest charges." In fact, the CARC used some of this money to pay 6 percent dividends to stockholders. Thus the shareholders received their splits before the growers got theirs. The contract dealt with the growers individually and unilaterally; it ran for a term of three years, and renewal was at the discretion of the CARC. With forty-four hundred of the state's sixty-five hundred growers under contract by the spring of 1913, the CARC controlled over 115,000 acres of vineyards, enabling it to offer an opening price of three and one-half cents, more money than the growers had seen since the Kearney organizations and enough to cover production costs.[12]

In addition to securing horizontal control over supply, the CARC began to integrate vertically. First, Giffen and Madison entered into exclusive contracts with sixteen of the valley's twenty-one dried fruit packers to process and pack raisins on the cooperative's terms. Because the packers could not collectively agree on what to charge the CARC for packing its raisins, Giffen and Madison proceeded to contract with the packers seriatim, dividing them as they scrambled to obtain the best possible terms for themselves.[13] The leading independent packers initially assumed that the CARC "was to be purely a farmers' cooperative organization dealing with the raw product only." After the CARC began operations, the "inside" packers—those under contract—continued to believe that its accumulated capital presented no serious threat. As the Bonner Packing Company informed its East Coast customers in 1913, "The money is not to be used in the purchase of or building of packinghouses but it is to be used as a working capital to enable the CARC to purchase outright from the growers and resell the raisins as the market requires them."[14] This "working capital" gave the CARC greater market power than any of its predecessors. Its monopoly gave it leverage over the inside packers, who ceded control over the prices they would receive, and the outside packers, who were forced to compete for the few raisins not under CARC control. An outside packer gloomily announced the new order to his customers: "The Association controls absolutely the Raisin industry in California, and are [sic] in position to set and control prices."[15]

These developments did not escape official notice. The packers' trade circulars stirred up deep animosity toward the CARC among brokers, jobbers, and wholesalers in the East. Their complaints reached Washington in the spring of 1913, months before the CARC marketed its first crop. An Illinois wholesale grocers' firm asserted that the CARC was "the most extreme airtight monopoly that we have ever come in contact with." A Massachusetts wholesaler

CARC treasurer James Madison (at left) and president Wylie M. Giffen (second from left) in the CARC offices, 1914. (Historical photographs from the C. "Pop" Laval Collection, Fresno, California)

called the CARC "one of the worst of all Combines or Trusts." The Arkansas Brokerage Company put the matter succinctly: "[The CARC] . . . is a combination in restraint of Interstate Commerce and in violation of the Sherman Anti-Trust law." By stirring up federal officials, these businesses hoped to make some kind of antitrust objection stick long enough to slow the CARC's monopoly-building.[16]

The U.S. Department of Justice (DOJ) immediately dispatched special agents to Fresno to interview growers and packers. Fortified by good crop prices, many growers expressed satisfaction with the CARC and its marketing plans. The general feeling among the growers was that the CARC provided "a feeling of security although it still means hard work and plenty of it to get results."[17] The special agent and the U.S. attorney in San Francisco were "not disposed to think that any immense trust exists." They reported that since the CARC's monopoly did not enter into interstate commerce, it did not raise federal legal issues. The DOJ's higher-ups thought otherwise. They instructed the San Francisco office of the Bureau of Antitrust Investigations to continue to monitor both the CARC and the "Packers' Combine." [18] Federal officials resolved to keep an eye on events in Fresno, a vigilance spurred by furious packers who continuously rained letters and wires on Washington. The appropriations rider did not revoke the DOJ's authority to investigate agricultural cooperatives that raised prices above "fair and reasonable," but despite the urging of the attorney general, U.S. Attorney Albert Schoonover put off an official investigation, pleading overwork from a heavy trial caseload. [19]

The CARC marketed the 1913 crop amid much self-congratulatory publicity. In addition to selling all surplus raisins from 1912, Giffen and Madison increased the CARC's authorized capital to $1.5 million. This money was used to lease a packing plant that processed part of the 1913 crop — the first outward sign of the CARC's ambition to integrate vertically. [20] The following year, the CARC continued to acquire packing plants through lease, purchase, or construction. Taking a cue from the citrus growers, the CARC established its own brand name for the raisins it marketed. Its eastern agents suggested "Sun-Made," to emphasize the unique California conditions under which the raisins were produced. A "happy thought" inspired the chief of the CARC's advertising department to combine natural imagery with feminine allure — a common practice in fruit and food promotion during the Progressive Era — and to devise the pun "Sun-Maid." Casting about for "a pretty girl" to symbolize the natural purity of sun-dried raisins, a Fresno artist plucked fifteen-year-old Loraine Collett from a Raisin Day parade to pose for a portrait in her white apron and red bonnet. The picture of Collett holding a tray of raisins against the backdrop of a radiant sun instantly became one of the most recognized food trademarks in the country, as well as one of the most valuable. For her work Collett earned $5. [21]

The Sun-Maid trademark was part and parcel of an innovative marketing strategy designed to familiarize American consumers with the growers' brand and create new outlets for the industry's rapidly increasing production. In 1915, Giffen and Madison deputized a battalion of sales agents to hawk Sun-Maid raisins directly to grocers in major cities. CARC salesmen became the only trade channel through which Sun-Maid raisins could be purchased, cutting the distribution network out of the picture. Coordinated from the CARC's Fresno offices, the national marketing strategy was to make raisins a year-round staple, instead of just a holiday-time specialty. An extensive advertising campaign in national newspapers and magazines increased home consumption dramatically. The pages of prominent women's magazines were soon filled with CARC ads pitching Sun-Maid raisins as healthy, nutritious, and tasty. By 1918, the annual advertising budget exceeded $400,000. The CARC's success in creating demand for raisins encouraged growers to expand

James Madison (third from right) with "Sun-Maid girls," CARC salesmen, and the Sun-Maid trademark carton, 1916. (Historical photographs from the C. "Pop" Laval Collection, Fresno, California)

production further and attracted thousands of newcomers to the industry. By 1915, raisins ranked second only to the citrus industry in importance to the state's prosperous fruit-growing economy.[22]

The CARC's impact registered beyond the state's boundaries. The *Country Gentleman* noticed the CARC's efficiency in clearing out the carryover while boosting production and sales at the same time: "Cooperative enterprise, high-class selling efficiency and scientific boosting have been chiefly responsible for this record fruit year. The Golden State has never had better proof that it pays to advertise."[23] An agricultural economist observed, "The raisin industry, flat on its back in 1911, had been made highly profitable by a cooperative organization which had employed a competent sales director, traffic manager, and other officials."[24] In just a few years, the CARC had created enormous demand for a product whose price and supply it nearly completely controlled. The CARC had transformed raisins from a rare luxury to an inexpensive staple that American housewives could not do without. As Kearney had urged in vain to an earlier generation, the growers' organization took a page from the trusts' handbook and made it work (Table 6.2).

Early CARC advertisement for Sun-Maid Raisins, *Ladies' Home Journal*, December 1915, inside back cover. (White House Antiques, Addison, Illinois)

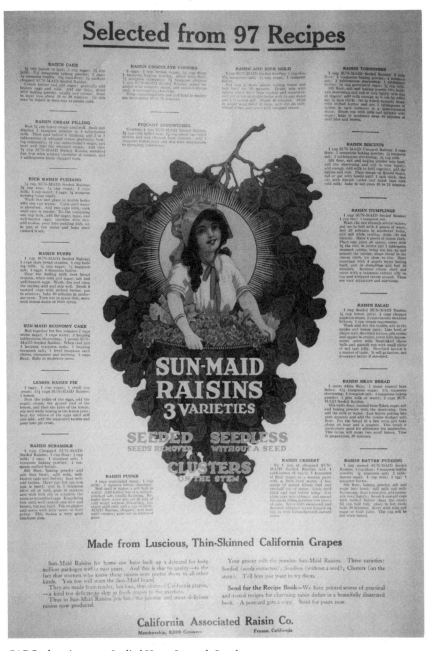

CARC advertisement, *Ladies' Home Journal*, October 1917.
(White House Antiques, Addison, Illinois)

California Associated Raisin Company brokers and salesmen, 1916.
(Historical photographs from the C. "Pop" Laval Collection, Fresno, California)

The CARC's effect on the market could no longer be ignored. In February 1916, two of the U.S. attorney's assistants filed a preliminary report on the CARC. They concluded that the CARC had the power to fix prices for the entire industry: "Working upon the basis of this plan of organization [the voting trust], the contracts with the growers, [and] contracts with packers and leases of their plants, the California Raisin CARC [*sic*] has absolutely fixed and determined the price of raisins." Having an effect on prices, however, was not sufficient to warrant prosecution; in fact, the agents conceded that the CARC had been able to reduce prices to the consumer by eliminating the profits of speculators and middle merchants. Nevertheless, they believed that the cooperative's economic control over the industry raised significant legal issues. DOJ attorneys in Washington agreed that the situation merited continuing supervision, but they decided to wait and see if the CARC committed a clear violation of the antitrust laws before proceeding to the courts. The DOJ should proceed with caution, the DOJ's special agents wrote, because the CARC's fortunes affected "probably fifty thousand people" and every commercial interest in the Central Valley: "All bankers, merchants and tradesmen, architects and builders, artisans, laborers and professional men within the raisin district are interested in the maintenance of the association because they believe that its

TABLE 6.2. Financial Condition of the California Associated Raisin Company, 1914–1920

| Year | Earnings[a] | Patronage Dividends[b] | Selling Expenses[c] | Operating Expenses[d] | Packing Costs and Packing Plants[e] |
|------|------------|------------------------|---------------------|-----------------------|--------------------------------------|
| 1914 | $146,818.32 | none paid | $117,452.64 | $1,268,131.72 | $1,359,882.07 |
| 1915 | 168,830.57 | $1,082,888.33 | n/a | 2,593,118.58 | 3,456,452.46 |
| 1916 | 385,854.69 | 1,739,503.70 | n/a | 3,040,263.95 | 3,381,105.52 |
| 1917 | 406,136.76 | 1,263,879.95 | 284,778.79 | 4,074,352.89 | 2,241,700.37 |
| 1918 | 406,540.95 | 1,370,196.86 | 291,756.44 | 5,657,645.71 | 1,603,423.85 |
| 1919 | 456,623.19 | 3,334,881.12 | 1,312,498.16 | 7,583,337.31 | 4,156,785.43 |
| 1920 | 522,096.15 | 7,338,906.02 | 1,832,446.45 | 13,819,932.97 | 3,401,734.85 |

*Source:* Manuscript financial statements of the California Associated Raisin Company, Box 18, Erdman Papers.

*Note:* n/a = data not available

[a] Reported variously as "undivided profits," "undivided profits and surplus," and "surplus"

[b] Year-end final settlements to growers; includes payment of all excess proceeds from sale of crop but does not include dividends paid on capital-common stock

[c] Includes all selling expenses (brokerage fees, salaries, advertising, miscellaneous expenses)

[d] Includes business, selling, and operating expenses; tallied on balance sheets as total current liabilities

[e] Total cost of packing raisins sold by the CARC, including in-house packing, and total amount invested in packing plants and equipment (through purchase and lease). For 1915, and 1916, figures include selling expenses (duplicating costs counted in previous column).

TABLE 6.3. Average Retail Prices of Raisins in the United States, in Cents per Pound, 1915–1928

| Year | Price | Year | Price |
|------|-------|------|-------|
| 1915 | 12.5 | 1922 | 23.0 |
| 1916 | 12.9 | 1923 | 17.6 |
| 1917 | 14.6 | 1924 | 15.4 |
| 1918 | 15.3 | 1925 | 14.5 |
| 1919 | 18.4 | 1926 | 14.7 |
| 1920 | 28.6 | 1927 | 14.2 |
| 1921 | 29.8 | 1928 | 13.2 |

Source: U.S. Department of Labor, Bureau of Labor Statistics, Retail Prices, 1890–1928, Bulletin of the Bureau of Labor Statistics, No. 495 (Washington: U.S. Government Printing Office, 1929), 50–51.

operations have increased and stabilized their own business and trades and increased the value of their properties." Finding an impartial jury should the government decide to prosecute would be difficult.[25]

The DOJ proceeded cautiously, but neither the special agents nor the staff attorneys in Washington doubted that the department had the authority to prosecute if it found that the CARC had raised prices above the "fair and reasonable" standard set out in the rider. Retail prices for raisins had increased in 1915, but to file suit, the DOJ would have to prove that the CARC caused the increase. The DOJ's California representatives saw little to implicate the CARC under the Sherman Act. Under the rule of reason, size alone did not violate the antitrust law. To this point, no one in the DOJ had determined whether the appropriations rider would protect a farmers' cooperative that violated the rule of reason or whether the CARC would qualify for the Clayton Act exemption as an agricultural cooperative. The CARC was neither in imminent danger nor completely safe, a position that permitted it to exploit the law's ambiguity in the short term (Table 6.3).

ENFORCING COOPERATION

In the absence of immediate prosecution, the CARC expanded its monopoly over the growers. The first crop contract expired with the 1915 crop, and the CARC had the option to renew the contract for two more years. Before exercising the option, Giffen and Madison decided to hold a referendum campaign to solidify grower support and to bring in new members. The CARC averaged an 83 percent share of the crop in its first three years, but that statistic was misleading. The CARC's control had been reduced each year for several reasons: breach of contract by members, the CARC's inability to sign up new growers, and the transfer of land from members to nonmembers. In addition, outside packers offered lucrative prices to lure CARC members away. The CARC accused several packers of inducing growers to break their contracts by keeping their plants open at night to enable growers to deliver under cover of darkness. The packers' prices created an incentive for some members to free ride, meaning that they delivered to the CARC when its prices were higher but jumped ship when outside packers offered better prices.[26] The CARC promptly sued growers it suspected of cheating in this way, and it won enough of these breach of contract cases to induce most growers to honor their contracts. These problems spurred the CARC to cast its net still wider (Table 6.4).[27]

On January 1, 1916, Giffen announced in the *Sun-Maid Herald* that the CARC planned to exercise the option to renew—but only if an additional fifteen thousand acres were pledged to the CARC by April 1. This announcement sounded the first note of a hotly contested campaign to extend the CARC's control over the crop. The CARC's house organ and the Fresno newspapers ran a virtual propaganda campaign urging growers to sign for the sake of the community: "Our appeal is a community appeal, as every person is either directly or indirectly prospered in proportion as the raisin industry prospers. . . . Especially do we appeal to the growers. First, to you who are on the outside that you quit yourselves like men and in the interest of the common good lay aside the slight advantage which you now have. And secondly, to you who are already members of this organization that you should for the next three months, forgetting everything else, spend your time and your influence towards the perpetration of the organization that has brought about your emancipation."[28] Any grower who refused to sign was, in effect, a traitor. The charged atmosphere of this first membership campaign presented an uncompromising choice to growers: sign up and share in the wealth or face the community's displeasure.

The 1916 campaign was the first test of the CARC's ability to link its existence to the survival of the raisin industry and the prosperity of the community. The CARC urged everyone to join its monopoly. It had an open membership policy out of economic necessity: the cooperative needed all growers, or at least as many as possible, so it could monopolize supply. The CARC's campaign propaganda declared that the cooperative both ensured the individual grower's economic freedom and preserved the political community. A 1916 advertisement in the *Fresno Morning Republican* analogized the CARC's fate

TABLE 6.4. The California Raisin Crop, the CARC's Share of the Raisin Crop, and Prices Paid to Growers, in Cents per Pound, 1912–1920

| | Tonnage | | | CARC Price | |
|---|---|---|---|---|---|
| Year | Statewide Crop | Marketed by CARC | % of Crop under CARC | Muscat (Seeded) | Thompson (Seedless) |
| 1912 | 31,426 | 24,512 | 71.2 | n/a | n/a |
| 1913 | 65,930 | 59,770 | 90.7 | 3.5 | 3.9 |
| 1914 | 91,000 | 73,660 | 80.9 | 3.3 | 4.6 |
| 1915 | 127,100 | 98,405 | 77.4 | 3.6 | 5.0 |
| 1916 | 132,000 | 105,100 | 79.6 | 4.2 | 6.6 |
| 1917 | 163,000 | 127,000 | 77.9 | 4.9 | 6.9 |
| 1918 | 167,000 | 149,710 | 89.6 | 5.3 | 6.9 |
| 1919 | 182,593 | 159,260 | 87.2 | 10.4 | 12.0 |
| 1920 | 173,528 | 152,499 | 87.9 | 11.2 | 14.8 |

*Sources*: R. B. Forrester, *Report upon Large Scale Co-operative Marketing in the United States of America* (London: H.M. Stationery Office, 1925), 141; "Annual Financial Report, California Associated Raisin Company," *Associated Grower* 1 (September 1921): following 26, at viii; ibid., 5 (January 1923): 16.
*Note*: n/a = data not available

to the nation's birth. Beneath a drawing of the signing of the Declaration of Independence, the ad read, "The Independence of the United States depended not only on co-operation, but ALL the States participating in the cooperation. So it is with the Raisin Growers. The Independence of the Raisin Growers depends on the co-operation of ALL the growers."[29] The CARC used moral and religious ideas to justify monopoly: "It was the world's greatest teacher who 2000 years ago said, 'Every kingdom divided against itself is brought to desolation;' and we say that this community should not be divided against itself."[30] Other voices within the larger agricultural economy opposed the extension of the CARC's monopoly. Early in the campaign, the *California Fruit News* expressed reservations about the CARC's methods: "Legally the Associated can accomplish what it tells the growers in its ultimatum it intends to do and a few other things as well. Morally, we think, they have no right to do it."[31]

The CARC and its campaigns created a new definition of community, to be sure, but this new definition could not mend the existing tears in the social fabric of early twentieth-century California. In fact, the CARC only rent the fabric further. Its economic imperative could not erase social prejudice and in fact was probably not designed to do so. The split that had always existed between the dominant economy run by white growers and packers and the subindustry headed by Armenian packers and minority growers with substantial acreages continued to widen after the organization of the CARC. Immigrants lacking strong economic ties to the Anglo-American community did not need them to prosper. As Armenian plants continued to thrive even after the CARC was formed, not all immigrants and minorities necessarily had to sell to the CARC. Cooperation did not erase racial or cultural distinctions; CARC loyalists believed it was more important to extend the monopoly than to undermine it by excluding growers, whatever their ethnicity.[32]

The subindustry's self-sufficiency also meant that immigrants devised varying strategies of resistance. The Germans tried to assimilate as quickly as possible, a task World War I made more difficult. The Japanese arrived in larger numbers after Chinese exclusion took effect in the 1880s; they began as contract laborers and steadily became landowners. They sought to be compatible members of the community, even as discrimination against them grew. As a Japanese-American farmer recalled in 1980, "We were just trying to get along the best we could and had no political power."[33] Because of their established stake in the industry, Armenians initially joined the groundswell of support for the cooperative. They may have felt socially or politically disfranchised in California, but minorities had more economic power than they realized. As a group they were a substantial proportion of the horticultural industries.[34]

The proclivity of ethnic and racial minorities to decline to sign CARC contracts or to cancel their memberships not only exacerbated underlying social tensions but also led to overt acts of violence. In the whipped-up fervor of the CARC's membership campaigns, Euro-American growers resorted to night riding both to reinforce the CARC's monopoly and to intimidate foreigners. These were collateral goals. While anyone—white or nonwhite—who refused to sign a CARC contract could expect a visit from the night riders, Japanese and especially Armenian growers, who were distinct by virtue of color and language, were signaled out most frequently, forced to sign up, and coerced into remaining CARC members when they wanted out. The violence of the night riders spoke volumes about an individual grower's ability to undermine the cooperative.

The campaign to exercise the option in 1916 was more than a test of growers' loyalty to industry and community; it was an exercise in cultural assimilation for racial and ethnic groups. As such, it was one of many manifestations of the ethnic and racial biases pervading California during the Progressive Era. The violence of the night riders is ironic only because it was conducted in the name of cooperation. A movement that began as a voluntaristic act of self-help and mutual benefit became transformed into an economic imperative from which no dissent could be tolerated.

Initially, the coercion took mild forms. Local merchants and banks agreed to boycott any grower not under contract. Out in the vineyards, the gloves came off, as roving mobs resorted to intimidation, threats, property damage, and in some cases physical injury. In one incident, "a grower was let out on a bridge over a canal with a rope around his neck and yanked into the water and then yanked out." In another case, the child of an outside grower suffered a beating at the hands of his peers while their schoolteacher stood by; the child's father promptly signed a CARC contract.[35]

Giffen publicly disavowed any knowledge of the activities of the night riders, but he did nothing to stop them before the campaign ended on April 1. Indeed, he claimed that the CARC was justified in preserving its monopoly: "We do not concede the moral right of most men to sell their raisins to the independent packers." Opposition packers spread word of the violence to state agricultural journals. The *California Fruit News* decried the CARC's use of "methods contrary to the spirit and intention of our trust and monopoly legislation." Giffen was undeterred. The problem, he believed, was not overzealous CARC members but uncooperative outside growers who should be made to bear the costs of their independence.[36]

The packers later acknowledged that CARC officers and directors did not participate in the violence and that the cooperative publicly sought to discourage it. Nevertheless, night riding was an effective recruiting tool for the CARC, helping to bring in growers on the margins who otherwise might have undermined the monopoly. The membership campaign put thirty-two thousand more acres under contract to the CARC—more than double its published goal—and increased its market control to 85 percent. The *Sun-Maid Herald* then published a token rebuke to the night riders: "We wish to say that from this time on no contracts will be taken by mob violence and even night riding will only be permitted in its milder forms." What those "milder forms" might be, Giffen did not specify. A month after the 1916 drive concluded, he told a DOJ investigator that during the campaign the CARC officers had "stepped in and warned [members] not to use such tactics, telling them the Association could not exist if such methods were used." But if the CARC did issue such warnings, it did not do so in print. More to the point, the cooperative refused to cancel the contracts of growers who claimed that their signatures had been obtained under duress. Public opinion in California seemed to agree with Giffen; the night riders were, perhaps, excessively zealous but tolerable if their presence helped to maintain cooperation in the raisin industry. According to the *California Citrograph*, "It would have been a tremendous blot to co-operative marketing, had this most successful association discontinued." [37]

## THE DEPARTMENT OF JUSTICE AND REASONABLE PRICES

The packers thought the CARC was too successful to continue. They accused the CARC of peremptory and arbitrary dealing with them and with the trade.[38] Prompted by the packers' complaints, DOJ agents resumed their investigation, and the U.S. attorney finally filed his report a year after the campaign ended. Although the Fresno newspapers reported little of the night riders' actions, DOJ special agents learned about the violence by interviewing victims and witnesses. Nevertheless, Schoonover found that no violation of the Sherman Act had occurred: "Any conduct of this kind on the part of any of the growers," Schoonover informed his superiors in 1917, "is wholly without the knowledge and sanction of the association." As long as the CARC used its market control to decrease prices to the consumer, Schoonover contended, federal officials had no reason to worry.[39]

The department's Washington staff disagreed. They now believed that the CARC was neither a nonstock, nonprofit cooperative under the Clayton Act nor an organization concerned with "obtain[ing] and maintain[ing] a fair and reasonable price" as the appropriations rider stipulated. In the department's view, the CARC was not a true growers' association because only 45 percent of the growers under contract actually held shares in the cooperative. The voting trust agreement further emphasized the corporate character of the organization. In the view of Lincoln Clark, a DOJ special assistant, the CARC "as now constituted and conducted is a combination in violation of the Sherman Law." Clark took the strong position that the exemption in the Clayton Act and the appropriations rider were unconstitutional. His assessment of the CARC invoked the reasoning of *Connolly v. Union Sewer Pipe Co.*: "I am unable to see any distinction between a pool of commodities which are the product of agriculture and a pool of commodities which are the result of manufacture." [40]

The DOJ concluded in 1917 that it could prosecute the CARC under the Sherman Act. Since *Connolly* was still good law, and since the CARC was not entitled to the Clayton Act exemption, the department believed it had legal grounds to file suit against the CARC at that time. Yet it hesitated to do so. The uncertain state of federal antitrust law made prosecution tactics difficult to gauge. According to Special Assistant Attorney General G. Carroll Todd, "the lower Federal courts have been thrown into such confusion regarding the law of restraint of trade" that the DOJ could not be sure of its chances for success against a farmers' organization. Until "some further deliverance by the Supreme Court" on the meaning of the rule of reason, Todd reluctantly concluded that no federal prosecution should be undertaken.[41]

Despite all the information it had collected in its investigation, the DOJ still could not prove conclusively that the rise in the retail price of raisins was caused by the CARC's monopoly and not other social or economic factors. World War I, for example, had raised prices for most food commodities. Political considerations weighed in the department's decision as well. Even though the rider carried no weight as substantive law, in this case it prevailed over the Antitrust Division's legal judgment. In a time of dwindling congressional commitment to enforcement of the antitrust laws, the DOJ could not risk losing its appropriations by prosecuting a farmers' association—even if that organization bore, in the words of the chair of the Federal Trade Commission, "all the elements of a ruthless trust monopoly." [42]

## "JESUS WOULD SIGN UP"

Meanwhile, the CARC confidently continued its vertical integration. In 1917, the cooperative built six new packing plants. Giffen also initiated two new selling practices with the national trade. Under the first, known as firm at opening price, the CARC took orders from wholesalers and jobbers in the spring of each year at prices to be named by the cooperative in late summer. Second, the CARC guaranteed this price against any decline from the date it was named until December 31, the end of the holiday trade. Through these two trade practices the CARC could hold buyers to their commitments until the size and quality of the crop could be determined. Giffen believed that such dealing would reduce the incidence of speculative purchases and ensure a profitable return to the growers.[43]

Of greater consequence was the crop contract. The option signed in 1916 expired in 1917. In 1918, the CARC ran a campaign to obtain signatures on a brand-new crop contract, one that differed from the first in several important respects. To put an end to the problem of losing vineyard acreage by sale or tenancy, this new contract ran with the land. Succeeding owners or tenants were bound to the contract just as if they had been parties to it. Thus the new contract imposed a servitude on the land rather than effecting a bargain between two parties. In addition, if the previous owner failed to notify the CARC of the transfer, that person was still personally bound by the contract and liable for damages. To discourage breach of contract, the agreement imposed liquidated damages of two cents for every pound of raisins sold outside the CARC. The liquidated damages clause was a California innovation; it

Women workers packing raisins in the CARC plant on G Street in Fresno, 1917.
(Historical photographs from the C. "Pop" Laval Collection, Fresno, California)

provided the strongest weapon against free riding that American agricultural cooperatives had ever seen. Finally, the CARC raised the guaranteed price paid to the growers to three and a half cents per pound. The new contract was expressly designed to correct the "defects" of the old one, which allowed too many vineyards to slip out of the cooperative's control.[44]

The tensions of the 1918 campaign were exacerbated by wartime inflation, social tension, and xenophobia. With overseas markets shut off by the European conflict, effective domestic marketing was all the more crucial, and Giffen argued that only the CARC could protect the growers from economic instability: "If the growers will sign up the new contract, the California Associated Raisin Company, their great co-operative organization, will become greater, more powerful, and more useful to the community, and continued prosperity will be assured. If the contracts are not signed the growers will lose control of their business and we must be prepared for a return to the era of ruinous competition and resultant low prices." The CARC shamelessly invoked the idea of the greater good in its campaign circulars: "The result of the campaign upon which we now enter determines not alone of the prosperity of ourselves, but in a much larger measure than we realize, affects the welfare of the state. The eyes of California are upon us today, and let no man falter!"[45]

This sort of exhortation encouraged night riders to resort to "strenuous methods" to obtain signatures. They threatened to have German growers ar-

Growers signing contracts in the CARC offices, 1917. William Saroyan wrote in 1980 that some of the men pictured here were Armenian. (Historical photographs from the C. "Pop" Laval Collection, Fresno, California)

rested as spies. They burned farm equipment. They organized boycotts, had people fired from their jobs, and induced growers to break contracts with other packers. They threatened to lynch people. The plight of an Armenian grower named Rustigian whose wooden raisin trays were destroyed by arsonists actually made the Fresno papers. The CARC reimbursed Rustigian for his loss, but the perpetrators were fined only $5 because of "the sentiment in the community."[46]

Frightened at the prospect of losing their land and perhaps their lives, Armenian growers reluctantly capitulated to the mobs. Charles Bonner, one of the leading outside packers, tried in vain to retain his growers' business. His Armenian neighbor, who had resolved not to join, confessed to Bonner that he capitulated after a mob paid him a visit: " 'The people came after me the other day, a lot of these men, and they came and they just scared me, make me scared of my life, and they make me sign the contract.' " Bonner then took a more active approach; when the Missikian brothers were threatened, Bonner took them to an attorney who drew up affidavits to serve on the CARC as proof of the coercion. But the brothers, afraid of being hounded out of the valley, refused to sign. Bonner then confronted Giffen himself, but the conversation infuriated the packer: " 'Why, of course everybody knows we don't approve of these methods,'" Giffen reportedly said, "'but it is beyond us to control it. There is a pretty strong sentiment in this community and it is probably surprising that there is not a little more violence and lawlessness; but there is not anything that we can really do to stop it.'"[47]

Toward the end of the campaign, the CARC published a full-page advertisement in the *Fresno Morning Republican* "counseling prudence" and exhorting growers to "make no threats, indulge in no abuse, coerce no one, and forget the boycott." Giffen did nothing further, however, and his position only reinforced the packers' suspicions that the CARC was involved in orchestrating the activities of the mobs. When an employee of the Alta Irrigation District was threatened with dismissal if he did not sign with the CARC, he "made some inquiries" and discovered that James Madison had appeared before the directors of the district "and had made the request of them that they put it up to him just in that way that he should join the Association or that he should lose his job."[48]

To the CARC and its devotees, the cooperative and the community were indistinguishable. Growers thought of cooperation in overtly religious terms. They circulated the story of a Fresno preacher who sermonized on what Jesus would do if offered an CARC contract. The moral of the sermon "was that Jesus would sign up." The mobs and their excursions became a central focus of community life in the raisin district, according to raisin packer E. L. Chaddock: "One day they declared a holiday in Fresno, shut up the stores, and all the merchants formed these caravans. There would be a dozen automobiles go to a man's place, get a man surrounded, and put all kinds of pressure and persuasive methods trying to compel that man against his will to sign this contract."[49]

The results of the 1918 campaign bore stark testimony to the new order. More than ninety-two hundred raisin growers, or 88 percent of the industry, were now under contract to the CARC.[50] With a secure, strong monopoly over supply, Giffen and Madison proceeded to complete the CARC's control over packing and processing. When the membership campaign ended, they notified the packers that the CARC would no longer sell them raisins to pack and sell to their own customers. As Giffen told one packer, "It may seem cold-blooded to say so, but . . . we have decided that it is to the interest of the growers to have the profit which you have been making, and . . . the easiest way to bring that result is to eliminate the packers." This decision stripped the packers of 75 percent of their business and gave the CARC unchecked authority over prices. The CARC could easily afford to cut the packers out of the raisin trade; the proceeds from the one-quarter-cent fund enabled the cooperative to build an impressive array of state-of-the-art packing plants with

An aerial view of the Sun-Maid plant at Hamilton and E Streets, Fresno, 1936.
(Historical photographs from the C. "Pop" Laval Collection, Fresno, California)

sufficient capacity to handle most of its business. The CARC retained the services of only one packer. In February 1918, the giant California Packing Company (CPC) received a preferential contract to pack the rest of the CARC's
raisins—a reward, the packers' attorney later charged, for the CPC's noninter-
ference in the 1918 crop contract campaign. Later that year, the CARC declared
a stock dividend of 8 percent.[51]

The year 1918 marked the pinnacle of the CARC's economic life. It had succeeded in ousting the packers from the raisin industry, a goal Kearney never
achieved, and through persuasion and force had secured a greater monopoly
over the growers than any other cooperative. In 1918, nearly 90 percent of the
crop was marketed through the CARC. Its prices for Muscat and Thompson
raisins, the two major varieties, steadily increased even as statewide production set new records every year. Few cared that the CARC had rewritten the
cooperative gospel in establishing its "good trust," except perhaps the victims
of the night riders. What mattered was that American farmers had finally tri-
umphed over their perennial foes, had won a just share of the fruits of their
labors.

The CARC's very success waved a red flag in front of DOJ attorneys. The
packers quickly circulated reports of night riding to Washington, thinking

that this evidence would finally launch a prosecution against the CARC. The federal government's continued inaction astounded the packers, the distribution firms, and the eastern traders. How long could farmers conduct a notorious monopoly with impunity? The entry of the United States into World War I delayed the legal showdown, but the war created the conditions that forced the CARC from its protected position in the legal shadows into the stark theater of Progressive Era antitrust sensibilities.

# 7 ∾ Busting the Raisin Trust?

*Call us a trust, if you want to, but we're a benevolent one. — Wylie M. Giffen, 1920*

The CARC's phenomenal growth did much more than increase the availability and price of raisins. It challenged the way Americans looked at farmers. By becoming, or at least appearing to become, as large and powerful as any trust in industry, the CARC sundered the picture of American farmers as scattered, individualistic, and dedicated to organizational inefficiency. In undermining the cultural ideal of American farming, the CARC polarized the debate about farmers' role in the new economy. The CARC's monolithic, corporate character did not provide a compelling substitute for the old ideal. Instead, it ignited a legal and political conflict over agriculture at a time when world war made it difficult to measure a cooperative's actual impact on prices.

When World War I made food prices a critical political issue, the CARC's opponents had the opportunity they needed to force the government to act. The CARC's success in expanding the market for raisins ironically led people to expect raisins to be both easily available and cheap. Immediately after the war, the CARC raised its prices by 100 percent over the government's wartime rate, at last giving the federal government grounds for a Sherman Act prosecution. The trigger of the antitrust gun, as it turned out, was price, the standard contained in the rider, and not organic structure, the standard in the Clayton Act. Price was the key to the DOJ's jurisdiction. Only when the war skewed perceptions of reasonable prices did the political and legal grounds for prosecution emerge. Only when the CARC's prices exceeded "fair and reasonable" could the DOJ step outside the restrictions in the rider.

The raisin antitrust case raised as many issues as it purported to settle about the legal status of cooperation, the cultural image of farming, and the farmer's place in the economy. As the raisin case dragged on in federal district court at Los Angeles, Congress was considering new legislation to clarify cooperatives' standing under the federal antitrust laws. The Capper-Volstead Act resulted from two contradictory impulses: first, that farmers should be protected from the antitrust laws, with the implication that farmers deserved special treatment *as farmers*; and second, that farmers could combine in cooperatives without significantly threatening free competition. The image of farming had significantly changed since 1900. Though farmers remained an important cultural symbol, they also had significant market interests to protect.

Congress's consideration of this law effectively undercut the government's prosecution. One month before the Capper-Volstead Act became law, the DOJ entered into a consent decree with the CARC that reflected how little antitrust standards had changed. The CARC agreed to refrain from market practices that constituted unfair competition, but its monopoly and organic structure were left intact. The legal order embraced the cultural preference for the older ideal of farming while recognizing that the new economic powers embodied in the CARC's model of cooperation were here to stay.

## WORLD WAR I AND AGRICULTURE

World War I transformed domestic agriculture. The war closed lucrative export markets to U.S. farmers, who responded by producing less. Staple crop production leveled off after 1914, leading to sporadic shortages of wheat, potatoes, dairy products, and meat after the United States declared war on Germany in April 1917. The resulting price inflation triggered an unusual exercise of the government's emergency powers. In August 1917 Congress passed one law to encourage food production and another to authorize the U.S. Food Administration to impose price controls for the duration of the war. Headed by Herbert Hoover, the Food Administration had broad, unprecedented authority to regulate agricultural markets, encourage farmers to organize marketing and distribution cooperatives, and promote farm conservation. Its activities pointed in the direction that agricultural policy would take during the 1920s, but in the short run its broad regulatory capacity helped relieve the economic and legal pressure on farmers and cooperatives exerted by both the war and the industrial conditions of the Progressive Era.[1]

The war also altered the public's views of the sanctity of farming. During the war, farmers' patriotic contributions to the war effort and to supplying the home front were roundly celebrated. When the government lifted price controls immediately after the war, the resulting surge in food prices marked farmers as crass profiteers and led to a backlash of renewed antitrust enforcement against cooperatives and their officers at both the state and federal levels. Then, agriculture's postwar tailspin shifted the onus to speculators. By mid-1919, it became imperative to ensure that farmers could get a living wage, remain in the business of farming, and feed the nation — goals that gained importance in the wake of the war and the devastation in Europe. Farmers had gone from patriots to profiteers to prey: "Nobody objects to dairymen's, or other farmers' or fruit growers' cooperative associations per se. What is objected to . . . is that Big Business should get hold of them and exploit them and the public by purely trust methods which not only work great injury to producer and consumer, but are against the laws of the land." [2]

The "farm crisis" that followed the war brought agriculture's economic distress back onto the center stage of national politics for the first time in nearly thirty years. Farmers suffered steep losses when commodity prices collapsed in 1919. A Virginia farmer succinctly described the crisis: "Many of us in the coming year, in spite of slogans such as 'Food Wins the War' or 'Feed the Nation' or 'Feed Starving Europe' will merely try to feed the farmer." Prices reached a nadir in 1921, but two years later farm prices remained much lower than retail prices for manufactured goods. Farmers sought assurance from the federal government that they could use marketing organizations to improve their prices without risking legal action against them. As historian James Shideler put it, "If governmental policy could change the conditions of agriculture during a war emergency, government might also be used on behalf of agriculture during an economic emergency." [3]

Some courts responded to the war's emergency by softening prior precedents against capital stock cooperatives. In 1919, a regularly incorporated cooperative persuaded the Alabama Supreme Court that its purpose was "mutual benefit through the application of cooperati[ve] . . . principles," a conclusion probably supported by the fact that the organization provided services to members at cost and paid no dividends to stockholders. [4] The same year, the New York Court of Appeal similarly construed the facts in two cases involving the assessment of liquidated damages by dairy cooperatives. Instead of regarding cooperatives as a threat to the market, the court focused on their purpose and function: "It cannot be said that an agreement such as this would tend to restrain trade or stifle competition; on the contrary . . . it encourages competition by bringing a new creamery into being." [5]

The willingness of a few state courts to break down what had been an inflexible link between capital stock and restraint of trade marked the first judicial acknowledgment of the new form of cooperation. As the war altered national economic conditions, it shifted attention from the goal of decentralizing market power by thwarting the accumulation of capital in corporate cooperatives, as was the courts' purpose before 1917, to a new priority that would dominate agricultural policy throughout the 1920s: stabilizing the farm sector against a relentless cycle of overproduction and low prices. The 1919 crash was so steep and severe that the likelihood that farmers would manipulate cooperative organizations to enrich themselves unjustly seemed greatly diminished. Indeed, the Alabama and New York courts endorsed key features of the new model of cooperation without reconciling the new results with earlier hostile decisions. The New York court justified liquidated damages—which earlier courts ruled were in restraint of trade—by noting the investment-backed expectations of dairy farmers in their organization: "It was quite necessary that they be assured of the continued patronage of their members in order to justify the association in going to the expense of acquiring and erecting a creamery."[6]

Though it came in the throes of postwar depression, the recognition that cooperatives could help farmers secure "a fair and reasonable price" for their products, as the federal rider mandated, was an essential prerequisite for accepting cooperation as a legitimate form of economic organization. That larger recognition would still be tied to legal expressions of cooperation, which in turn continued to be understood in the context of antitrust law, monopoly, and restraint of trade. The Alabama and New York decisions, which arose out of private disputes between cooperatives and their members, were not as influential in shaping the new legal conception of cooperation as public prosecutions of cooperatives by states and the federal government. Antitrust prosecutions not only raised the issue of the legality of cooperatives' market practices, they also forced a reconsideration of the underlying issue of whether farmers deserved to be—or could constitutionally be—treated separately from other entrepreneurs under the law. The Sherman Act suit against the CARC, which involved an actual monopoly, united these issues and brought into the open the latent contradiction between the idealization of farming and Americans' expectations for a strong, self-sufficient farm economy.

## THE PACKERS STRIKE BACK

The cultural image of farming had little relevance to the CARC's opponents. They viewed the CARC as an impediment to free trade in raisins not because they opposed cooperation as such but because they believed the CARC did not play fair. Their aim was not to destroy the CARC, because the CARC had helped the industry by expanding demand; rather, they sought only to peel away some of its market control so they could obtain more raisins to sell. Their lawyer said as much in 1921: "What we are seeking to do is to cut down the percentage of control of the raisin production held by this corporation to such an extent that the independent dealers may have some portion of the crop for their trade, and thereby compete with [the CARC]."[7]

The war's end found the raisin packers in as desperate a position as they had ever been. The results of the 1918 contract campaign and the CARC's decision to cut commercial packers out of the business completed the cooperative's consolidation of the industry. After the CARC took delivery of the 1918 crop, which totaled 167,000 tons, fewer than 20,000 tons were free for "six or seven" packers to purchase. Most of the packers responded by "paying practically any price necessary" to get raisins. The heightened competition increased the already ample incentive for growers to break their agreements—and for the packers to induce breach: "[Charles] Bonner," one packer wrote to another, "is lending money in vast quantities to every scalawag who wants to borrow it, providing the aforesaid scalawag will agree to deliver a few raisins to him. The scalawag usually hails from Armenia and is full of the wiles of the average Turk." The slur demonstrated the low regard in which some packers held Armenian growers. It infuriated the packers to have to pay outrageous sums to growers. These firms could not both buy raisins for a price higher than the CARC's and sell them to consumers for less than what the CARC charged. Franklin P. Nutting's American Seedless Raisin Company (ASRC) was forced to "borrow heavily" to meet its expenses while marketing only 2,000 tons of raisins in 1919.[8]

The packers tried extreme measures to preserve their businesses. They encouraged growers to test the legality of cooperative crop contracts, but the California courts upheld these agreements.[9] Then the ASRC's Nutting took the CARC to court himself, contesting the CARC's accounting for business it transacted with his company before the CARC began dealing exclusively with the California Packing Company. Nutting won the right to inspect the CARC's books, but Wylie Giffen stalled for months while negotiating a settlement.[10] Since a state prosecution seemed unlikely in view of California's lenient antitrust provisions, the packers focused on federal remedies. In 1918, even before the conclusion of the contract campaign, the packers stepped up the legal assault on the CARC that they had initiated before the war in the halls of Congress, the Department of Justice, and the Federal Trade Commission.

To take the packers' case before the federal government, Nutting, the son of former CARC director William R. Nutting, organized the "Big Five" packing firms into a "raisin committee" to finance federal legal action against the CARC. This group included the ASRC, the J. B. Inderrieden Company, Bonner Packing Company, Chaddock and Company, and Guggenheim and Company. When the DOJ's 1916 investigation of the CARC did not result in prosecution, the Big Five turned to the Federal Trade Commission (FTC), which eventually agreed to conduct its own investigation.[11] It summoned Giffen to Washington to appear in person at hearings held in August 1918; shortly thereafter, the commission issued its findings.[12]

The FTC's interpretation of the CARC's legal character agreed with that of the packers. "It appears that a corporate device is being employed at Fresno, California, which is clearly intended to appear as a co-operative institution but which is really a private corporation with all the powers and privileges of a private corporation. . . . it is a weapon too powerful to be permitted to exist for it is quite conceivable that a single man or a single large interest might obtain all of the capital stock or a majority of it and turn the corporation into an instrument of oppression as to the farmers and extortion as to the consumers." The FTC proposed two remedies: outright dissolution of the CARC, which the commission was reluctant to order, or substantial alterations "that would end the unlawful conduct of the corporation and yet preserve its power for good."[13]

This, however, was as far as the FTC could go. Under the Federal Trade Commission Act of 1914, the FTC had no judicial powers of its own; additionally, the FTC assumed that the CARC operated in interstate commerce, but the DOJ would have to prove it. Moreover, the FTC was in a politically precarious position. Its investigation into the meatpacking industry in 1917 won support from farmers and consumers. The National Farmers' Organization quickly and publicly allied itself with the FTC. After a few months in wartime Washington, Nutting realized that the Big Five's attack on the CARC could not have been more poorly timed: "Now we are asking the Commission to attack organized farmers. The atmosphere around the Commission's offices seemed just a little blue this morning when I went to see them."[14] The FTC turned its findings over to the DOJ, but as Nutting knew, the attorney general's staff was hamstrung by the appropriations rider.[15] By the end of 1918, all that the complaints, investigations, and reports had produced was a legal impasse: the FTC was convinced of the illegality of the CARC's structure and market behavior but had no authority to sue; the DOJ had the authority to sue but was reluctant to use it without convincing evidence of the CARC's impact on price.

In the wake of the FTC's report, the packers debated strategy. Nutting wanted the DOJ to file suit for dissolution, despite the government's belief that a criminal conviction would be nearly impossible to obtain in Fresno. Charles Bonner thought the packers had a better chance of winning a case that attacked only the CARC's preferential contract with the California Packing Company. Rather than call for the CARC's dissolution, he counseled, the packers should milk local sympathy for their position: "Business men in Fresno and . . . independent and Association growers alike . . . nearly all regard this discrimination against us as improper, unfair, and we could go into the Courts here with a great deal of sentiment in our favor if our claim was only for equal rights."[16]

Bonner had a point. The federal government was unwilling to challenge the existence of farmers' cooperatives during the Progressive Era. The U.S. Food Administration's wartime regulations made it difficult to prove that cooperatives had unreasonably increased prices. Agricultural leaders publicly proclaimed the farm sector's patriotic sacrifices in supplying the nation and the Allies during the war. Furthermore, the CARC had powerful political allies, as Bonner informed Nutting: "[Giffen and William Sutherland] are going to work through [Colonel Harris] Weinstock, the Market Director in California, who is daffy on the subject of co-operation and who holds the confidence of Senator [Hiram] Johnson." Nutting was unmoved. To him, the CARC was a monopoly, a trust, and a capital-stock corporation. As such, it could make no claim to protection under the Clayton Act, and so, he thought, the government could at least order the CARC "to give up the monopoly."[17]

## THE CHANGED POSTWAR CLIMATE

The government declined to issue any such orders during the war. The DOJ did not have the political capital to spend on a prosecution of a farmers' cooperative while the nation's attention remained fixed on the European conflict. The armistice, however, changed everything. Politically, it permitted Congress to return to domestic policy; economically, the repeal of price controls sent retail prices for food and finished goods spinning upward. Accusations of profiteering abounded as inflation infected every sector of the economy. Farmers were no longer immune to legal attacks. Beginning in 1917 and lasting for the better part of two decades, the DOJ began Sherman Act investigations of the organizations of dairy farmers in several major cities. In an effort to force urban milk distributors to pay better prices, dairy cooperatives staged "milk strikes" in Chicago and New York. They fixed prices, boycotted milk distributors, cornered supplies, diverted shipments from their intended destinations, and "us[ed] economic sanctions to keep members (and nonmembers) 'in line.'" State officials quickly filed suit to enjoin the cooperatives from further obstruction of trade. In 1920, U.S. attorneys procured indictments — later quashed — against the milk producers at New Orleans.[18]

These events outlined some limits to agriculture's favored status. At President Wilson's behest, the House was considering amendments to the food control law "to enable the government to get at profiteers." According to California Republican representative J. A. Elston, this measure would not substantially curb the "farmer trusts": "At the present time Congress seems to be very partial to the farmer and is willing to permit him to commit individually and collectively almost any kind of a crime on the calendar. . . . We are up against an absolute stone wall so far as getting any congressional favor for those who oppose farmer monopolies."[19] Since neither the DOJ nor the FTC shared Congress's favoritism toward farmers, antitrust enforcement would depend on what happened in the market. Changes in price would shape the government's interpretation of the law.

American farmers saw the high prices of the postwar era as a sign of increasing demand for their crops resulting from the reopening of international markets. For raisin growers, the perceived upswing had other causes: the advent of Prohibition increased demand for raisins as "thousands of people began experimenting in the production of homebrew."[20] Raisin prices after the war, like everything else, would be measured against prices during the war. The 1918 crop contract specified a minimum price of four cents per pound. The Food Administration set a 1918 price of five and one-half cents. Giffen and CARC general manager Fred A. Seymour expected that the government's price would yield sufficient profit to cover the increased costs of labor and "everything used in the vineyard." The weather did not cooperate. Heavy rainfall during the drying season destroyed thirty thousand tons of raisins and left most of the rest of the crop too wet and moldy to pack. The short supply not only reduced growers' income—at this time the market was not permitted to respond to the decrease in supply—but also meant that few raisins from the 1918 crop would be carried over into 1919.[21]

Normally the absence of a carryover would be good news, but 1919 proved to be anything but a normal year. During the spring months, independent packers sold their 1919 raisins on contracts for delivery at a future date, as was their custom, at a record price of ten cents per pound. Demand was skyrocketing, and the packers, desperate to find raisins still in growers' hands, circulated rumors that the CARC's as yet unannounced 1919 prices would be no higher than 1918 levels. They were gambling that growers would still profit from selling outside the CARC even after paying liquidated damages. Giffen took to the pages of the *Sun-Maid Herald* to deny the rumors and prod growers to honor their contracts when the CARC named its prices in August. Giffen predicted that the packers' ten-cent price would not hold up for long: "These prices are admittedly a result of a very unique situation as far as market conditions are concerned and though it may be proper to take advantage of this situation, we do not believe that these prices can be maintained for a long period of time."[22]

Giffen guessed wrong. Without imports to supplement the California crop, the CARC's raisins had to satisfy a level of demand pushed even higher by Prohibition: "Raisins have become our national drink." On August 22, the CARC's

TABLE 7.1. CARC Prices to Growers and Wholesalers for Muscat Raisins, in Cents per Pound, 1914–1920

| Year | Price to Growers | Price to Trade | Difference (CARC Profit)[a] |
|---|---|---|---|
| 1914 | 3.3 | 6.9 | 3.6 |
| 1915 | 3.6 | 7.0 | 3.4 |
| 1916 | 4.2 | 7.0 | 2.8 |
| 1917 | 4.9 | 9.0 | 4.1 |
| 1918 | 5.5 | 9.6 | 4.1 |
| 1919 | 10.0 | 15.0 | 5.0 |
| 1920 | 15.0 | 21.0 | 6.0 |

*Source*: Testimony of Wylie M. Giffen, U.S. Congress, Senate, Committee on the Judiciary, *Hearings Authorizing Associations of Producers of Agricultural Products*, 67th Cong., 1st sess., June 1921 (Washington, D.C.: U.S. Government Printing Office, 1921), 113.

[a] This difference is what the CARC claimed it returned to growers as patronage dividends and what the packers argued the CARC paid out as earnings on capital stock.

board of directors issued a price schedule that shocked the national trade: Sun-Maid raisins would retail in 1919 for fifteen cents per pound. Bowing to pressure from growers who expected the CARC at least to meet the packers' prices, the CARC fixed prices to the growers at ten and one-half cents per pound, double the previous year's war-controlled prices. Still, the CARC claimed that its price was lower than the independents', and brokers took advantage of the CARC's price increase to add a thirteen-cent cushion of their own (Table 7.1).[23]

It was the CARC's price that grabbed all the attention, however. Temporary market conditions notwithstanding, the packers were furious; the CARC's price would bring between $8 and $10 million more to the industry that year but, with no raisins to sell, they would not share in the bounty. The wholesalers and retail grocers fumed because firm-at-opening-price contracts compelled them to commit to their purchases before the price was named, usually months later. Now they were legally bound to take delivery of raisins that were going to be considerably more expensive than they expected. Few retailers believed that consumers would buy raisins at CARC prices. Supplied with inside information by the packers, eastern trade newspapers accused the CARC of profiteering and branded the CARC "the Raisin Trust." One indignant Pennsylvania grocer swept all California cooperatives into the same indictment:

"The California people seem to get away with things that no other state attempts."[24] Giffen argued that the CARC's price was dictated by the packers' own high prices the preceding spring, but people outside the industry remained unconvinced that free competition actually existed.[25]

The changed national legal and political climate gave the raisin packers an opening in which to wedge their own grievance. Early in 1919, the Big Five retained Breed, Abbott & Morgan, the New York law firm whose clients included the National Wholesale Grocers' Association, and Joseph E. Davies, former chair of the U.S. Bureau of Corporations, whose Washington, D.C., practice often took him before federal regulatory agencies. The packers' New York lawyers reported to their clients that the DOJ had reopened its investigation "with a view to prosecution if the facts warrant."[26] On September 23, however, the House passed another annual appropriations rider; the packer Alex Goldstein noted that Senate passage would bar the attorney general from taking action against the CARC. Before the Senate could act on the appropriations bill, Attorney General A. Mitchell Palmer launched a preemptive strike. On September 30, invoking his authority under Section 6(e) of the Federal Trade Commission Act for the first time, Palmer formally requested the FTC to investigate charges of exorbitant prices in the raisin industry. This unprecedented maneuver served Palmer's presidential ambitions brilliantly. It conveyed the impression that he was actively moving against profiteers; at the same time, it protected him by making the FTC, not the DOJ, investigate a farmers' organization. At the same time, by delaying a decision to prosecute the CARC for the third time that decade, Palmer threw the political hot potato back at Congress.[27]

## THE FTC SUMMONS

Palmer's order detonated political cannons in the Central Valley congressional delegation. Senators Hiram Johnson and James Phelan took personal charge of the appropriations bill to ensure its speedy passage. A Washington, D.C., rally of the National Bureau of Farm Organizations in September induced the House to reverse a vote to kill the rider. House Democrats Henry E. Barbour of Fresno and Hugh S. Hersman of Gilroy presented a bill exempting farmers' organizations from the federal antitrust laws to the House Judiciary Committee in October, "asking that it should be rushed through because the Department of Justice and Federal Trade Commission have threatened to prosecute the Raisin Growers' Association." Representative Elston sent the packers a behind-the-scenes report: "It is my belief that Hersman and Barbour, with the aid of Phelan, are endeavoring to obtain either delay or milder treatment of your Fresno friends." The Senate's passage of the appropriations bill on November 4 seemed to guarantee the latter.[28]

The California congressional delegation, however, could not delay the FTC hearing. In October, the FTC summoned all parties to Washington to present evidence. Both sides brought big legal guns. John W. Preston, a former assistant U.S. attorney from San Francisco, led the packers' charge against the CARC. By his own admission, Preston had never set foot on a raisin vineyard, but he fought the packers' cause as if it were his own. Preston's co-counsel, the firm of Breed, Abbott, & Morgan, was experienced in defending firms in Sherman Act cases and now asserted its clients' interests in the CARC proceedings. Joseph E. Davies also signed the packers' briefs; his role at the hearings was minor. Counsel for the CARC was former California state assemblyman William A. Sutherland, a Fresno attorney who had served as legal adviser to the CARC since its inception. Before Palmer's order, Sutherland worked furiously to avoid federal proceedings; after the FTC scheduled its hearings, he put it to the DOJ that no farmers' cooperative could prevent speculation in foodstuffs: "Unless an effort is made . . . to curb this profiteering by the middleman, either jobber or retailer, the disruption of the producers' organization will accomplish no good so far as the consumer is concerned."[29]

The packers were unimpressed with the CARC's attempts to sway the federal government. Nutting boasted to his subordinates: "Our attorneys are far and away superior to the Association's attorneys." He was ambivalent, however, about his goals: "On the one hand, I would like to see the Raisin Trust thoroughly regulated by the Government in such a way as to compel it to maintain competition in the selling of goods. . . . On the other hand, we are among the largest growers of raisins, and thoroughly interested now in the maintenance of at least reasonably high prices." In fact, the CARC was a convenient scapegoat for high postwar prices, as George H. McCandless, ASRC general manager, admitted: "We are not in any position to condemn anybody for profiteering and in consequence of this we have been keeping very quiet on the price question here. Some of our jobbers occasionally accuse us and we try to throw the burden on the Associated making it necessary to pay the farmers so much."[30]

What, then, did the packers hope to achieve with an adversarial proceeding before the FTC? One thing is clear: they did not seek the dissolution of the CARC. Rather, they objected to the CARC's horizontal and vertical integration. As McCandless put it, "The thing that I have been hoping for and I think what the wholesale grocers associations are working for is to get the Government to force the Associated to give up control of enough of the total crop to come within the limits that have been generally fixed for a monopoly." This position required the packers to sustain two related arguments: first, that the CARC was not a true farmers' cooperative; and second, that a farmers' cooperative organized along the lines of the CARC constituted an illegal monopoly. In a telegram to their New York attorneys, the packers outlined their position: "The association is not a cooperative company but is a stock corporation, makes profits, and pays dividends and therefore its operations should be regulated and its control restricted."[31]

As they headed for the hearings in Washington, both sides knew that neither could do business without the other. The CARC could not pack all the raisins its growers produced, and the packers could not pack and sell raisins if the growers could not afford to grow them. Ignoring their mutual dependence, both sides headed to Washington embittered and angry—the packers at their heavy losses, the CARC at the packers for the concerted attack on "the pride of the Valley."[32]

## "THESE GENTLEMEN ARE NOT LONG-HAIRED, DOWN-TRODDEN FARMERS"

The FTC's hearings began on November 20, 1919. An investigation that the commissioners expected to take three or four days stretched into two weeks as the commission and attorneys for the DOJ, the CARC, the packers, and the wholesale grocers locked horns over the legality of the CARC's selling practices, its control of the growers, and its corporate structure. The CARC's fate depended on two issues: whether the CARC's 1919 price violated the "fair and reasonable" standard of the Sherman Act, and, if so, whether the CARC's identity as a cooperative would bring it under the Clayton Act's exemption for nonstock associations. If the CARC could prove that it met the legal definition of a cooperative, it might mitigate any adverse findings the FTC might make as to price.

Preston saw the hearings as his best opportunity to secure a ruling that would limit the ability of farmers' cooperatives to control commodity prices: "The California Associated Raisin Company is now and during its entire existence has been operating in violation of the anti-trust laws of the United States, [and] it is not . . . under any exemption from the operations of that law." The CARC had initially held itself out to be merely the growers' selling agent, but once it obtained a monopoly over supply, it went "into the commercial end of the business." This deception, together with the monopoly, Preston argued, enabled the CARC to set unreasonably high prices in 1919, and the only way to ensure that it did not do so again was to dissolve the monopoly.[33]

To buttress this argument, Preston claimed that the crop contract, stock-holding pattern, and the one-quarter-cent fund all furnished evidence of the CARC's corporate identity. The 1918 crop contract gave growers little freedom; if they breached, the CARC could claim liens on vineyards as well as damages. The CARC's shareholding was trickier to interpret. In 1919, 2,454 growers held 7,765 shares and 853 nongrowers held 2,554 shares. In fact, these figures were not out of proportion with each other. Growers, who represented about 75 percent of the shareholders, owned about 75 percent of the stock, but since the CARC had more than 9,000 growers under contract, that meant that most growers held no shares. Preston noted that by 1919, the CARC had a surplus of $323,000 even after paying stock dividends, an amount that protected the CARC against possible losses if market prices dropped below the growers' contractually guaranteed price.[34]

As much as he detested the CARC's monopoly, Preston understood that price was the real issue. He knew he had to persuade the FTC that the CARC's prices were arbitrary and not the product of reasonably different calculations of costs. Both sides agreed that it was nearly impossible to compare the prices of all brands of raisins sold in the United States because they differed in size, quality, and packing. The packer Charles Bonner estimated a reasonable price for 1919 at six and one-half cents per pound; the one-cent increase over 1918 covered increases in the costs of packing and growing.[35] In response, Sutherland produced figures showing that the CARC was underselling the ASRC on eleven-ounce packs by eighty-four cents, while saving consumers hundreds of thousands of dollars. He claimed the CARC's 1919 price was reasonable given the growers' rising production costs and the small carryover from 1918. And Sutherland got Bonner to admit under cross-examination that the packers had not included in their estimates some costs accruing to the growers.[36]

When the packers concluded their presentation, Sutherland's first move took everyone by surprise. On the third day of the hearings, he made a bold concession: "I wish to say . . . there is no contention, never has been and never will be any contention that this concern is within the proviso of Section 6 of the Clayton Act." What he gave with one hand he retook with the other: "(Barring the mere fact that our company is corporate in form) the California Associated Raisin Company is in fact a cooperative concern in the highest sense and is at least embraced by the Congressional intent that farmers' organizations shall be exempted from the operation of the anti-trust laws."[37] The CARC's attorney sought to exploit the Clayton Act's ambiguity on the legal status of cooperatives by persuading the FTC that the CARC's corporate structure came within "congressional intent." If the FTC bought that argument, then a conflicting but equally "reasonable" calculation of price would undermine the packers' claim that the 1919 price was illegal on its face, and the case against the CARC would collapse.

That was exactly what Preston feared most. Casting the CARC as a powerful, financially independent corporation, he painted a picture of the cooperative ideal gone bad: "[The CARC] has obtained and does now maintain a complete, thorough and air tight monopoly on the production of raisins and upon the marketing of the same. . . . These gentlemen are not long-haired, down-trodden farmers." Preston urged the commission to recognize what he saw as the true nature of the CARC: "You cannot endorse a monopoly such as this. This concern is admittedly a profit-sharing stock corporation. So long as it remains such, and monopolizes as it admits it does more than 90 per cent of a product, both in production and distribution, you cannot endorse it. If you put the stamp of approval upon an organization of that character, the injury that would result to the nation would be without any method of reckoning."[38]

The FTC determined that it had the authority only to judge the legality of the CARC's market behavior. It declined to address the broader question of the CARC's legal status. In June 1920, it reported its findings to the attorney general, focusing on price as the linchpin of the case and rejecting the CARC's rationale for the 1919 price: "In the absence of a showing of a greater increase in the cost of production . . . after considering the diminishing purchasing power of the dollar, our conclusion is that the price fixed by the Raisin Company for the 1919 crop was in excess of a fair and reasonable price." As the DOJ had done in prosecuting the milk producers immediately after the war, the FTC applied a price standard it did not quantify, relied on only one factor (cost of production), and declined to specify the price that would meet its standard.[39]

Ruling on the legality of the CARC's market practices, the FTC declared that firm-at-opening-price contracts restrained trade, that the preferential contract with the California Packing Company violated the antitrust laws, and that some grower contracts had been obtained through coercion. Authorized by the Federal Trade Commission Act to issue cease and desist orders in cases of unfair competition, the commission offered the CARC a choice: either divest itself of its corporate attributes — the capital stock, for-profit operations, and nongrower stockholding — or retain its corporate identity and cease all market practices prohibited under the Sherman Act — monopoly control over the growers, price-fixing, firm-at-opening-price contracts, and excessive vertical integration. The CARC's structure did not meet the Clayton Act's definition of a legitimate cooperative; thus its conduct came under the Sherman Act's prohibitions on regular corporations. Although the FTC's orders were "fully subject to judicial review," the attorney general was free to use the results of its investigations as he saw fit, including using them as the basis of judicial proceedings.[40]

The FTC's conclusions should have surprised no one. It had already ruled in 1917 that the CARC was "a ruthless trust monopoly."[41] At that time it expressed grave reservations about what would happen if the CARC continued its course unchecked. Three years later, the commissioners were unpersuaded that the CARC's price was based on a reasonable valuation of the growers' labor and other factors that the packers should have counted. The FTC's report showed no sentimental attachment to farming; indeed, it treated the raisin industry as a business and the CARC as a firm. The report conveyed the suspicion that such a powerful corporation, even when its constituents were agricultural producers, would not act in the public interest.

### "IMMEDIATELY FILE BILL AS DIRECTED"

The FTC's findings provided Attorney General Palmer with the legal basis for superseding the appropriations rider and proceeding against the CARC in federal court. The FTC's conclusion that the CARC did not come within the Clayton Act exemption left the CARC "open to challenge under [the] monopoly section [of the] Sherman law." In July 1920, Assistant Attorney General Henry S. Mitchell notified Sutherland that the DOJ regarded the FTC's findings on price "abundantly substantiated." Although the DOJ had not yet determined whether the reforms suggested by the FTC would be an "adequate remedy," Mitchell threatened immediate prosecution if the CARC extended the existing monopoly or charged prices that were "unreasonably high." In any case, unless the CARC cut its prices and reduced its control to the 60 percent level established as legal by lower federal courts, the government would file suit.[42]

Giffen simply did not believe that the department would proceed against an agricultural cooperative. It was getting late in the season, and the upcoming harvest put pressure on the CARC's officers to decide quickly. With Sutherland counseling that the rider would insulate the CARC from prosecution, Giffen and the trustees decided to call the government's bluff. Giffen told the attorney general, "Of course, we have hoped and prayed that a prosecution might be avoided, but when confronted with taking the chance of a prosecution or accepting your suggestions we have decided to take the former if you still feel that is necessary. As we see it there is at least some hope that we can win in a law suit, and there is no hope at all for the continuance of an organization like ours under the suggestions that you lay down."[43] As a show of goodwill, Giffen made three significant concessions for the 1920 season. The CARC dropped the firm-at-opening-price contracts, released 20 percent of the crop under its control to the packers, and auctioned thirty thousand tons of raisins to New York jobbers on August 1. To Giffen, a federal prosecution was unimaginable. As a witness to the industry's turbulent development, he took it as gospel that the CARC had improved the growers' lot while saving consumers money. What the government was contemplating, he argued to FTC commissioner Victor Murdock, would destroy the CARC, throw the competitive advantage to the packers, and impose higher prices on consumers: "How can an organization like ours continue to exist if the Government should say to us that we would not be allowed to pay over 10¢ for raisins and at the same time put no restrictions on competitive bidding of the independent packers?" [44]

Giffen and Sutherland badly misjudged the government's thinking. By prosecuting the nation's top meatpackers in 1919, Palmer had already demonstrated his willingness to use Sherman Act suits to exploit public anger over the high cost of living. While that case was pending, eastern trade papers used information leaked to them by the raisin packers for a series of editorials critical of Palmer's slowness in acting in the raisin case. For example, the *New York Journal of Commerce and Commercial Bulletin*, which covered the nation's processing, distribution, and wholesale industries with a decided East Coast bias, labeled the August raisin auction a sham: "The whole affair indicates a total lack of appreciation on the part of the Raisin Trust of its weak position under the law, let alone public sentiment." Adding fuel to the fire, Preston wired the attorney general that the raisins sold at auction were almost certainly going to be used "for the illegal manufacture of wines and liquors." Palmer was generally disinclined to prosecute businesses under the Sherman Act, but food prices were such a hot political issue that he made exceptions in the meat and raisin cases. The DOJ notified the CARC that its concessions did not bring it into compliance with the law and began preparing its complaint. [45]

Palmer believed his case was airtight. Determined now to bring the CARC into court as soon as possible, he sent special agents to California to collect evidence. On September 5, 1920, he personally ordered the U.S. attorney at Los Angeles, Robert O'Connor, to file suit. O'Connor asked for a delay and, in an encoded telegram, explained that local political and economic backlash was sure to be severe: "Filing bill at this time would array every person interested california fruits not merely members this association and would positively mean losses twenty five thousand votes to Democratic Party in November making certain defeat Senator Phelan and losses this state for Governor Cox." Palmer wired back: "Please immediately file bill as directed." The complaint was filed in district court the next day. [46]

Initially, Palmer was confident of the government's chances for success. The lawsuit had been filed in equity, which meant that the government could seek injunctive relief against the CARC. In contrast to the FTC proceedings, which focused on the Clayton Act exemption, the DOJ pleaded directly to issues raised by the Sherman Act: the degree to which the CARC conducted a monopoly of the raisin trade, whether that monopoly operated in restraint of trade, and whether the CARC's control of the crop affected interstate commerce. The legal and political stakes were high. A loss in the federal government's first attempt to prosecute a cooperative under the Sherman Act would not only be embarrassing but also leave the antitrust liability of farmers' cooperatives even more difficult to establish.

The filing of the suit shook Fresno to its foundations. Desperate to avoid dissolution, Giffen changed attorneys and proffered settlement overtures, which the DOJ rebuffed. Meanwhile, the California press heaped hostile editorial comment on the government: "Does the Federal Administration propose to bring ruin to a GREAT CO-OPERATIVE GROUP OF INDUSTRIOUS AND PROGRESSIVE TILLERS OF THE SOIL?"[47] In interviews with agricultural journals, Giffen argued that the corporate aspects of the CARC were "essential" to fulfilling the goals of a traditional cooperative. He played on farmers' abiding hatred of middle merchants: " 'Our right to organize and fix prices is attacked, just as the milk producers' right to organize has brought farmers into court all over the United States. Cooperative marketing associations have their existence at stake in this fight, which is engineered by the speculative interests to destroy us.' "[48]

To make matters worse, the packers, whom the DOJ had excluded from all discussions after the FTC released its report, were "getting cold feet." The lawsuit, filed at the height of the harvest, threatened to destroy more than the CARC's monopoly. A victory for the government would cause raisin prices to collapse, and the packers would be unable to recover the high prices they had paid to growers in the spring of 1920. Unable to control the substance of the litigation or predict its outcome, the packers were nearly as unsettled by the legal process as the CARC itself. Thousands of growers, Armenians among them, were reportedly ready to sign affidavits that their signatures had not been coerced. No one in California, not even the packers who had instigated the suit, wanted the government to obtain an injunction against the CARC while the crop was still being shipped to its eastern purchasers. The DOJ was flooded with wires and letters from outraged Californians protesting the lawsuit, not the least of which was Senator Phelan's defense of "one of the worthiest cooperative associations in the United States."[49]

The district court in Los Angeles ruled against the government on several key motions during the fall of 1920. In September, the court denied the government's motion for an immediate injunction against the CARC. District Judge Benjamin F. Bledsoe said in open court that an injunction, if granted, would have produced "chaos and consternation" in the California economy.[50] On October 15, the judge heard arguments on the CARC's motion to dismiss the suit for lack of federal jurisdiction. The CARC argued that its control over the growers was wholly intrastate and thus its monopoly did not operate in interstate commerce. The judge agreed. Ruling the complaint "defective," Judge Bledsoe informed the U.S. attorney that the weight of legal authority lay with the defendants: "I cannot come to the conclusion that the Knight case . . . has been overruled or departed from by the United States Supreme Court. . . . Whatever else may have been done by these defendants in the matter of securing a *monopoly of the manufacture of raisins in California*, unless the defendants by themselves or through their chosen emissaries are *actually engaged in interstate commerce*, this proceeding will not lie under the authority conferred by the Sherman Anti-Trust Law." If the federal government chose not to amend its complaint, Bledsoe indicated that he would grant the CARC's motion to dismiss.[51] In disgust, an attorney with Breed, Abbott & Morgan commented to Nutting: "The Knight case is one of those ghosts that never die. . . . It seems to us that the Court may now fairly be said to have overruled it. Yet, naturally enough, it is continually cited when it seem[s] to serve either party to a pending case." The attorney had a point. Here the government was alleging the existence of a monopoly over the production of a crop grown entirely within the state of California. Bledsoe's reliance on the old production/commerce distinction, which Supreme Court rulings beginning with the *Swift & Co.* decision in 1905 had largely scuttled, signaled that the government's task would be a difficult one.[52]

Bledsoe's ruling gave the CARC an opening. It would take the government weeks to amend its complaint, and the suit would not come to trial for months after that. In December, the CARC requested that it be released from a stipulation barring it from exercising the option in the 1918 crop contract so that it could secure control over the 1921 crop that would almost certainly be marketed before the case was decided. A top DOJ official, still working on the government's amended complaint, told the U.S. attorney to "oppose it [the CARC's] motion in every way possible." But the DOJ was losing its grip on the case. In January 1921, Bledsoe ruled that the CARC could offer growers a new fifteen-year contract approved by the court. The CARC promptly announced that a new sign-up campaign would begin on February 1, 1921.[53]

Bledsoe's approval of the new contract for such a long term took much of the steam out of the government, the packers, and their attorneys. Preston was convinced that the judge's ruling on the contract was incontrovertible evidence of bias: "Judge approved informally the new fifteen year contract and otherwise showed that he was absolutely against us." The DOJ knew that the crop contract lay at the heart of the monopoly; if the CARC could evade liability for the 1918 contract by canceling it and offering a new one, nothing would bar the cooperative from maintaining its monopoly and its power to set prices. The DOJ's staff attorneys felt deceived by the judge's action, but Palmer had no intention of abandoning the lawsuit, even though his departure from office was imminent.[54]

The CARC's 1921 contract campaign, which even the packers admitted was conducted "in a careful and conscientious way," brought between twelve and thirteen thousand growers under contract, increasing its monopoly from 88 to 92 percent.[55] This achievement precipitated another round of furious correspondence between John Preston and the new attorney general, Harry M. Daugherty: "This corporation and its officers and trustees affect to believe that they have full legal right to control the entire output of raisins in the United States." Months later, Preston was still urging the DOJ to act: "We fear . . . that in view of the pronounced favoritism of the Judge to this Defendant, that he may peremptorily dismiss this case." He reminded the DOJ that a bill was coming up in Congress that was "intended by its framers to legalize" the CARC, arguing that "it was not the intention of Congress to in any wise interfere with legal proceedings such as are pending in the above matter, or to in any wise legalize monopolies of this character." But the new administration froze him out.[56]

The DOJ's commitment to prosecuting the case flagged after the change in presidential administrations. Daugherty had little confidence in the government's case and even less dedication to prosecuting trusts. The district court's insistence on *E. C. Knight* as the controlling precedent required the government to prove the existence of an interstate monopoly in order to establish jurisdiction; the 1920 Supreme Court decision in *U.S. v. U.S. Steel* made it impossible to dissolve the monopoly just because it was large.[57] The government believed that the CARC's admission of monopolistic behavior brought it directly under the Sherman Act, but the judge ruled that the CARC had entered into no interstate agreements to control sales. The chronic lack of funds for enforcement seriously hampered the government's ability to prosecute antitrust cases, and the new administration's dismissal of all DOJ staff attorneys undermined the CARC case. In view of the government's flagging commitment to antitrust enforcement during this time, what is remarkable is not that the prosecution failed but that the DOJ came as close to breaking up the CARC as it did.[58]

Politically, too, much had changed in the year since the suit was filed. Versions of legislation to amend the Clayton Act were making their way through House and Senate committees. The Capper-Hersman bill, introduced in 1920 in response to both the raisin case and prosecutions of dairy farmers, extended the Clayton Act exemption to all "associations, corporate or otherwise, with or without capital stock . . . any law to the contrary notwithstanding." This bill died in late 1920 when the House and Senate could not agree on who should enforce the law.[59] The ongoing legislative debates were a sign of increasing support for a broader antitrust immunity for farmers and the changing legal order's growing acceptance of the use of law to govern market behavior.

A new version, introduced in April 1921, showed that the farm groups were holding their ground. H.R. 2372, the Capper-Volstead bill, permitted cooperatives to handle commodities for nonmembers while maintaining antitrust immunity. A second alteration tossed a bone to antimonopoly hounds in the Senate. The sweeping antitrust exemption ("any law to the contrary notwithstanding") was dropped. Instead, the exemption was to be inferred rather than explicit; the bill authorized the existence of all associations that operated for the "mutual benefit of the members," adhered to the one-person, one-vote principle, and paid no more than 8 percent dividends on its capital. To come within the exemption, cooperatives could conduct no more than half their business with nonmembers. Enforcement was the responsibility of the secretary of agriculture, who was to determine if any association "restrains trade or lessens competition to such an extent that the price of any agricultural products is, or is about to become, unduly enhanced." Upon such a finding, after notification and hearings, the secretary could order the association to "cease and desist" from price-enhancing practices. If the association refused to comply, the secretary could file suit in federal district court, and the DOJ would enforce all court orders. The newly organized congressional Farm Bloc made the cooperative marketing bill its top priority.[60]

Farm groups were instrumental in securing these changes. The American Farm Bureau Federation, the National Milk Producers' Federation, the Farmers' National Council, and the National Board of Farm Organizations used the raisin and milk prosecutions to show that the DOJ did not understand the economic and legal nature of marketing cooperatives. The real problem was that the rider worked both for and against farmers. As the farm groups saw it, when cooperatives succeeded in exerting any influence at all over prices, the DOJ deemed that influence a violation of the antitrust laws. The rider, however, did not permit the DOJ to use any measure other than prices to judge cooperatives' impact on the market. As a result, farm leaders viewed the DOJ's enforcement decisions as arbitrary. Milo Campbell, president of the milk producers' association, made that point in 1921: "All that our Department of Justice has done with any of our great trusts was to try to dissolve them; and they charged even more after they were dissolved than before. Nobody was ever put in jail, except a farmer, under the antitrust laws."[61]

The milk producers, who were leading the charge for extending farmers' antitrust immunity, believed that neither the Clayton Act nor the annual appropriations riders sufficiently deflected prosecutorial attention because neither guaranteed farmers complete protection from antitrust liability.[62] The milk producers believed farmers should have the freedom to cooperate in the form and manner they deemed appropriate, with capital stock and dominant market control if they so chose, free from DOJ "harass[ment]." The farm interests dismissed accusations of monopoly with the claim that cooperatives placed "reasonable organizational restraints" on competition for the legitimate purpose of securing better prices for agricultural commodities.[63] Coming as it did in the middle of their antitrust prosecutions, this argument struck federal officials and some members of Congress as self-serving.

The DOJ's antitrust enforcement policy at this time was selective but not arbitrary, shaped by political and legal considerations rather than the concerted aim of persecuting farmers. Before the war, the number of Sherman Act lawsuits had decreased sharply; according to Morton Keller, the conditions of the postwar economy called for a more energetic approach: "One of the few areas in which the Justice Department showed some vitality was the politically sensitive one of market control in the food industry." The government's targets were mostly chain stores, food manufacturers and processors, and meatpackers; the producers were merely one piece of a larger battery of interests and firms controlling the nation's food supply and food prices. The farm lobby, however, was adept at creating the perception that farmers were being singled out, and that context controlled the legislative debates.[64]

Accordingly, the bills loomed over the CARC lawsuit like storm clouds. "The whole Bill appears to be framed up to suit the operating conditions of the Raisin Association," Charles Bonner complained. That it took the bill the better part of two congressional sessions to become law did not slow farmers' momentum with Congress; in December 1920, Fresno representative Barbour arranged for the CARC's lawyers to meet personally with the attorney general.[65]

The House quickly passed the bill in 1921, leaving the meat of the debate to the Senate. Many prominent farm leaders believed that an adverse decision in the raisin case would stunt the growth of the cooperative movement. In response to their pressure, the Senate Judiciary Subcommittee summoned some of them to appear at its June 1921 hearings, including defendants in both the raisin and milk cases. Only one of the eleven witnesses opposed the bill, and that was John W. Preston. His appearance reinforced the impression, as Nutting sardonically observed, "that no one is opposing [the CARC], excepting five packers, and that all the rest of the United States seems to be favorable to them."[66]

The hearing was hardly a one-sided affair. CARC officials were in danger of doing the bill more harm than good. Senator Thomas J. Walsh, a Montana Democrat, professed to sympathize with farmers but indicated he did not countenance monopolies in farm commodities. To him, the CARC represented everything the antitrust laws were intended to prevent: "one single association . . . [with] a complete monopoly of the raisins produced in the country." Gray Silver, president of the American Farm Bureau Federation, and Carl Lindsay, attorney for the CARC, sought to distance the raisin organization's troubles from the legislation under consideration. Lindsay gave the senators to believe he was in town to negotiate with the DOJ and had only that day discovered that hearings on Capper-Volstead were occurring. "We have never taken any particular interest in the passage of this law," Lindsay informed the committee. He maintained that the bill, even as amended by the Senate to prohibit monopolies, would not affect the CARC's operations: "I do not think that we will fall within the provisions of the bill, and I am quite sure that it will give to us no greater protection than we have at the present time, under the law." Dismissal of the indictment against the milk producers in 1921 left the CARC as the only farmers' cooperative still being prosecuted, however halfheartedly, while the bill was under consideration.[67]

The legal issues raised in the hearings reflected the ongoing fluctuations in Progressive Era antitrust law. The basic questions—what exactly was a monopoly, and when did prices become unfair and unreasonable—had been complicated by the rise of new legal forms and organizations in the market. Courts and legislatures were struggling to understand the legal implications of these changes. Walsh noted that the bill was not intended to permit farmers to engage in monopolies, but he seemed unsure of whether marketing cooperatives led to or inherently constituted monopolies. The secretary of agriculture's insistence that the bill would leave farmers subject to the Sherman Act was partly responsible for this uncertainty. "The bill . . . would not authorize an association of agricultural producers to engage in unlawful restraints upon trade or to form monopolies or to do any other things forbidden by the antitrust laws."[68] Secretary Henry C. Wallace's lukewarm support for the bill fed Senate skepticism about farmers' motives in seeking a "reform" that did not substantially change their antitrust liability. Walsh opposed the bill because he thought it did not sufficiently guard against monopoly. In his view, the CARC supplied the strongest argument against a bill that would permit farmers to conduct monopolies under the guise of cooperative marketing.[69]

Troubled by the bill's vague language on monopolistic behavior and pricing, the Senate Judiciary Committee deleted the entire enforcement section and substituted a provision that forbade the "creation of or attempt to create a monopoly" or any behavior prohibited under the Federal Trade Commission Act of 1914. The amendments reflected the conviction that despite the claims of the farm groups, there was no need to protect farmers from the antitrust laws. When they did violate the law, this belief ran, they deserved to be prosecuted. Walsh, for example, was "frightened" by the "image of 85,000 dairymen from six states bargaining collectively with New York milk dealers." The CARC, embroiled in a Sherman Act lawsuit that challenged its existence, was repudiated as "a confessed monopoly," a worst-case scenario come true. Yet even such fierce opponents of monopoly as Walsh believed those cases were exceptions; in general, he believed, farmers were incapable of conducting monopolies. Walsh contended that the amendments were sufficient to remove the "peril, if there be peril, of prosecution" under Section 1 of the Sherman Act.[70]

The farm lobbyists were horrified at Walsh's changes. By default, they argued, these amendments left all farmers, not just the milk producers and raisin growers, just where they were—at the mercy of the Justice Department. To the farm groups, this result was unacceptable. Cooperative organizer Aaron Sapiro informed Senator Capper, "The Walsh amendment . . . will not give to the growers of the United States any important single thing which they do not have now in their rights, and will do definite harm to cooperation. . . . The farmers' organizations would rather see no legislation at all than such an act."[71]

Such an outcome was hardly likely in view of the widespread political support for agricultural relief generated by the farm organizations and the congressional Farm Bloc.[72] President Warren G. Harding endorsed the bill in the midst of Senate debate in January 1922. Aside from the antimonopoly forces in the Senate, the only support for the Judiciary Committee's version of the bill came from Secretary Wallace. Since the farm groups denied any intent to monopolize or "unduly enhance" prices, he said, the amended version of the bill would provide farmers with all the protection they needed. The Senate rejected this argument. The war and the emerging political acceptance of collective economic action, with its anticompetitive effects, had changed the regulatory culture in Congress and the courts. Nebraska senator George Norris commented during the last day of debate, "There is no use trying to dodge the proposition that it is a price-fixing proposition, for it is." The bill passed the Senate by a margin of fifty-eight to one on February 8 and became law ten days later.[73]

## MONOPOLY PRESERVED

The impending passage of the Capper-Volstead bill spurred rumors that the administration was losing interest in the CARC suit and would agree to settle the case on weaker terms than the packers wanted. Congress's attention to the economic plight of farmers emphasized the continuing political conviction that farmers occupied a distinct place in the market and deserved special protection from economic forces that overwhelmed them and undermined their ability to feed the nation. The law's earlier insistence on equality within the market for all entrepreneurs was giving way to the recognition that formal equality could blight economic growth. This recognition pushed the liberty of contract ideal aside, relaxed the equal protection sensibility that denied farmers antitrust immunity, and ultimately forced the DOJ to concede that it could not win the CARC case.[74]

In December 1921, Daugherty ordered the case settled. The consent decree, filed on January 18, 1922, voided the 1921 crop contract and permanently enjoined the CARC from engaging in market practices prohibited by the Sherman Act. These practices included obtaining crop contracts through coercion; using sales contracts in which prices were made contingent on future market conditions; purchasing the plants and businesses of competing packing firms; dealing exclusively or discriminating in any way among purchasers of raisins; using the packers as the CARC's selling agents; buying raisins for the purpose of fixing prices; or restricting production in any way.[75]

The decree said nothing, however, about the legal issue of the CARC's corporate identity. It did not indicate how much market control a farmers' cooperative could legally attain. Consistent with the Clayton and Capper-Volstead Acts, the decree focused on market behavior rather than organizational structure, on business dealings rather than the existence of a monopoly, and left unanswered the question of whether a cooperative that resembled a regular corporation constituted an abuse of market power. In effect, the CARC was barred from further vertical and horizontal integration, but its monopoly was left intact. For Preston, whose goal all along had been the dissolution of the monopoly, the decree was a bitter pill: "I know that it will not in any way provide for adequate relief."[76] Congress's ongoing deliberations on the Capper-Volstead Act were an undercurrent throughout the litigation, constantly eroding the legal basis for prosecution. The DOJ threw in the towel and signed the consent decree barely one month before President Harding signed the act into law.

The settlement in the raisin case provided the nation's farmers with specific guidelines on legal market practices. It also furnished policy makers with a model of cooperation that at once accommodated the commercial nature of agriculture and incorporated its legal innovations. Congress had concluded that cooperatives could properly borrow from the corporate form and still comply with the nation's policy on free, competitive markets. As such, cooperatives could boldly claim the modern trust as their own and embellish it with a special claim to serve the public interest. This insight was the key to changing cooperation's legal status. As Wylie Giffen asserted in defense of the organization that "saved" the raisin industry while cutting the excess profits of commercial packers, middle merchants, and speculators, "Call us a trust, if you want to, but we're a benevolent one."[77]

The legal and intellectual roots of the Capper-Volstead Act's new conception of cooperation lay in the CARC and the willingness of its leaders to flout the law. CARC leaders followed the precedent Kearney laid down twenty years before, but they packaged his organizational innovations differently. Instead of relying on analogies to industrial trusts, modern corporate forms, and the robber barons, Giffen asserted that the CARC fulfilled the cooperative ideal. Changes in the law of equal protection, federal antitrust law, and enforcement policy between 1902 and 1922 were essential in helping the CARC succeed where Kearney had failed. The CARC, in achieving a far greater measure of integrated market control than Kearney's organizations ever managed, continually asserted that in spirit it was an association of growers—and refused to concede the point throughout three years of federal investigation and litigation. In the end, because of the CARC's distinctive economic circumstances and its leaders' willingness to push beyond the law's existing categories, their conviction that monopoly and the cooperative ideal were thoroughly compatible became incorporated into federal law.

The litigation did take a toll on the CARC, and it worsened after the case finally came to an end. The CARC built its monopoly using a single-minded legal and economic strategy; now the issue was whether the consent decree would materially alter that strategy. The decree's prohibitions made it clear that the DOJ would base future enforcement decisions on the CARC's effect on price. Market control such as the CARC had enjoyed between 1916 and 1921 was illegal precisely because it enabled the cooperative to dominate the price-setting function.[78] The decree permitted collective action by farmers as long as price—the best measure of competitive freedom and the DOJ's only real enforcement tool—remained ultimately unaffected by that collective action.

Having striven to impart to growers a theory of cooperation in which collective action gave them more control over the prices they received, the CARC and its officers were caught between the law's insistence on prices that were not unreasonably affected by cooperative activity and the CARC's economic imperative. According to that imperative, the participation of everyone in the industry could be compelled by force when necessary. The cooperative regime had yielded such high rewards that the industry's appetite for an effective monopoly only increased after 1922. But the decree's restraints undermined the organization's ability to hold a monopoly together, even as the legal order's indulgence of cooperative monopoly — the "benevolent trust" of which Giffen boasted — reached new heights.

# 8 ∾ Decline of the Benevolent Trust

*The recent raisin drive here was a reign of terror, a blot on the good name*

*of the San Joaquin Valley, and, more still, a disgrace to civilization.*

—*Raisin grower to the Department of Justice, 1923*

*We have taken the law into our own hands and you will have to sign.*

—*One grower to another, 1925*

The CARC antitrust case and the Capper-Volstead Act ratified cooperation's new legal form. The consent decree in the CARC case set out a series of rules to govern market behavior; the new statute laid out a new conception of cooperatives' structure. Farmers welcomed both developments. Most in Congress considered the issue of monopolies in agriculture settled. The CARC's president, officers, and loyalists all believed that the new law swept away whatever uncertainty the consent decree might have imposed on the raisin industry.[1]

As it turned out, however, neither the consent decree nor Capper-Volstead had the expected stabilizing effect on the agricultural economy. The fundamental question of what was a reasonable price and the USDA's difficulties with administrative enforcement opened new avenues for contesting cooperatives' impact on the market. As it had been during the formation of both the law and the consent decree, the CARC remained at the center of a fierce struggle, this time between two arms of the federal executive branch. The issue was not merely on whose regulatory turf cooperatives belonged but also who got to define legitimate cooperative practice.

Both the DOJ and the USDA were well armed in the fight over the authority to regulate cooperatives. The DOJ had no intention of relinquishing its authority to enforce the decree. The USDA, although ambivalent about its responsibilities under Capper-Volstead, wanted to protect cooperatives from prosecution. The CARC, whose case had played a pivotal role in shaping the Capper-Volstead Act, provided the test case for determining who held federal regulatory authority over cooperatives. The antitrust liability of farmers' organizations under the new statute would be defined against the backdrop of the consent decree issued in the only federal antitrust prosecution of an agricultural cooperative.

The two departments were also at war over the meaning of cooperatives' responsibilities under the new law. The consent decree not only assumed that farmers were capable of monopoly and restraint of trade, it also prohibited unfair trade practices and excessive market control. In contrast, the Capper-Volstead Act gave cooperatives a broad blanket of immunity and provided only the most general enforcement trigger. The struggle for administrative jurisdiction in the raisin case, then, was not just a struggle over the legality of the CARC's affairs; it was a battle to determine how the contradiction between the cultural ideal of the farmer, which continued to permeate federal law, and the very real market power of cooperatives would be resolved. Even when presented with stark, irrefutable evidence of a monopoly predicated on violence and coercion, the USDA abided by Capper-Volstead's indulgent terms and declined to hold the CARC responsible.

The USDA's beneficent interpretation of the Capper-Volstead Act, however, had far less impact on the CARC's destiny than the CARC's own lack of control over supply. No federal or state law gave cooperatives the power to limit production. During the 1920s, agriculture's productive capacity—stoked by World War I—became farmers' great burden. Annual surpluses piled up, pushing prices down and sinking agriculture into a prolonged depression. The CARC's downward slide mirrored the experience of many American farmers during the 1920s. The raisin growers' resort to increasingly violent means to prevent the slide, however, did not.

RENAMING AND REORGANIZING THE CARC

When the consent decree was signed in January 1922, Wylie M. Giffen publicly exulted. The raisin growers' association, he said, was essentially left intact: "We . . . are only enjoined from doing thing[s] that we have never done and never expect to do." Privately, though, Giffen knew that the cooperative faced significant economic difficulties. Although he declared publicly that the consent decree would not affect the cooperative's business operations, he was unsure as to how the settlement would affect the cooperative's monopoly over the growers or its ability to market the crop.[2]

The consent decree worked definite alterations on the CARC, the most obvious of which was the cooperative's name. One day before Capper-Volstead became law, the CARC legally changed its corporate name to the Sun-Maid Raisin Growers of California. CARC officials wanted to make as obvious a break with the past as possible while insisting that the raisin growers' organization remained essentially unchanged. The *Associated Grower* magazine hastened to point out that the association was the "same in every way" as the old one. The new name identified the association with its well-known brand name; it also emphasized the organization's cooperative character rather than its corporate structure.[3]

Sun-Maid planned to market the 1922 crop under the terms of the 1921 contract, even though the consent decree declared this contract no longer in force. When the 1922 crop came in, Sun-Maid realized it had an economic disaster on its hands. The 1921 crop had been so large that for the first time Sun-Maid had been unable to sell it all. Making things worse, as harvest began in 1922, "frenzied ranchers were mortgaging themselves up to the eyes to set out new miles of vines." All this new planting boded ill for the future; Sun-Maid had no way of knowing the market conditions in three years, when the new vines came into bearing. And though retail prices had dropped somewhat since 1920, they remained high. The results were a saturated market, a sharp decline in the prices Sun-Maid could obtain from the wholesale trade, and a $4.5 million loss on the 1921 crop.[4] During the winter of 1922, Sun-Maid officials could not secure financing for the 1923 crop and were forced to pay growers in demand notes that Fresno banks would not cash. These notes were "put out in such quantities as to bring about an ever-present danger of bankruptcy should any number of these notes be presented for payment."[5] Not surprisingly, "the bottom dropped out of the raisin market."[6] Under intense pressure from the financial establishment, Sun-Maid's board of directors concluded that only new management could save the cooperative from bankruptcy. They chose Ralph P. Merritt to assume the new position of managing director, with full executive authority. Wylie Giffen remained president—he was far too sentimentally important to the growers to ditch unceremoniously—but in two more years that title, too, would belong to Merritt (Tables 8.1 and 8.2).[7]

Merritt came to Sun-Maid with ties to the financial community and California agriculture but none to the raisin industry. A University of California graduate, he had worked for the Miller and Lux cattle firm, served as comptroller of the University of California, and was appointed food administrator

TABLE 8.1. Sun-Maid's Share of the California Raisin Crop, in Tons, 1921–1929

| Year | Statewide Production | Sun-Maid's Total Receipts from Growers | Sun-Maid's Share of Crop (%) | Industry Carryover |
|------|---------------------|----------------------------------------|------------------------------|---------------------|
| 1921 | 177,000 | 152,497 | 86.2 | 36,000 |
| 1922 | 145,000 | 123,665 | 85.3 | 34,000 |
| 1923 | 237,000 | 204,652 | 86.4 | 86,000 |
| 1924 | 290,000 | 246,373 | 84.9 | 186,000 |
| 1925 | 170,000 | 129,853 | 76.4 | 67,000 |
| 1926 | 200,000 | 136,309 | 68.2 | 59,000 |
| 1927 | 272,000 | 177,332 | 65.2 | 108,000 |
| 1928 | 285,000 | 144,068 | 50.6 | 124,000 |
| 1929 | 261,000 | 82,303 | 31.5 | n/a |

*Source*: Henry E. Erdman, "Sun-Maid—and How It Got That Way," unpublished manuscript, Aug. 16, 1935, p. 42 (total receipts, citing data from Sun-Maid Statistical Department); p. 49 (California production; based on data from G. M. Peterson and S. W. Shear, "The California Muscat Grape Outlook"), Box 18, Erdman Papers; S. W. Shear and R. M. Howe, "Factors Affecting California Raisin Sales and Prices, 1922–1929," *Hilgardia* 6 (Sept. 1931): 74–75.

*Note*: n/a = data not available

for California by Herbert Hoover during World War I. In 1921, Sacramento bankers turned to Merritt to solve the surplus problems in the California rice industry; he organized the California Rice Growers' Association and sold the entire forty-million-pound surplus to Japan, "which had never imported a ton of rice in its history."[8] To the banks, Merritt seemed ideal for the job of reorganizing Sun-Maid, securing its finances, and eliminating the "liability" of the crop advance. It would not be an easy task. The "necessity of reducing costs while increasing the volume of sales" was, in Merritt's words, like " 'carry[ing] a ten ton load on a one ton truck.' "[9]

After taking office in January 1923, Merritt planned to reorganize Sun-Maid into two separate companies. One, the Sun-Maid Raisin Growers of California, was organized under the California nonstock cooperative law. It adhered to the one-person, one-vote principle and limited membership to producers. This organization, Merritt said, would "become a marketing association owned and managed exclusively by growers of raisins."[10] The new cooperative reverted to a more traditional form to comply unambiguously with Capper-Volstead.

TABLE 8.2. Sun-Maid's Prices to Growers, in Cents per Pound, 1920–1929

| Year | Muscats | Thompsons | Retail Price |
|------|---------|-----------|--------------|
| 1920 | 11.2 | 14.8 | 26.2 |
| 1921 | 7.0 | 8.1 | 31.2 |
| 1922 | 5.8 | 6.5 | 24.6 |
| 1923 | 2.5 | 2.4 | 18.2 |
| 1924 | 3.0 | 3.1 | 15.7 |
| 1925 | 4.0 | 3.6 | 14.6 |
| 1926 | 2.8 | 2.6 | 14.6 |
| 1927 | 2.2 | 2.4 | 14.4 |
| 1928 | 1.9 | 1.8 | 13.6 |
| 1929 | 2.0 | 3.2 | 11.7 |

*Sources*: Sun-Maid prices: S. W. Shear and R. M. Howe, "Factors Affecting California Raisin Sales and Prices, 1922–1929," *Hilgardia* 6 (Sept. 1931): 74–75; Annual Financial Report, California Associated Raisin Company, *AG* 1 (Sept. 1920): following page 26, at viii; Wylie M. Giffen, "Analysis of 1921 Raisin Crop Returns," *AG* 5 (Jan. 1923): 12; Giffen, "The 1922 Raisin Prices," *AG* 4 (Sept. 1922): 5; Henry E. Erdman, "Sun-Maid—and How It Got That Way," unpublished manuscript, Aug. 16, 1935, p. 51, Box 18, Erdman Papers (citing data from Sun-Maid Accounting Division, August 3, 1935). Retail prices: Erdman, "Sun-Maid—and How It Got That Way," p. 40a (data taken from U.S. Bureau of Labor Statistics, "Retail Prices, 1890–1918," *Bulletin* 495 [Washington, D.C.: U.S. Government Printing Office, 1919], 26; U.S. Bureau of Labor Statistics, "Retail Prices, 1923–1936," *Bulletin* 635 [Washington, D.C.: U.S. Government Printing Office, 1937], 196). Retail prices recorded by Erdman are primarily for fifteen-ounce packages of seeded (Muscat) raisins.

To handle the capital-intensive processing and packing operations, a second company, the Sun-Maid Raisin Growers' Association, was incorporated under Delaware law. The Delaware Sun-Maid took title to all assets of the old Sun-Maid, including its plants, equipment, and $3.7 million in capital stock. Any investor could purchase nonvoting preferred stock in the Delaware company. The subsidiary was free from the limits on capital and membership incumbent on cooperatives. Merritt did not care that in separating the marketing and processing functions he was conceding the point that M. Theo Kearney and Wylie Giffen had insisted upon: the growers' cooperative should handle both the pooling and processing of the crop. What did not change was that the most capital-intensive parts of the raisin business were controlled by a corporation that nominally answered to the growers but operated according to the imperatives of the financial community (Map 4).[11]

To recover from the losses incurred in 1921 and 1922, Merritt sought to increase the Delaware company's capital to $8 million. Since "the raisin growers had no money," this new capital would have to come from the investment community. The banks would support this plan only if the minimum price guarantee were eliminated from the crop contract, so Merritt called for a two-faceted campaign to secure stock subscriptions for the Delaware company and procure signatures on a new crop contract without a guaranteed minimum price.[12] As it turned out, the 1921 contract was jettisoned not because the consent decree required it but because it had become a financial liability.

If what the banks wanted was a complete shake-up of the cooperative, they could not have chosen a person more intent upon that task. Merritt's hierarchical management style replaced whatever attachment existed between growers and the cooperative with "high power business methods." That style, together with the overall economic uncertainty in the industry, produced disaffection among growers. On a visit to the Sun-Maid offices, University of California economist Henry E. Erdman observed irritated growers waiting to receive their advances: "Almost without exception there was criticism of the association," he observed.[13] Giffen had appealed to community values, the industry's historical struggles, and the growers' shared suffering. In contrast, Merritt framed his plan as a business proposition to be judged on its profitability alone.[14]

## THE 1923 CAMPAIGNS

As a business proposition, Merritt's plan had a gaping hole. Although no cooperative in the raisin industry had ever succeeded in attracting members on the strength of ideology alone, the CARC had always combined cooperative ideology and economic prosperity in its appeal to growers. Merritt's relative lack of interest in cooperative ideals emphasized the all-or-nothing tenor of Sun-Maid's post-1922 crop contract campaigns and gave zealots even greater license to use violence. The willingness of local legal officials to look the other way when victims lodged complaints made it seem, at least for a time, that nothing could hold Sun-Maid back.

The campaigns for new stock subscriptions and the new contract began on April 2, 1923. In many respects, this new contract was harsher than the CARC's previous crop contracts. Liquidated damages rose from two and one-half to three cents per pound. Growers were to bear the loss incurred in marketing the 1922 crop, from which a carryover of seventy-five to eighty thousand tons remained. The lien on the land was also retained. Most important, the new contract promised only to make "as large an advance per pound as in [Sun-Maid's] judgment financial and market conditions will justify." Whatever pay-

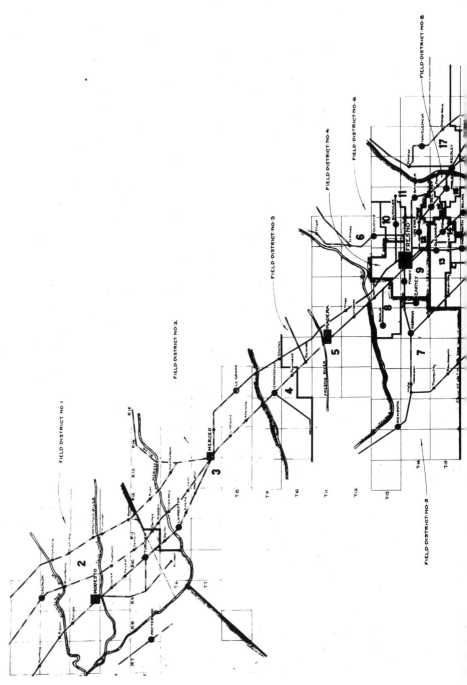

Map 4. Sun-Maid plants, receiving stations, and unit divisions, 1923. (Henry E. Erdman Papers Relating to Agricultural Economics and Cooperatives [BANC MSS 74/161c], Bancroft Library, University of California at Berkeley)

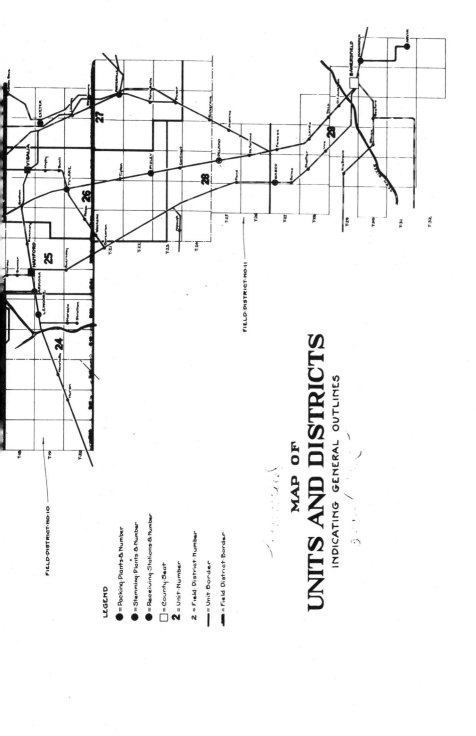

MAP OF
# UNITS AND DISTRICTS
INDICATING GENERAL OUTLINES

LEGEND

● = Packing Plants & Number
● = Stemming Plants & Number
● = Receiving Stations & Number
☐ = County Seat
**2** = Unit Number

**2** = Field District Number

= Unit Border

━ = Field District Border

FIELD-DISTRICT-NO-10

FIELD-DISTRICT-NO-11

ment the growers received for their raisins would depend on the cooperative's financial ability to borrow. In one notable way, however, the consent decree influenced the terms of this new contract. Growers now had the right to withdraw from the cooperative every two years. Sun-Maid's board of directors announced on March 27 that 85 percent of the industry's acreage had to be signed for Sun-Maid to continue in business. The monopoly imperative continued unabated.[15]

The stock campaign jumped off to a fast start. The bankers, whom a DOJ agent described as "the men behind the guns," immediately started a public crusade. They proclaimed their support in full-page advertisements and provided growers with blank contracts and subscription forms.[16] Aided by a timely visit to Fresno by Secretary of Agriculture Henry C. Wallace, during which he endorsed Merritt's reorganization plan, Sun-Maid raised $1,834,900 by April 17. One Fresno banker summarized the financial community's position: "If this thing falls through it will put the whole of the San Joaquin Valley back twenty years. . . . And the whole cooperative idea throughout the United States will get a black eye." On April 26, the day of the industry's Raisin Day festival, Merritt announced that the stock subscription drive had met its goal.[17]

The crop contract had not fared nearly as well, despite the efforts of a coordinated recruitment force under the leadership of Sun-Maid publicity director Al C. Joy. By Raisin Day, only 60 percent of the acreage was signed, well short of the 85 percent goal.[18] Disgruntled by the loss of the guaranteed price, many growers were refusing to sign. At a Raisin Day mass meeting, Wylie Giffen—a figurehead symbolically leading the charge on the contract campaign—told the growers that the guarantee had to go: " 'A guarantee we make to ourselves is absolutely worthless. . . . The only price we can obtain is the price the crop will bring.' " Merritt gave Sun-Maid loyalists until May 5 to bring in the other 25 percent.[19]

Sun-Maid leaders instructed the solicitation teams not to resort to violence or coercion. "We have heard of coercion," Merritt said. "We do not want this drive put over by any other than lawful and proper means." And from Giffen came words that would later prove ironic: "Let us do nothing in this campaign that would leave a blot on it." A group of prominent Armenian growers formed an Armenian Drive Committee to bring all unsigned Armenians into the Sun-Maid fold. Prosperous leaders of the Armenian community were now starting to sound like the industry establishment. Committee chair Arzapat Setrakian declared, "No man who expect[s] to live on this side of the ocean or whom [sic] want[s] to be an American citizen should stay out of this organization."[20]

It soon became obvious that the Raisin Day speeches, rather than fostering restraint, only encouraged violence. The next day, a mob numbering between seventy-five and one hundred men called at the home of Nazaret and Ashken Der Torosian around midnight. The Der Torosians were leasing their ranch from Abraham Kulekjian, who had gone to Los Angeles to "escape signing the raisin association contract." His tenants were not as fortunate. The mob surrounded the house and shouted to the frightened inhabitants, "Come out, you black Turks, your house is on fire." As the mob began to break windows and doors, someone in the house found a rifle. Two shots were fired. One hit William L. Osterhaudt, a Sun-Maid employee, in the right leg and right arm. The bullet shattered his knee, and the leg was later amputated. The Der Torosians were arrested on charges of attempted murder. The police arrested them both because, due to Der Torosian's inability to speak English, they were not sure who fired the gun. His wife claimed responsibility, but the sheriff and district attorney thought that she was only protecting her husband. The night riders swore that "not one member of the party was armed" and no one had threatened the Der Torosians before the shooting. The leader of the local recruitment committee told the local newspaper, "I know all in the party to be level headed men, and after learning that no return of the shooting had been made, felt satisfied that not one was armed." The men explained that they came at midnight because no one had been home during the day. These defenses, however, did not, explain the presence of two different caliber bullets found at the scene—the 30-30 shells from the Der Torosian rifle and a handgun bullet. The Der Torosians hired an Armenian attorney—the uncle of the writer William Saroyan—and were eventually acquitted.[21]

The shooting at the Der Torosian ranch was far from an isolated event. A reporter said that Fresno during the final week of the 1923 campaign "was like [a city] besieged, so tense the excitement, so poignant [the] realization of the issue hanging in the balance." Disavowing knowledge of the night riding, Merritt announced that if the contract campaign failed, Sun-Maid would immediately file for bankruptcy, bringing "ruin" and "demoralization" on the growers. Merritt reminded members that Sun-Maid was "still operating under the court consent decree" and that no coercion could be used to maintain the organization.[22]

Nevertheless, it was the night riders who forced across the last signatures in the final days. The night riders employed their usual tricks of causing property damage and personal injury, but this time, unlike previous campaigns, unwilling growers fought back just as violently. The Fresno County grand jury was impaneled to probe the night riding for the first time and began to investigate the Der Torosian shooting and other allegations of coercion. Fresno district attorney George R. Lovejoy urged all growers with any complaints of violence, property destruction, or coercion to come forward and give evidence to the grand jury. Anticipating "raisin riots," the grand jury geared up for a long session.[23]

When the campaign ended on May 5, all admissions of violence and coercion were silenced by official expressions of victory and success. More than fourteen thousand growers, representing 270,000 acres, signed the new contract. Merritt proclaimed the campaign a "great victory" marked by "comparatively few cases of violence." Giffen was more circumspect; he noted that while the successful drive would enable Sun-Maid to weather the storm of readjustment, prices would probably never return to their previous levels. The *Fresno Morning Republican*, apparently forgetting all it had published during the previous week, boasted, "Class feeling between American, Armenian, Italian and Russian faded away into insignificance on the last few days as the need for unity became apparent. Americans and Armenians worked side by side for the common good which the association will produce for all." [24]

Testimony presented to the grand jury belied this peaceful image. A Del Ray couple told of the destruction on their ranch: about an acre of raisin vines chopped down, several windows broken, and their house damaged. M. Samgochian said that a mob of "marauders" chopped down 375 vines on his ranch and that members of the Sanger Chamber of Commerce then told him "he had better look at his vineyard and then sign the contract before he got into more trouble." In fact, Samgochian had already signed with Sun-Maid. The district attorney belatedly condemned the night riders and expressed sympathy for the victims: "Just because a campaign is on is no excuse to go into a neighbor's home and shoot him and his family, burn his house, and otherwise destroy his property." Lovejoy promised swift prosecutions of all indicted persons.[25]

The grand jury's action prompted a response from Sun-Maid. The day after Lovejoy's statement, Merritt published a notice in valley newspapers addressed to growers "who may claim to have signed the Sun-Maid crop contract against their will or under coercion." Anyone who had been victimized by the night riders could submit an affidavit describing the violence and identifying the culprits. Sun-Maid would investigate and cancel all contracts where the accusations were substantiated. But many growers feared reprisals if they accepted Sun-Maid's offer. According to the cooperative, only thirty-four growers asked for cancellation. Sun-Maid postponed action on these contracts until after the 1923 harvest.[26]

Sun-Maid's announcement seemed to absolve local officials from their responsibility to prosecute. The following day, the grand jury declined to issue indictments in the four cases it had heard, suspended all other investigations, and adjourned indefinitely. The foreman reported that the grand jury had no grounds to indict anyone: "Virtually all of the evidence offered . . . was either hearsay, indirect, or not positive identification to warrant the prosecution of any persons." Part of the problem lay with the local police; among those who complained, "word was that if one called the Sheriff for help the reply was apt to be 'Sign up.' " The biggest problem was the district attorney himself. Lovejoy's commitment to prosecuting night riders flagged after the campaign closed. He promised to present all complaints and evidence to the grand jury, but since the date of the grand jury's next session was not known, growers had little reason to believe their complaints would be taken seriously. The DOJ later discovered that the district attorney himself had advised dissenting growers "to 'sign up' and avoid further trouble." Lovejoy admitted that he personally stood to lose $12,000 had the campaign failed. By the end of May, both he and the grand jury had washed their hands of the matter.[27] Local legal officials were too closely connected to the Sun-Maid establishment to enforce the law. Further, the DOJ, after learning that Wallace had approved the reorganization plan, took no steps to enforce the consent decree while the campaign was taking place. Whether Sun-Maid directly connived with legal officials to support the campaign cannot be proven, but if the district attorney's action was any indication, Sun-Maid knew it could count on favorable treatment from law enforcement in the Central Valley.

## THE PROBLEM OF CONTROLLING SUPPLY

Sun-Maid's triumph would prove short-lived. Its crop contracts gave it control over the crop only after it had been produced; it had no power to limit production. To maintain prices, much less increase them, Sun-Maid had to continue to do what the CARC had done effectively for ten years: expand demand as supply increased. Even after the postwar farm crisis ended, however, American agriculture continued to be plagued by excess production and depressed prices. The raisin industry was caught in that cycle.

Annual surpluses in the raisin industry began in 1921 and became increasingly larger. The 1923 and 1924 crops produced carryovers totaling 272,000 tons — nearly as large as the 1924 crop alone. Freed from the four-cent guarantee, Sun-Maid lowered its 1923 prices to growers by over three cents per pound for Muscats and over four cents per pound for Thompsons. These prices were the lowest since 1912; as if that were not enough, production costs had skyrocketed.[28] When Sun-Maid announced that it had lost $4.25 million on the 1922 crop, some commercial packers began to dream of ending Sun-Maid's reign: "California Packing Corporation and its allies are watching and hoping for an opportunity to buy the plants and equipment of the new organization and to put cooperation in the raisin industry out of the running."[29]

Some growers were thinking much the same thing. Discontent raged among those who felt they had been coerced to sign the 1923 contract and others who refused to sign only to find that Sun-Maid still considered them bound to deliver under the 1921 contract. Realizing that Sun-Maid could not possibly recover its 1922 losses from the growers, Merritt decided to put Sun-Maid through bankruptcy. The directors promised dollar-for-dollar liquidation of the assets so that all debts, including growers' notes for 1922, would be paid in full. Losses on the 1920 and 1921 crops, however, would be billed back to the growers if Sun-Maid's control dropped below 85 percent, and anyone leaving the cooperative would have to pay the billback. Grower J. T. Turner recognized that Sun-Maid was forcing the growers to absorb its losses: "It is . . . queer and dishonorable to flatly refuse to pay demand notes to the growers who have loaned their money to carry the association along." The cooperative ideal had disappeared.[30]

The underlying problem was that the industry had gotten too large for Sun-Maid to manage profitably. Because they could not control production, Sun-Maid officials had no choice but to expand consumption. Merritt doggedly worked to enlarge raisin markets overseas, find new domestic outlets for the large crops, and increase the efficiency of Sun-Maid's business operations.[31] These solutions, however, were stopgaps at best. The impact of the industry's excess production had not even begun to hit the market, and Sun-Maid had a more pressing problem on its hands: the growers were in open revolt. The withdrawal clause in the 1923 contract forced the cooperative to supervise growers vigilantly. In 1924, Sun-Maid hired private police to ensure delivery of the crop. The directors urged "loyal members" to help obtain complete compliance by reporting all "suspicious movement of raisins, night running of trucks, [or] statements of intention to deliver Sun-Maid raisins anywhere else than to a Sun-Maid receiving station." Growers demanding their final payments for the 1923 crops were asked to sign statements that they had delivered their 1924 raisins to the cooperative. Finally, Sun-Maid attempted to head off the exodus by asking growers to waive their right to withdraw during the first withdrawal period of December 21 to 31, 1924.[32]

Growers now saw Sun-Maid as openly coercive. Grower Hans Dahl informed Merritt, "You have as yet got the chance to regain and earn our confidence but you will not have it forever . . . there is a limit to the amount of coercion we can stand." Some growers informed Sun-Maid that they planned to withdraw. J. T. Turner warned Sun-Maid, "I intend to live up to my contract until the 24th of Dec. After that—no more Association for me until they can or will in my opinion do what is right—and just by the growers." The withdrawal period made the banks nervous. Officers of the Bank of Italy in Merced told University of California economist Henry Erdman that the bank planned to withhold loans to Sun-Maid until after the withdrawal period. Erdman predicted that the bank's "fears on this ground are probably ill-founded because the contract should hold."[33]

Judging by local accounts, Erdman's forecast appeared correct. When the waiver period concluded on December 31, Sun-Maid and the local press proclaimed that the cooperative had maintained its monopoly. The *Fresno Morning Republican* announced that about 300 of Sun-Maid's 16,500 members, holding about 15,000 acres, filed withdrawal notices. Most of the withdrawals came from Fresno County, where 234 growers with an average vineyard size of 35 acres withdrew. The newspaper estimated that the withdrawals dropped the association's control from 87.5 to 82 percent. According to the press, this loss was less than expected; Merritt's stump speeches during the withdrawal period were credited with reconverting dubious growers. Others reported a less sanguine outcome. Fresno attorney and Sun-Maid director Merris B. Harris privately estimated that Sun-Maid's control was reduced "considerably below 85 percent and possibly as low as 65 percent." The DOJ noted that "a large number of growers had availed themselves" of the right to withdraw.[34]

The withdrawals hurt Sun-Maid more than it was willing to admit publicly. Immediately Merritt moved to get the withdrawing growers back in the fold. The directors scheduled a "referendum period" for January 26 to February 14, for the purpose of giving "the community and the membership" the chance to engage in "a thorough consideration of the problems involved" if Sun-Maid ceased doing business as a cooperative and dropped all attempts at monopoly control. As the *Fresno Morning Republican* described it, the referendum period would permit "the community to invite desirable growers who have withdrawn or who have not been members to join." If the referendum period failed to maintain the monopoly, Sun-Maid announced that it would "weed out" members "who by temperament are not suited to co-operative effort." The unsuitable included "those malcontents who by habit have indicated that they cannot be expected to live up to their contract[s]" and "that portion of the alien element who do not understand American principles and who persistently refuse to co-operate upon American lines." This language was loaded with double meanings. The "outsiders" were not only those who refused to join the cooperatives but also anyone who could not cooperate along "American lines."[35] Either way, these people were socially and economically unfit.

It was easy to see just who Sun-Maid and the Fresno establishment had in

TABLE 8.3. Armenian Growers Withdrawing from Sun-Maid, by County, 1924

| County | Acreage Withdrawn | Armenians Withdrawn | Percentage of Total Withdrawals (N) |
|---|---|---|---|
| Fresno | 10,000 | 112 | 48 (234)[a] |
| Kern | 400 | 8 | 25 (8) |
| Kings | 410 | 1 | 10 (10) |
| Madera | 280 | 0 | 0 (2)[a] |
| Merced | 1,100 | 0 | 0 (1) |
| Stanislaus | 0 | 0 | 0 (1) |

*Source*: "Small Percentage of Raisin Growers Leave Association," *Fresno Morning Republican*, Jan. 4, 1925, p. 1; "Sun-Maid Raisin Withdrawals in County Total 234," *Fresno Morning Republican*, Jan. 5, 1925, pp. 1, 5.

[a] Includes one Japanese grower

mind. Armenians constituted over 40 percent of the withdrawn growers. The referendum campaign offered an opportunity to make an example of these "traitors." Although Sun-Maid publicly stated that the reason for the new campaign was to "invite desirable growers who have withdrawn or who have not been members to join," the cooperative did not intend to let any member leave without a fight. The *Fresno Morning Republican* published the names of the 234 growers who withdrew, essentially inviting the night riders to call on them during the referendum.[36] The monopoly imperative required that everyone belong, and in securing the monopoly, Sun-Maid enabled the zealous to act on their ethnic prejudices (Table 8.3).

Changes in the broader legal environment helped to insulate Sun-Maid from effective state or federal accountability at this time. The sweeping movement among the states to pass laws permitting cooperatives to restrain trade and fix prices, discussed in Chapter 9, had already begun by the time Sun-Maid ran its 1923 crop contract campaign. By 1925, state supreme courts had handed down important decisions upholding these laws and the larger policy they represented. Notably, California was one of the few states to limit cooperatives' powers to force members to abide by crop contracts or to obtain liquidated damages in case of breach. At the time of the referendum, state officials had made no move to check Sun-Maid's expansion; the actions of the USDA and the DOJ in monitoring the situation, which are described after the account of the referendum, turned on the issue of whether the DOJ could apply the consent decree to the new Sun-Maid. As the agencies argued over that question, Sun-Maid ran its last campaign to keep its monopoly and the night riders made their last stand.

## THE LAST RIDE OF THE NIGHT RIDERS

Sun-Maid's relations with disgruntled growers deteriorated as it scrambled to keep them under contract. Merritt warned that "those who are not with us are against us," setting an ominous tone for the referendum. Instead of directly supervising the campaign, as Sun-Maid had always done, Merritt delegated this job to the San Joaquin Valley Progress Committee, an unofficial body of Sun-Maid boosters that included former CARC attorney William Sutherland, state senator and Sun-Maid director Merris Harris, and the heads of several local chambers of commerce. Sun-Maid officials and employees were still intimately involved with the campaign; some participated in the violence against withdrawing growers.[37]

The 1925 referendum campaign left no doubt that Sun-Maid ran a company town. The business community publicly resolved to withhold "any credit or business courtesy" from all withdrawn Sun-Maid members and to honor boycotts organized to ostracize outsiders. Four banks stood ready to foreclose on those refusing to rejoin. Sun-Maid pulled its insurance business from a broker who withdrew his twenty acres. The banks became the arm of the association, denying credit to nonmembers, assigning Sun-Maid crop payments to growers' mortgages, and tricking a grower who did not read English into assigning his 1925 crop to the bank, which then delivered it to Sun-Maid. Members known to be "troublemakers" were slapped with suits for nondelivery when they attempted to sell elsewhere; those delivering to Sun-Maid had their raisins graded as "hog feed," which was assessed a value of $10 per ton.[38]

Sun-Maid all but turned the night riders loose that February. Despite a Sun-Maid official's bold assertion that "the time of the night-rider is past," allegations of violence and coercion surfaced during the referendum period.[39] The pattern was the same as before: all outside growers, regardless of race, were pressured, but minority growers—Armenians, German-Russians, Italians, and Japanese—got the worst of it.[40] The *Pacific Rural Press* described events in the raisin industry as "virtually a race conflict" directed against the "aliens" who made up the majority of the "holdouts." According to the writer, the violence was more localized this time, less pervasive, but more intense, and it expressed personal "grudges" and "feuds." Indeed, a majority of both the victims and the men they identified as their assailants resided in the Fresno County town of Kerman, where the chamber of commerce had been especially vocal in support of Sun-Maid. Thirty-four victims, nearly half of them Armenian, identified forty night riders. The mobs embodied the interrelatedness of rural and urban interests in the valley. Among the aggressors were ranchers, merchants, constables, a physician, a high school principal, a postmaster, and a "constable." At least one was identified as a "driver for the Raisin Association"; still other victims knew some members of the mobs to be Sun-Maid employees.[41]

Again, growers sought legal protection and redress for the crimes committed by the night riders. As before, police and local law enforcement officials refused to help. The county sheriff refused to investigate victims' complaints, even in cases of serious personal injuries. The district attorney told one grower he was "too busy to bother with such small matters." Newspaper reports of the violence finally pressured him into reconvening the grand jury. No doubt remembering the results of past investigations—and knowing where Lovejoy's sympathies lay—some complaining growers declined to press charges and instead sought to secure the return of their contracts. Lovejoy advised them to take that matter up with Sun-Maid. From all indications, the state of California never intended to enforce the state antitrust law against Sun-Maid, and if it had, local officials probably would not have offered much assistance.[42]

Sun-Maid made token efforts to show that it was complying with the consent decree's prohibition on the taking of contracts under duress. "Members who were forced to sign," Merritt said, "do not make desirable members." Of course, Sun-Maid did not care whether they signed willingly or not; it needed all the growers to maintain its monopoly. Merritt turned the tables on the victims, blaming the "un-American" growers who heaped "filthy abuse" on the Sun-Maid solicitors for the injuries they received. Sun-Maid invited growers to report all incidents of coercion, violence, and destruction of property to the local chambers of commerce—an exercise in futility in Kerman, where the chamber of commerce was complicit in the violence; elsewhere, Sun-Maid's own employees were the ones taking the complaints under an agreement reached in 1923 by the USDA and the DOJ, described below.[43]

Faced with obvious collusion between legal officials and Sun-Maid, growers turned to local attorneys for help. Several Fresno lawyers, some in private practice, others on retainer to independent packers, prevailed on Sun-Maid to return the contracts of growers who claimed coercion. This approach was not always successful, but nine growers reported to the DOJ that they obtained their contracts in 1925.[44] In a tactic that was proving increasingly effective with every Sun-Maid contract campaign, growers were taking the cooperative to court and winning judgments releasing them from all contractual obligations. Four growers alleging coercion and duress in the 1923 and 1925 campaigns won suits in which Sun-Maid "was adjudged as conniving at the actions of the mob that forced the execution of the waiver." These results were consistent with California Supreme Court rulings denying cooperatives the remedy of specific performance handed down during this time, discussed in Chapter 9. These cases showed that while cooperatives could win liquidated damages after the breach took place, the courts' willingness to void contracts obtained through violence gave growers a fairly reliable remedy.[45]

Some growers, intent on finding more immediate help than litigation would provide, wrote to the federal government, describing the illegal actions of the mobs and the connivance of the entire Fresno business community in the 1925 referendum. A local newspaper editor's plea to the DOJ implicated the law as handmaiden to the mobs: "Judges and officers of the Court were intimidated until law became a travesty and its executives became co-conspirators with the crooked officials of the Raisin association who were inciting criminal acts, coercion, violence and bloodshed to force signatures to an illegal contract." One grower's wire to the secretary of agriculture described "the worst type of coercion being used to force raisin growers signing fifteen-year contract today about fifty men came to my ranch dragged me from disc destroyed vines threatened death burn buildings cut all vines isolate my family if I do not sign sunmaid contract what relief has grower?" Wallace offered him none, advising him to swear out a complaint with the local police.[46]

Wallace's response was disingenuous. The violence in the raisin vineyards had always attracted federal attention. Although the DOJ could not prosecute local crimes, its officials used incidents of night riding to bolster the 1920 antitrust prosecution. After the lawsuit the raisin cooperative came under federal scrutiny by dint of either the consent decree or the Capper-Volstead Act or both. If the consent decree applied to the reorganized Sun-Maid, then the DOJ retained jurisdiction; the USDA's authority under Capper-Volstead was less clear but ostensibly embraced the regulation of such patently illegal acts as night riding. The jurisdictional tangle over Sun-Maid involving the DOJ, USDA, state courts, and local law enforcement resulted from overlapping but uncoordinated sources of legal authority over the raisin industry. Each level of government had its own institutional, political, and legal concerns to pursue, concerns that often diverted officials' focus away from the violence associated with Sun-Maid's monopoly. As the federal agencies negotiated the issues of authority and jurisdiction, Sun-Maid exploited the uncertainty created by the conflict between the consent decree and Capper-Volstead, waging its campaigns with essentially no supervision and evading accountability for the night riding.

The Sun-Maid leaders knew they could count on the USDA as a reliable ally. Immediately upon taking office in 1923, Ralph Merritt used the Capper-Volstead Act to distance Sun-Maid from the consent decree and DOJ oversight by seeking USDA approval for Sun-Maid's reorganization plan. Secretary Wallace endorsed the plan: "I am much pleased with the way you are proceeding with the reorganization of this association to meet the requirements of the Capper-Volstead Act." He added only one caveat: "It is understood that all steps taken by the Sun-Maid Raisin Growers of California with respect to the contemplated reorganization and all methods of operation which will be employed by it will be consistent with the terms of the consent decree entered against it in 1920 [sic]." Unsure of the extent of his authority under Capper-Volstead, Wallace assumed that the new Sun-Maid, as successor to the old, was bound by the consent decree.[47]

After the 1923 campaign, the independent packers decided to reinvolve the DOJ in the industry's affairs, protesting that coercion had been employed in the drive to gather signatures. The U.S. attorney promptly dispatched a special agent to determine whether Sun-Maid had violated the consent decree's prohibitions against the use of "coercion, intimidation or duress" in sign-up campaigns.[48] A week spent interviewing growers convinced the special agent, Ralph Colvin, that there were "literally hundreds of cases" of violence and coercion. To maintain the monopoly, he reported, Sun-Maid management and the banks had condoned "outrages, personal violations, destruction of property, etc., perpetrated by 'committees' of volunteer workers." Colvin noted that "many individuals" seconded grower Augustus F. Jewett's characterization of the campaign as a "reign of terror" conducted by the "Raisin Ku Klux Klan." Serious violations of the consent decree had taken place, he concluded.[49] In view of the continued harassment that dissenting growers were likely to suffer, from which the district attorney was willing to provide "damned little" protection, Colvin recommended that the DOJ determine "a policy to be pursued in this case."[50] The special agent assumed that the DOJ's policy was to enforce the consent decree.

Six weeks after Colvin filed his report, Fresno was rife with rumors that another Sherman Act suit was in the works. "If the government can extend forgiveness to wealthy manufacturers' combines in the East," the *Fresno Bee* commented, "it is to be wondered why the farmers are to be harassed through the courts with a policy of vengeance rather than justice."[51] Chances of new litigation were actually slim. Even as the U.S. attorney in San Francisco was laying down an ultimatum to Sun-Maid, he was moving to resolve the conflict in a way that would reflect positively on his boss, Attorney General Harry M. Daugherty.[52] After declaring publicly that "coercion had been resorted to in some instances to compel growers to sign the new contracts," the DOJ conceded that the crop contract and its provisions belonged "under the jurisdiction of the Secretary of Agriculture." In late 1923, Assistant Attorney General William D. Riter released the newly reorganized Sun-Maid from the consent degree, publicly assuring Merritt that the cooperative fell "entirely within the Capper-Volstead Act and need fear no prosecution from the Federal Government under the anti-trust statutes."[53]

The action of federal officials in 1923 freed Sun-Maid from the restraints of the consent decree. Except for the special agent, whose duty was merely to investigate, no local or federal officials deemed the charges of coercion worthy of investigation or prosecution. The subsidiary marketing corporation with nonproducer stockholders had won the approval of two federal executive departments. Small wonder that Merritt could express gratification at the "friendly spirit" of the DOJ and the "helpful and constructive attitude" of Secretary Wallace. The legal apparatus of the benevolent trust seemed complete. As a result of the DOJ's decision to release the reorganized Sun-Maid from the consent decree, Merritt and the directors assumed that they "were starting anew and were operating solely within the terms of the Capper-Volstead Act, and, as a natural consequence, do not consider themselves bound by the decree, and, in fact, have operated without any reference to said decree."[54]

Unwilling to take an activist role in applying federal antitrust law to farmers under Capper-Volstead, the USDA continued to rubber-stamp Sun-Maid's dealings with its members. In 1924, Merritt sought and received Wallace's cautious approval for the withdrawal campaign. At first, Wallace's staff intended to tell Merritt that the secretary had no jurisdiction over the "manner in which [cooperatives] organize or with the terms of their contracts" and that Merritt should take his inquiry to the attorney general. At the insistence of Lloyd S. Tenny, chief of the USDA's Bureau of Agricultural Economics (BAE) and a personal friend of Merritt, the letter was rewritten to give Merritt the ringing endorsement he sought: "No reason is apparent to this Department, from the standpoint of the Capper-Volstead Act, why you may not . . . obtain waivers from those [growers] who desire to execute them."[55]

When the referendum produced fresh complaints of violence and coercion, new DOJ attorneys on the West Coast decided to revisit the jurisdiction question. Frustrated by the inaction of the Fresno grand jury, many growers pressured the DOJ to enforce the law. Michael Jensen issued an "appeal for justice" from a situation in which "the criminal decides if he has committed a crime." In November 1925, H. H. Atkinson, an assistant U.S. attorney in the Western

Antitrust Section, forwarded these complaints to Washington and requested permission to investigate possible violations of the consent decree during the February referendum.[56]

By this time, scandal had forced Daugherty from his post, and the new attorney general, John G. Sargent, thought it advisable to distance the DOJ from the deal Daugherty struck with Sun-Maid and the USDA in 1923. Daugherty neglected to formalize his release of Sun-Maid from the consent decree; thus the DOJ retained the authority to enforce the decree through further orders by the federal court. To lay the groundwork for future court orders, Sargent agreed to Atkinson's request and dispatched three special agents to Fresno in early December. In two weeks these agents interviewed nearly one hundred people, including the referee in Sun-Maid's bankruptcy case, a member of the Fresno grand jury, several lawyers, Merritt, and scores of growers. The agents found that the referendum period was a time of unchecked mob rule, during which the night riders flouted the law in the name of Sun-Maid. Frank Hamm told the agents that the president of the Dinuba Chamber of Commerce had told him, " 'We go and get the contracts no matter how we get them.' " One solicitor admitted Sun-Maid's involvement to Henry Gussman: " 'I am put on by the association to make the growers sign the contracts.' "[57]

The agents collected substantial evidence of the extent of the damage and the identities of the victims and night riders. Out of a total of thirty-seven complaints, fifteen came from U.S. citizens and thirteen from Armenians and Armenian-Americans. The rest were six Germans and one Italian and one Dane. The victims were typical smallholders; they owned vineyards with a median size of twenty acres, and all told they lost about twenty-five thousand grapevines to the mobs. The agents identified eleven growers who claimed they signed contracts under duress in 1923, and eight of these subsequently obtained the return of their contracts from Sun-Maid. The most valuable information in the agents' report, from a historical perspective, is the identification of some night riders by occupation. Of those named, twenty-four were landowners (and presumably full-time growers), six were members of professions, two were merchants, one was a local police official, and three were Sun-Maid employees.[58]

The agents had nearly completed their work when, in late December, the *Fresno Morning Republican* broke the news of the investigation. It reported that a broad, far-reaching investigation of Sun-Maid was inquiring not only into possible consent decree violations but also the bankruptcy of the old association and Sun-Maid's responsibility for the damages suffered by growers who withdrew in 1924. Contacted by the Associated Press, Atkinson confirmed that an investigation was in progress but provided no other details. The story ruined the DOJ's plans. Attorney General Sargent demanded to know who had leaked the story: "That investigation should have been made quickly and without publicity." Once the news was out, Sargent told Atkinson, the DOJ's position was compromised: "Unauthorized publicity has embarrassed the Secretary of Agriculture in his plans for development of cooperative marketing associations." Atkinson replied that the agents had conducted their investigation as discreetly as possible and still managed to turn up plenty of evidence of violations of the consent decree. The DOJ had the legal authority to enforce the decree; politically, however, it was on less certain ground. When William M. Jardine succeeded Henry C. Wallace as secretary of agriculture in early 1925, he inherited Wallace's problematic relationship with Commerce Secretary Herbert Hoover, who had designs on bringing cooperative marketing into his department. Not only was Jardine willing to take a more activist role in determining cooperatives' liabilities under Capper-Volstead, he also was determined to use the law to fend off Hoover's encroachment.[59]

Thus the politics of the situation had everything to do with the USDA's perception of its position in relation to its farmer constituents. Merritt immediately pressured Jardine to intervene, asking him to prevail on the attorney general to issue an "unqualified statement that cooperatives are not harassed by his department and [the] Sun-Maid case is not actually directly or indirectly before him." On December 23, Merritt received assurance that the matter "will be settled by conference" between the USDA and DOJ that day; meanwhile, the special agents continued to press Merritt for information, an act which Merritt believed "violates understanding and agreement for administering cooperative matters thru Department Agriculture." BAE chief Tenny assured Merritt that the problem was being dealt with at the highest levels of government. In Tenny's opinion, the investigation had been initiated by West Coast Justice attorneys, not by anyone in Washington; he believed the entire affair was an internal DOJ problem, unrelated to federal policy regarding cooperatives.[60]

The evidence uncovered by the DOJ agents could not overcome the political pressure on the attorney general to relinquish oversight of Sun-Maid to the USDA. After Sargent and Jardine met on December 24, both departments issued statements that agreed in only two particulars: the press reports of the investigation were "wholly unauthorized and unfounded," and Sun-Maid was "operating under the provisions of the Capper-Volstead Act, under the jurisdiction of the Department of Agriculture." On December 28, Sargent ordered Atkinson to cancel the investigation and refer all inquiries to Washington.

Sargent tendered Jardine a copy of the special agents' report and asked him to resume the inquiry, since the "Secretary probably has [the] power under Capper-Volstead Act to afford [a] remedy if necessary."[61]

The USDA had that authority but was not about to use it against Sun-Maid. A USDA legal expert interpreted the consent decree to apply only to the CARC, which was no longer in existence: "The 'consent decree' rendered against the old company would be without force as to the new." In the view of USDA attorneys, the DOJ's acquiescence in the legality of Sun-Maid's membership and financial affairs in 1923 imposed a kind of estoppel on investigations related to the consent decree: "Only the highest and most imperative reasons would justify a departure from the conclusions reached in 1923 that the affairs of the Sun-Maid were in conformity with the law." No such reasons apparently existed. The consent decree, which provided the DOJ with the legal basis for retaining jurisdiction, made Sun-Maid an exceptional case, and USDA officials did not want this distinctive case to set the dangerous precedent of narrowing the secretary's authority under Capper-Volstead. As for the night riding—the sole infringement of the consent decree for which there was actual evidence—the USDA retreated behind the curtain of dual federalism. Any "questions of violence or intimidation . . . would be dealt with by the local authorities"—presumably those officials who had already shown how keen they were to defend growers' rights. Matters involving "restraint of trade resulting in undue enhancement of prices" remained the responsibility of the USDA. As Tenny intended, this decision left little room for the DOJ to become involved in the affairs of cooperatives. Jardine endorsed Tenny's view of the case, and to Merritt's satisfaction Tenny closed the file.[62]

The idea that prosecuting acts of coercion and violence should be left to local authorities gained acceptance in the DOJ as well. DOJ staff attorneys decided they had no legal basis for holding Sun-Maid to the terms of the consent decree. The CARC had reorganized in conformity with Capper-Volstead, and the new Sun-Maid purchased the assets of the bankrupt Sun-Maid free and clear of all liabilities. While the decree technically remained in force against the CARC, the DOJ could not use it to restrain Sun-Maid. Further, the DOJ had no authority to prosecute such local crimes as assault and malicious mischief. The special agents' evidence that Sun-Maid employees were involved in the night riding was not enough to link the cooperative to the illegal activities. "Outrageous as the situation in the raisin growing country of California is," an assistant attorney general noted, "it is nevertheless outside Federal jurisdiction."[63]

The DOJ and the USDA came to their conclusions with particular aims in mind. The DOJ was hamstrung by its approval of Sun-Maid's reorganization and conduct during the 1923 campaign. Once Daugherty agreed that Sun-Maid was operating under Capper-Volstead, Sargent had no basis to reverse that judgment, and Jardine prevented him from trying. Although the USDA had yet to develop a coherent enforcement policy, Jardine acted to protect the secretary's statutory authority over cooperatives not just from the DOJ but from Commerce as well. The night riding presented a sticky problem that both departments preferred to avoid. They disposed of it by declaring that only local authorities—who refused to prosecute in 1923 and 1925—had the authority to enforce law and order in the raisin-growing districts. Growers who wanted out after 1925 found that their most effective means of escape was a suit for the return of contracts obtained under duress. Within two years, moreover, a more permanent solution became available.

## GROWERS BREAK THE TRUST

The Capper-Volstead Act, as interpreted by executive branch departments, relieved Sun-Maid from liability for the activities of the night riders. The USDA's timely intervention permanently suspended the 1925 DOJ investigation and settled the issue of which federal agency would administer the enforcement powers the statute conferred. The USDA approved not only Sun-Maid's new legal structure but also the recruitment methods of its members and employees. Nothing that the USDA did, however, could give Sun-Maid the legal power to do the one thing essential for its survival: control production. The unchecked increase in the production of raisins after 1921 proved more effective than anything dissenting growers could have done to undermine the "benevolent trust."

Although the federal government refused to hold Sun-Maid accountable for the violence and local officials refused to prosecute the individuals involved, growers resorted to other remedies. Some simply breached their contracts and delivered to other packers, incurring breach-of-contract suits. Of the thirty-two suits Sun-Maid filed in Fresno County Superior Court between 1921 and 1925, twenty-seven were dismissed or settled. One grower won outright. Twenty-three of the defendants were Armenians and two were Japanese, reflecting Armenians' dissatisfaction with Sun-Maid.[64]

Other growers directly challenged the legality of the contracts. Krikor Arakelian, who had refused to pay liquidated damages in 1924, and Paul Mosesian both won relief from the contract from a state appeals court in 1928. In both cases, the court upheld the legality of the liquidated damages provision and carefully noted that neither case "involve[d] either the moral or prestige of a cooperative association." Sun-Maid lost because neither defendant belonged to "a cooperative association." Both had signed with the CARC in 1921, and the old Sun-Maid assigned those contracts to the new Sun-Maid in 1923. The court held that the assignment did not bind the defendants to Sun-Maid: "This action . . . is simply one where a co-operative association is seeking damages from an outsider for refusing to be bound by the terms of a contract made with a corporation organized for profit, of which contract the plaintiff has become the assignee."[65] Interestingly, the court saw the relationship between the CARC and Sun-Maid just as the USDA had: Sun-Maid was the successor to the CARC. In these cases, however, disassociating CARC from Sun-Maid made the successor organization worse off.

Downplaying the rulings, the *Sun-Maid Business* declared that "the Mosesian and Arakelian decisions have no significance in cooperative law and no bearing on the relations between Sun-Maid and its members." It was wrong on both counts. The cases voided the 1921 contracts six years after the consent decree did so. Moreover, the decisions reinforced the legal significance of a cooperative association: "There is a marked distinction between cases such as the one with which we are dealing," the court said in *Mosesian*, "where the defendant entered into a contract for the sale of grapes with a corporation organized for profit, and one where the contract exists between members of a co-operative organization."[66]

As it turned out, few growers required the remedy provided by the *Mosesian* and *Arakelian* cases. Market conditions did not bear out Merritt's confident assertion that Sun-Maid could prosper as a competitive business. The withdrawals in 1924 and 1925 reduced the cooperative's control to around 60 percent and crippled Sun-Maid's ability to maintain prices. Even with the complicity of the banks, which denied credit to former Sun-Maid members after 1925, more growers sold to independent packers. As surpluses continued to accumulate, Merritt could only strongly recommend to growers that they plant no more vines and produce fewer raisins.[67]

Few of them listened. The immediate postwar boom in planting raisin vines had a rebound effect that lasted for most of the decade. Each year, the total surplus grew. The eighty-six-thousand-ton carryover from the 1923 crop took almost three years to sell; although smaller carryovers were produced in the following three years, the cumulative effect was demoralizing. On top of that burden, growers had to bear the costs of the advertising and marketing programs, which each year ate up $2.5 million out of about $30 million in total sales. Despite Sun-Maid's program of financial austerity—for which Merritt unfailingly claimed credit in the *Sun-Maid Business*—growers' returns decreased relative to retail prices. As prices returned to pre-CARC lows, many growers lost their vineyards to foreclosure. One observer estimated that nearly eighty thousand acres reverted to the banks.[68]

The 1927 season marked an especially bad year for American agriculture and the beginning of the end for Sun-Maid. Prices slumped, demand was sluggish, and Merritt accused the independent packers of "ruthless" price cutting at Sun-Maid's expense. The real cause of the problem, as Merritt knew, was that "Sun-Maid has no control of raw material and therefore of markets." Merritt was forced to borrow heavily to cover Sun-Maid's mounting debts, "but the final crash of prices in 1928 all but wrecked the association." The banks and investors insisted on reorganization. The secured creditors "resorted to remedies available to them" and took over the Delaware corporation and its assets.[69] Six months later, the banks fired Merritt, the cooperative filed for bankruptcy, and a mere 32 percent of the state's raisin growers counted themselves as Sun-Maid members. Not until the Federal Farm Board paid off the banks with federal loans in 1931 did control of the Delaware corporation revert to the remaining members.[70]

Contemporaries knew what had caused Sun-Maid's failure. An Australian observer noted in 1929 that "Sun-Maid is criticized from end to end of the Valley, and the lesson of its fall should not be lost." The lesson was that "the only thing which could have saved the situation would have been authority to destroy the carryover of 100,000 tons, or the loss of a season's crop."[71] In short, the marketing monopoly was necessary but not sufficient for success. Organized, profitable marketing was impossible as long as control of the supply was left to the individual decisions of sixteen thousand growers. The economic power that Sun-Maid lacked — the ability to control supply through production controls or destruction of the surplus — the law did not sanction. As a result, even the sympathetic invocation of the Capper-Volstead Act failed to prolong the heyday of the benevolent trust. The decline of Sun-Maid indicated that the marriage of cooperation and monopoly required more than an indulgent body of law. In the end, it required precisely the cooperative spirit that Sun-Maid boosters proclaimed with their words and undermined with their deeds.

But while it lasted, Sun-Maid's ride was an impressive one. The formal legal culture of the Central Valley so indulged commercial agriculture that it gave license to extralegal uses of law to force minorities to conform. As Chapter 9 illustrates, the 1920s marked the ascendance of cooperation in national politics, and the strength of agricultural influence blinded local legal officials and the USDA to Sun-Maid's continuing violations of the antitrust laws. The night riders' use of force, however, was ultimately trumped by constitutional restrictions on cooperatives' power to control supply. As much as the law now encouraged collective action, it still withheld from economic associations the ability to control production. Perversely, in agriculture, this meant that even the strongest of monopolies could still be undermined by the uncoordinated economic decisions of individual producers.

Sun-Maid collapsed in spite of the efforts by zealous mobs to maintain it; more surprising is the fact that it collapsed in spite of the elaborate legal shelter erected to protect it. Administrators in both the USDA and DOJ effectively gutted the consent decree as a mechanism for regulating Sun-Maid's activities, without ever obtaining federal judicial termination of the decree. Yet, curiously, Sun-Maid continued to offer growers the right to withdraw from their crop contracts—a clause that significantly contributed to the cooperative's downfall. And as much as Ralph Merritt proclaimed Sun-Maid's autonomy from the decree between 1923 and 1925, thereafter he unhesitatingly blamed it for Sun-Maid's problems: "The consent decree put the Sun-Maid organization in a disadvantageous position in further competition."[72] Making scapegoats out of the consent decree and the growers it was meant to protect was Sun-Maid's only weapon against what it could not control: the stampede to plant raisins immediately after the war and the resulting avalanche that cascaded into the cooperative's packing plants. The consent decree forced Sun-Maid to change its legal relationship to the growers, but it had little impact on the way it conducted business and none at all on the dynamics of supply and demand.

In the end, Sun-Maid's "benevolent trust" theory of cooperation turned out to be an all-or-nothing proposition. Sun-Maid needed to maintain a monopoly on supply in order to control prices, not so much to profiteer as to keep the industry from collapsing under the weight of short-run overproduction. Yet Sun-Maid's own high prices induced growers to plant more than the cooperative could profitably sell. Anything less than a 100 percent monopoly created the destructive competitive conditions Sun-Maid's presence was supposed to prevent. The violence of the night riders, their disdain for the law, and their contempt for "aliens" and "outsiders" ironically hastened Sun-Maid's downfall.

Yet for all its distinctiveness, Sun-Maid created a legal form for cooperatives that was adopted in scores of states even as Sun-Maid was collapsing. Sun-Maid's lasting contribution was to disassociate Rochdale values from the economic function of marketing. As appealing as Rochdale was, it was designed for collective buying; American farmers required an organization legally equipped to perform collective selling. In giving their sweeping endorsement to the California model of cooperative marketing, lawmakers ignored the fundamental question of cooperatives' impact on price, and the courts, by deferring to legislative judgment on that issue, completed the legal recognition of the benevolent trust just as the Great Depression rendered it essentially irrelevant.

PART FOUR

Cooperation in the Industrial

Economy

1920–1945

# 9 ∾ Associationalism and Regulation

*The only people in the United States who may and who have the power to organize*

*without limitation are the farmers of the United States. No industry can do what you*

*may do if you will by organization. Only the farmer can have a complete unlimited*

*monopoly and still be in any measure within the law. —Aaron Sapiro, 1926*

The CARC and Sun-Maid showed the nation two images of agricultural co-operation. One depicted a unified industry of growers improving their position in the market through the application of corporate business practices. The other raised the specter of a monopoly that fixed prices and eliminated competition. Both images helped to shape the direction of state and federal policy during the 1920s. Cooperation's new framework for coordinated economic action nicely complemented associationalism, Herbert Hoover's approach to the regulation of markets that became the decade's watchword for government-business relations. The Janus-faced character of cooperation fed both the associationalist impulse, with its emphasis on self-regulation, and the enduring individualism of American capitalism, which helped to bring about the Great Depression.[1]

It was that continuing dedication to individualism and competition, how-ever tempered by associationalism, that made people uneasy about monopoly in agriculture. From the perspective of eastern observers, the California co-operatives gave farmers too much power over too many citizens. A columnist summarized this fear in 1925: "Seven million five hundred thousand farmers and their families cannot combine to hold up the other two-thirds of the populace, who could be made the victims of extortion beyond that forced by owners and miners upon the consumers of coal. Nothing more cruel could be devised than an agrarian monopoly."[2] Farmers' monopolies seemed cruel because they exacerbated urban dependence on the rural sector. Yet its promoters argued that cooperation was the best bet for preventing agricultural instability.

The Capper-Volstead Act signaled that agrarian monopolies, though distasteful, were now legally acceptable because they helped farmers to help themselves. While the federal statute implicitly acknowledged that cooperation might lead to monopoly, it established nominal restrictions on cooperatives' business practices to prevent them from charging unfairly high prices. Congress knew its standards were vague, but the bill's supporters assumed that the administrative state would step in when agricultural monopolies abused their market power. What Congress did not anticipate was that rather than monitor prices to make sure they stayed within the bounds of fair and reasonable, the USDA would follow much the same pattern established in the Sun-Maid case and confine its oversight of the market to advising farmers whether their organizations met the standards set out in Capper-Volstead. The USDA was vitally interested in making the agricultural economy more efficient and more competitive, but throughout the 1920s the department saw its role as furnishing expertise to farmers "through a cooperative planning system relying on education and voluntary action." [3]

A more definitive policy on agrarian monopoly came from the states. Beginning even before the passage of Capper-Volstead, a wave of legislation flowed from the states to license cooperatives to do business along the lines established by the California model. The model cooperative marketing statute was drafted by a California attorney, Aaron Sapiro, who had formed tens of cooperatives in California and several hundred more across the rest of the country and Canada. Sapiro's model statute not only recognized that cooperatives might monopolize, it also declared that the goal of public policy should be to encourage them to do so. This idea marked the high point of the cooperative movement's legal transformation. Thirty-eight states adopted this law or versions of it by 1928. The Supreme Court endorsed the general policy without determining whether cooperatives' restraint of trade exceeded statutory bounds. By the end of the decade, the Supreme Court was expressing reservations about whether cooperatives needed preferential treatment in the market, ruling that states could not set up one set of regulations for corporations and a less stringent regime for cooperatives.

Though the state marketing acts went further than any previous court decision or federal law in freeing farmers from the antitrust laws, they stopped short of giving cooperatives the power to act as true monopolies: they could not limit output. Without that power, cooperatives were unable to prevent the reckless accumulation of agricultural surpluses during the postwar decade.

The nearly universal judicial validation of the new statutes, ironically, came too late substantially to affect the economy's direction.

## COOPERATIVES' LEGAL STATUS UNDER CAPPER-VOLSTEAD

Far from marking "the beginnings of a legislative definition of a true cooperative," the Capper-Volstead Act represented a stage in the continuing evolution of cooperatives' legal status. The law recognized the overlap between cooperative and corporation that had arisen in economic practice. As a USDA official proclaimed, "This legislation merely tends to grant to cooperative organizations some of the privileges of the business organization." At the same time, the law acknowledged the distinctive character of associations formed "for the mutual benefit of the members thereof."[4]

Cooperatives' legal structure had hardly entered into the congressional debates. Both houses focused entirely on monopoly as manifested in price control. Supporters maintained that farmers would almost never be able to charge "unduly enhanced" prices: "The farmer is not in a position to do as a manufacturer does. He can not control his markets and he can not make his own prices, and he never ought to have been made subject to the provisions of the antitrust law." Opponents were concerned with the prospect of illegal prices to the exclusion of everything else. One eastern senator predicted that the nation's farmers, planters, ranchmen, dairymen, nut and fruit growers "may increase prices as much as they please . . . until the Secretary of Agriculture— whose appointment is so largely controlled and influenced by the very classes of people named in this bill—can be persuaded that they are asking an unconscionable price." The legislation passed because a majority agreed that the Clayton Act exemption should cover all cooperatives. As long as cooperatives observed the principle of democratic voting, they could resemble and behave as much like regular corporations as they pleased. Edwin G. Nourse, the leading expert of his generation on cooperation, wrote in 1928 that "the law is rather ambiguous as an expression of cooperative doctrine."[5]

In ensuing years, that ambiguity proved a handicap to the USDA. It was not price but legal form on which farmers demanded the USDA's interpretive judgment. Requests came to the USDA from farmers and cooperative organizers seeking to determine how to qualify for statutory immunity. The USDA consistently informed inquiring correspondents that "all associations desirous of obtaining [the law's] benefits must have only producers as stockholders." Following Congress's intent that Capper-Volstead would apply only to associations doing business in interstate commerce, Secretary Wallace issued the statement that it "does not change or supersede laws of the various states affecting or relating to the regulation of cooperative associations." Except in the Sun-Maid case, where the federal consent decree undermined its exclusive regulatory oversight, the USDA rendered only cursory advice to farmers on how to form associations that qualified for the exemption.[6]

The Capper-Volstead Act granted farmers a privilege that no other group of entrepreneurs received for more than a decade: recognition of the legality of collective action in the market. Labor unions did not attain this privilege until the Wagner Act of 1935.[7] Just how farmers would enjoy this advantage was less than clear. The law did not codify the appropriations rider; it did not permit cooperatives to engage in practices that violated the antitrust laws without threat of prosecution. The legality of their activities would be determined only if the USDA elected to perform its statutory duties.

As had been the case in the Progressive Era, federal policy had less impact on farmers' economic lives than state law. State laws authorizing the formation of cooperatives had largely forced changes in farmers' antitrust liability at the federal level before 1922. After the Capper-Volstead Act, farmers continued to make antitrust liability a political issue at the state level, in response to state prosecutions of farmers' cooperatives "with never a mention of corporate dealers who were doubtless more influential as price makers than were the farmers."[8] California-style cooperation provided the template for state legislation after 1922 because of the single-handed efforts of Aaron Sapiro.

## AARON SAPIRO AND THE ADMINISTRATION
## OF AGRICULTURAL MARKETS

Aaron Sapiro was cooperation's proselytizer to the nation during the 1920s. His cross-country crusades made him famous; his quarrels with national farm leaders and politicians made him notorious. Agricultural historians long credited him with originating the form of cooperation that he called "commodity cooperative marketing." He was, however, only the messenger — if a creative and forceful one — and not the source. The CFGE and CARC, which supplied the legal models for his program, were up and running concerns long before Sapiro began working in the field. As Grace Larsen and Henry Erdman wrote in the early 1960s, "Sapiro's achievement consisted in presenting his interpretation of the experiences of California co-operatives to a national audience in an attractive package, and in making a quick and extensive sale of his ideas." The model cooperative marketing acts were the centerpiece of his campaign to spread cooperation to farmers in every region of the country in every commodity.[9]

Sapiro began his career during the Progressive Era, when he served as legal counsel to the California State Market Commission and its director, Harris Weinstock. Weinstock was a wealthy Sacramento merchant who had hitched his political star to the Progressives. In 1915, Governor Hiram Johnson appointed him to run the California State Commission Market, which held the power both to set the price for and to sell, on commission, all food products produced in the state.[10] Weinstock proceeded to operate the Commission Market according to his personal vision of market reform: "organiz[ing] the state's farmers into state-wide marketing associations, industry by industry . . . until California farmers became, as he expressed it, the most effectively organized producers in America." With the wide discretionary powers conferred by the statute, Weinstock established the Market Commission to set and fix prices paid to farmers for their crops and hired Sapiro to organize farmers into cooperative marketing associations. By 1919, when a disagreement with Weinstock induced Sapiro to return to private practice, the latter had organized between twenty and thirty cooperatives in fresh and dried fruit, poultry and eggs, and dairy.[11]

Weinstock and Sapiro both took an entrepreneurial approach to the public administration of commodity markets. They believed that government should do more than encourage farmers to market their products more efficiently; government should show producers how to do so and, when necessary, should advise farmers about the proper legal forms to use. In adopting as official state policy the goals and tools of cooperative marketing in California, Weinstock and Sapiro not only made powerful enemies among commercial traders and food processors but also steered the state into unchartered regulatory territory. By establishing a market commission that regulated distribution, supply, and demand and by encouraging cooperative associations to fix prices, Weinstock made the California State Market Commission a market participant. The idea that the market should operate autonomously had been thrown over by a politically opportunistic Progressive reformer whose conception of the public interest in fair commodity prices entailed direct public supervision of private exchange. The State Market Commission's open intervention in the market drew sarcasm from Carl Plehn, a contemporary economist: "It is obvious that a commission market is just as different from a market commission as a chestnut horse is from a horse chestnut."[12]

The California experiment did not last long. Under heavy pressure from consumers' groups, urban mayors, and the fishing industry and beset by poor health, Weinstock retired from office in 1920. The Market Commission's public support had eroded. Legal hostility ended the roguish attempts by Weinstock and Sapiro to enlist administrative agencies in the service of agricultural producers. Echoing the conservative impulse of the postwar era, Plehn concluded, "[There is no] inherent reason why the gains and savings brought about by government action in the cases under review could not have been attained entirely by private initiative." After Weinstock's departure, the California Division of Markets continued to encourage the establishment of public markets at the municipal level, but it pared back the cooperative organizing work.[13]

Sapiro was convinced that farmers needed effective legal leadership if they were to form cooperatives of any economic consequence, and he believed no one was better equipped for that role than he. He may have been alone in that view, however. He had no farming background, and all he knew about the law of cooperative organizations he learned during his four years with Weinstock. He drew up model bylaws and articles of incorporation, "borrow[ing] freely from the documents of existing co-operatives [and] adapting them to the peculiar needs of each commodity." Leaving California for the South, he burst onto the national scene at the 1920 meeting of the American Cotton Association in Montgomery, Alabama. The delegates were electrified by his two-hour speech. One witness wrote in 1929, "The whole direction of the movement toward a new control of the cotton industry was changed by one man."[14]

Sapiro's ambition to organize American farmers according to the California plan of commodity marketing was more than a personal crusade. To him, it was the only way to assure the economic preservation of agriculture. His approach had a forked effect on national policy. On the one hand, he forced agricultural traditionalists to confront the implications of the California model for American agriculture. On the other, he exposed the divisions between national farm leadership and policy makers on the one hand and rank-and-file farmers on the other. These groups were divided not only by separate political allegiances and incompatible approaches to the farm problem of the 1920s but also by different understandings of the cooperative gospel. As members of Congress asserted that farmers could never monopolize the sale and distribution of their goods, Sapiro proclaimed monopoly to be the farmers' right: "The only people in the United States who may and who have the power to organize without limitation are the farmers of the United States. . . . Only the farmer can have a complete unlimited monopoly and still be in any measure within the law."[15]

Sapiro's declarations frustrated agricultural traditionalists such as Secretary of Agriculture Henry C. Wallace. Sapiro was adept at speaking the language of traditionalism while his model of cooperation expanded on the California organizations' most corporate features. Early in his rise to national prominence he stressed the values of unity and mutuality: "The cooperative associations of California have unified communities. They have brought all classes together. They have shown how the growers can help and serve each other." At the same time, Sapiro's recipe for commodity marketing stressed "the best established commercial principles" in recreating the relationships among farmers, their marketing organizations, and the larger economy. He was enough of a lawyer to know, however, that he could not cross the constitutional line separating cooperative marketing from the coordinated restriction of production: "The one great aim of cooperative marketing is to abolish the individual dumping of farm products and to substitute for it the merchandising of farm products, the controlled movement of crops into the markets of the world at such times and in such quantities that these markets can absorb the crops at fair prices."[16]

Sapiro dropped all pretense of asking individual farmers to forsake the advantages of free riding for the good of the community. He simply aimed to eliminate all cheating. He imposed inflexible legal ties between cooperatives and members, first by requiring a "conditional minimum" — the amount of market control without which the cooperative would not do business — and second by using an "iron-clad contract" to secure the participation of all producers. Although the consent decree in the CARC case had prohibited both these practices and guaranteed growers the right to withdraw from their crop contracts, Sapiro was intent on tighter, rather than looser, connections between organizations and producers. He publicly bragged that his crop contracts tied "those growers . . . to each other under as tight a contract as can be drawn." As in the raisin industry, Sapiro's crop contracts bound growers for a long term of years, specified liquidated damages in case of breach, and reserved the equitable remedy of specific performance — compelling the breaching party to perform as promised.[17] The lack of enforcement in the raisin case and the ambiguity in the Capper-Volstead Act enabled Sapiro to emphasize more rather than less monopoly, more rather than less coercive legal relations between organizations and farmers.

Sapiro promoted this plan with such intensity and dedication that the cooperative movement of the early 1920s became a virtual personality cult. By 1921, commodity marketing became a fixture in cotton, tobacco, and wheat, the nation's largest agricultural crops. At least fifty-five cooperative associations were organized under the Sapiro plan by 1923. That year, Sapiro helped launch the Canadian wheat pools. He became a regular figure in high agricultural policy circles, advising the American Farm Bureau Federation (AFBF) and James R. Howard, its president, on the relationship between business and agriculture. Sapiro also "assisted prominently" in the formation of the National Council of Farmers' Cooperative Marketing Associations (NCFCMA), an organization devoted specifically to promoting Sapiro's commodity marketing program. By 1926, one-tenth of American farms marketed their production through 8,135 different cooperatives, doing an aggregate business in excess of $1 billion in over one hundred different commodities.[18]

The market practices of commodity marketing raised many of the same legal and economic questions as the CARC and Sun-Maid had. Sapiro's insistence on strict adherence to his model of commodity marketing alienated national agricultural leaders who wanted to employ more traditional cooperative concepts. At the height of his popularity, opponents alleged that his law firm overcharged cooperatives for representing them and that Sapiro was abrasive and even unethical in his dealings with farmers. Not surprisingly, he incurred the enmity of the middle distributors, the scapegoats of his advocacy for the farmers. In 1924, one shipping agent complained to Secretary Wallace, "We people who are in this business think it advisable that the United States Department of Agriculture investigate the manipulations of this Sapiro at once, for the protection of all concerned."[19] The firms and corporations that marketed, distributed, and sold agricultural products considered themselves to be USDA constituents, and they pressured the USDA to keep federal help to cooperatives to a minimum.

The USDA was torn between serving its various constituents, abiding by its own policies about cooperation, and paying attention to the larger national political context. Wallace sought to keep a distance from Sapiro, who he believed exacerbated tensions between producers and distributors, and he also agreed with critics that Sapiro cooperatives ignored "the importance of the individual farmer." At the same time, Wallace saw cooperation's inability to unite farmers around a common goal as its greatest weakness.[20]

Wallace's discomfort with the way Sapiro was transforming marketing had less to do with his concern for farmers than with his ambivalence about government's regulatory role. Unable to mount a convincing economic case against commodity marketing, he contrasted the farmer's cultural image with the image of commodity marketing, which he saw as incompatible with traditional rural values. Just as critically, he refused to recognize that the USDA's position on cooperation carried important consequences in the marketplace. Wallace's equivocation left all sides dissatisfied and disinclined to rally behind him. Until he died in October 1924, he downplayed the marketing problem and insisted that the central goal of commodity marketing was unfeasible: "I doubt whether we will ever come to the time when we can exercise very direct control of production. That would involve either arbitrary government control or cooperative organizations strong enough to assert it. Direct government control is, I think out of the question. Cooperative control of production can come only after a very long period of education and experience in cooperation."[21]

Wallace's successor, William M. Jardine, more actively promoted cooperation and was more skilled at mollifying the USDA's combative constituents. Like Wallace, he disdained bold new initiatives, refrained from invoking the secretary's powers under Capper-Volstead, and opposed both the creation of a national cooperative marketing board and federal incorporation of cooperatives. Instead, he focused on expanding the administrative capacities of the Bureau of Agricultural Economics, the successor agency to the Bureau of Markets. He pushed hard for the passage of the Cooperative Marketing Act of 1925, which created a division of cooperative marketing within the BAE. Sounding much like Wallace, Jardine declared that the division's duties would be confined to "research, educational, and service work." The act, Jardine claimed, "does not provide for Government control or supervision, nor for the subsidizing of cooperatives."[22]

Jardine introduced a new wrinkle into the debate over cooperatives by suggesting that farmers did not want to be treated differently from other commercial concerns. Whereas Wallace had argued that farmers were distinctive and thus warranted special legal privileges, Jardine laid the status argument aside in favor of claims grounded in equity. If farmers were regulated or supervised in burdensome ways not incumbent on other businesses, they would be unable to compete. According to Jardine, the best way to assist the cooperative movement was to remove all impediments to efficient marketing and let the farmers take over from there. Hence Jardine promoted legislation to exempt cooperatives from paying corporate income taxes, a policy that became law in 1926. Jardine, as had Wallace, staunchly opposed all proposals to provide for federal incorporation of cooperatives and for a national cooperative marketing board. No proposal for federal incorporation made it out of Congress during the 1920s, but after President Hoover took office the Farm Board was established to encourage the formation and development of "effective" cooperative associations.[23] Throughout the 1920s, the federal government assisted cooperatives only indirectly, disseminating market information and publishing bulletins outlining various kinds of cooperatives.

The lack of direct assistance probably made little difference in the fate of the Sapiro cooperatives. The crown jewels of his work, the Burley Tobacco Cooperative in Kentucky and the Tri-State Tobacco Growers' Cooperative Marketing Association of North Carolina, South Carolina, and Virginia, began to fail in 1923. In addition to the perennial problem of overproduction, both were having trouble with disloyal growers; despite the "iron-clad" contracts, private warehouses lured producers away with promises of higher prices. As the tobacco cooperatives litigated breach-of-contract cases, they and other Sapiro cooperatives fell into receivership. The associations suffered from inexperienced management and sloppy organization. The geographic monopolies that made the California commodity marketing plan work so well were not present in the staple crops Sapiro sought to organize in the South and Midwest.[24]

As the cooperatives declined, the fervor Sapiro had created over the possibilities of commodity marketing gradually died down. New officers in the AFBF, mostly from the Midwest and South, were far less enamored with commodity marketing than their predecessors. They doubted—correctly, as it turned out—that the achievements of the California fruit and nut cooperatives could be duplicated in tobacco, cotton, and livestock because these commodities were produced over much more extensive areas and marketed in entirely different ways. When Sapiro and the AFBF failed to resolve their philosophical and personal differences, he was forced to resign in early 1924. The split nearly wrecked the AFBF—many of its rank-and-file members were staunch believers in Sapiroism—and it distanced Sapiro from national agricultural policy. The AFBF left the business of cooperative promotion to its less influential rival, the NCFCMA. Wallace and the USDA drew away from Sapiro after his quarrel with the AFBF, despite the charges of farm leaders that the BAE was failing to assist farmers with their marketing problems.[25]

As Sapiro's star faded, the farm groups threw their support behind the McNary-Haugen "dumping" legislation, a plan to dispose of agricultural surpluses by selling them overseas at rock-bottom prices. A series of McNary-Haugen bills repeatedly came up in Congress during the 1920s; President Calvin Coolidge vetoed the legislation in 1927. The legislation called for "publicly sanctioned market controls for the producers of staple commodities." Commerce Secretary Hoover and Secretary of Agriculture Jardine both opposed mandatory controls of any kind, as did the president. The polarizing effects of 1920s farm politics placed supporters of cooperation and McNary-Haugen at odds with each other. To try to win over the president, later versions of the bill relied more on cooperatives to implement voluntary production limitations, but the collapse of Sapiro's largest centralized cooperatives made that approach unfeasible, and the president's opposition to the legislation never wavered. The federal government produced no meaningful policy initiatives for agriculture until the Federal Farm Board Act of 1929. Although Sapiro had little impact on federal policy and his "days as an effective promoter of cooperatives were already over" by 1926, by then he had effectively made state law the vehicle for disseminating his vision of agricultural prosperity.[26]

The movement toward a uniform state law to regulate farmers' cooperatives continued the battle over the meaning and practice of cooperation that had begun in the nineteenth century. Drafted in 1921 by Sapiro and Kentucky state judge and tobacco grower Robert Bingham and circulated to state legislators by AFBF members, the uniform state cooperative marketing law codified antitrust exemptions for cooperatives, sanctioned the remedies of liquidated damages and specific performance, and enumerated economic powers for cooperatives that Capper-Volstead only hinted at. Unlike Capper-Volstead, the model law permitted farmers' cooperatives to use anticompetitive trade practices, and because cooperatives remained creatures of state law after Capper-Volstead, the model laws had the potential to affect the market in drastic ways. Thirty-eight states passed versions of the model cooperative marketing act (CMA) between 1921 and 1926; seven others passed cooperative marketing laws that differed in several particulars but echoed the fundamental ideal of Sapiro's statute.[27] A legal revolution in cooperative marketing took place during the 1920s, and it happened in the states.

The CMA invoked the ideal of economic efficiency to justify monopoly in agricultural marketing. Kentucky's version of the statute declared that it was in the public interest to permit growers to organize in large combinations to ensure agricultural prosperity: "The public interest urgently needs to prevent the migration from the farm to the city in order to keep up farm production and to preserve the agricultural supply of the nation; and the public interest demands that the farmer be encouraged to attain a superior and more direct system of marketing in the substitution of merchandising for the blind, unscientific and speculative selling of crops." The Kentucky law's policy goal was "to promote, foster and encourage the intelligent and orderly marketing of agricultural products through cooperation; and to eliminate speculation and waste; and to make the distribution of agricultural products between producer and consumer as direct as can be efficiently done; and to stabilize the marketing of agricultural products."[28] This declaration linked cooperation to the accomplishment of specific economic goals: agricultural stabilization and orderly marketing.

The CMA reflected Sapiro's theory about the relationship of cooperatives to the market and to the state. Rather than setting farmers apart from other entrepreneurs under the general antitrust policy, the CMA declared that state antitrust laws no longer applied to cooperative associations operating under the statute. Cooperatives earned this extraordinary privilege because they served an important public purpose. The difference between the earlier anti-trust exemptions and the CMA was that the latter explicitly stated this public purpose, thus making agricultural marketing ripe for state regulation under the "affected with a public purpose" doctrine of *Munn v. Illinois*.[29] Contemporary lawyers noticed that the model law was intended to resolve the equal protection issue that had hindered the movement since the late nineteenth century. The public interest rendered it "reasonable to make a distinct class of these co-operative associations of producers." According to California lawyer Mathew O. Tobriner, the constitutionality of the cooperative marketing laws rested less on the distinctive nature of cooperatives: "Farmers are not distinguished from bankers, business men, or laborers but cooperative associations for 'orderly marketing' [are distinguished] from illegal schemes of monopoly. Whereas one class of persons cannot be privileged to evade the law, special immunity may be conferred upon a well-defined type of organization, governmentally supervised." Another legal commentator resorted to a familiar analogy: "Agricultural associations not organized for profit are . . . in the situation of labor unions, and although their purpose is to establish a monopoly they are not for that reason illegal." Thus cooperatives could engage in pooling and bind members to deliver through tying contracts; but the statutes, as academic lawyers were relieved to discover, did not legalize coercion, secondary boycotts, or other activities prohibited by the Clayton Act.[30]

The CMAs specifically legalized iron-clad contracts and insulated them from interference or breach induced by competing processors, packing firms, and commodity warehouses. The cooperative had sole power to market the crop; members would receive the retail price in proportion to their contribution, minus marketing and overhead costs. Most states adopted the Kentucky section on marketing contracts, which was based on the 1918 CARC crop contract, with the addition of giving growers the right to withdraw each year.[31] The real teeth of the CMAs were the equitable remedies of specific performance and injunction. These remedies empowered cooperatives not only to put a stop to the tricks played by competing firms to induce breach of contract but also to force growers to fulfill their agreements even after a breach occurred. Further, the statute instructed that "such clauses providing for liquidated damages shall be enforceable as such and shall not be regarded as penalties." In one sentence the model law overturned the Iowa and Colorado courts' decisions in *Reeves v. Decorah Farmers' Co-operative Society* and *Burns v. Wray Farmers' Grain Co*.[32] Finally, the law made it a misdemeanor punishable by both fines and civil damages for third parties such as warehouses to induce a member to breach his marketing contract. In his speeches, Sapiro boasted of the strength of these provisions: " 'We regard cooperative contracts as more sacred than the bonds of matrimony; we permit no divorce. And if [farmers] fail to live up to their contracts we go at them like a ton of bricks.' "[33]

By the mid-1920s, a majority of state legislatures had decided that strong cooperatives were preferable to a weak agricultural economy. Effective enforcement of contracts was essential to successful cooperative marketing, so existing legal rules could no longer obstruct larger policy goals: "The need of encouragement of, and a strong public policy in favor of, such co-operative marketing organizations, justify the relaxation of the letter of the old technical rules and the granting of specific performance." What the state and the public got in return were nominal commitments to public accountability and a vague guarantee from Sapiro that cooperative monopolies would be held in check by the laws of supply and demand. If at any time prices became excessive, new producers would rush to enter the industry, and the resulting overproduction would bring prices back into line. Thus, Sapiro declared, no cooperative could monopolize indefinitely: "There is an automatic protection to the public and a constant threat to the growers themselves if they abuse their power through organization."[34]

In all its iterations, the CMA granted farmers legal privileges to combine in the market that no other category of entrepreneur then enjoyed. Sapiro boasted in the pages of a law review, "Farmers are the only great producers of actual commodities who may form the tightest type of combination, for both domestic and foreign trade, without interference of law by reason of the form or fact of organization." The CARC/Sun-Maid model was now legally available to the nation's farmers. Farmers had struggled for decades to balance the law's preference for the Rochdale model with the economic needs of marketing organizations; now, state law gave cooperatives the weapons California cooperative leaders contended were imperative to their economic survival.[35] How the courts would interpret the CMAs in light of existing precedent, the postwar economic climate, and their generally conservative posture during the 1920s posed the next challenge for the cooperative movement.

## THE COOPERATIVE MARKETING ACTS IN THE COURTS

Judicial reaction to the new special legal status of cooperatives was both immediate and extensive. As Sapiro predicted, the overwhelming legislative acceptance of the new policy provoked a cascade of lawsuits testing the constitutionality of the acts. As state courts upheld the CMAs in many significant respects, it became clear that the cutting-edge issue was no longer monopoly or restraint of trade but the validity of legislative assumptions about the public purpose of cooperation.[36] The overwhelming endorsement of the CMAs in the context of prolonged agricultural recession, moreover, helps to explain why the USDA indulged Sun-Maid during the 1920s and why the DOJ did not press ahead with enforcing the consent decree.

State courts upheld CMA provisions on contracts largely because they deferred to the legislatures' judgment that cooperatives did not unreasonably monopolize or restrain trade. This deference was so strong that the high courts of Texas and Oregon, first to rule on the constitutionality of the CMAs, avoided the hostile precedent in *Connolly v. Union Sewer Pipe Co.*, upheld the statutes as a valid exercise of legislative authority, and declined to distinguish the new result from the prior rule. The broadest ruling on the constitutionality of the CMAs and the legality of the crop contracts and damages for breach came down in North Carolina in 1923. In *Tobacco Growers' Co-operative Association v. Jones*, the state supreme court held that the legislature's purpose in enacting the law was to protect the public by stabilizing agricultural markets; it also said this purpose was justified by the postwar economic crisis. According to the court, the statute enabled farmers to help themselves without burdening the state or society: "[The producers] are taking all the risks. They are asking no assistance from the public treasury. They are forcing no one to join, and they are exacting no inordinate prices for their product. They are associating themselves as authorized by the statue, like other persons, and they have signed mutual and fair agreements among themselves which will be futile unless those who have signed such agreements can be held to abide by the terms of their contracts." The court lauded the cooperative movement as "the most hopeful movement ever inaugurated to obtain justice for and improve the financial condition of farmers and laborers."[37]

The North Carolina court's acceptance of the CMA's legal and economic assumptions drew fire from legal and economic critics.[38] Nevertheless, the decision paved the way for numerous other state courts to find that cooperatives "never can become a monopoly" because of the safeguards in the statute.[39] All the courts agreed that cooperatives organized under and operated according to the CMA presented no threat to free competition in the marketplace. It was, as the Wisconsin high court wrote in *Northern Wisconsin Co-operative Tobacco Pool v. Bekkedal*, simply a matter of adjusting the law to meet changed circumstances: "If in the course of time . . . such combinations . . . should come to be regarded as beneficial rather than injurious to the public interest, there is no doubt of the power of the legislature to completely reverse the public policy of the state with reference to such combinations and agreements and to promote rather than suppress the same."[40] No court save one reconciled the earlier rulings in *Ford v. Chicago Milk Association*, *People v. Milk Exchange*, or *Georgia Fruit Exchange v. Turnipseed*. The legal order's perspective had changed substantially in twenty-five years. Most courts seconded the sentiments of their colleagues in Minnesota: "The citizendom of the state is not complaining."[41]

Judicial ratification of the CMAs was overwhelming but not unlimited. The decisions effectively confined the legal status of cooperatives to the new statutes alone. As long as cooperatives abided by the statutory requirements, they did not restrain trade; if cooperatives were not organized under the new laws, they could be held to violate the antitrust laws. The statute permitted this result by encouraging cooperatives to reorganize under its aegis. Any cooperative activity that fell outside the scope of the statute, then, became tainted.[42] In effect, Sapiro-style cooperation had become the only kind of "true cooperative marketing association" permissible under the law. Although capital stock cooperatives incorporated under the old laws were still legally valid organizations, the CMAs did not protect them from antitrust liability. Some courts avoided sweeping rulings by construing the remedies narrowly and according to common law rules rather than the statutory goals. In the hands of the courts, the CMAs were something less than the full charter of economic freedom that Sapiro had intended.[43]

Endorsing the broad policy aspirations of the CMA was one thing; ascertaining the scope of the remedies it provided was quite another. In California, for example, since the turn of the century the courts had given a less than unqualified endorsement to cooperative crop contracts. Cooperatives had won the right to liquidated damages but not specific performance. Even though the CMA provided both remedies, the courts stood firm and continued to withhold specific performance from cooperatives. In its suits against growers, Sun-Maid sought liquidated damages rather than specific performance and still lost several key decisions. A trial court decision in 1924 followed the pre-CMA precedents instead of the statute, dissolved a temporary restraining order against three prune growers, and left the cooperative to seek damages at common law. By the time the appeal reached the California Supreme Court, the defendants had sold not only all their fruit but also their orchards. The court recognized the larger policy questions in the case but ruled that a mooted action was not the proper case in which to decide those questions.[44] No proper case arose and the California judiciary never ruled on the legality of equitable remedies or the CMA's constitutionality.[45] Sapiro, who had argued several of these cases, was unable to get the California courts to uphold the remedy of specific performance, the provision that in his view gave the CMA its teeth.

Other state courts hedged on the extent to which they awarded both remedies. In 1924 New York injected a new standard into the liquidated damages rule: the liquidated damages had to be reasonably related to the cooperatives' actual losses. In a case where a dairy cooperative's liquidated damages exceeded its marketing costs by over $1,100, the court held that the liquidated damages operated as a penalty. The court recognized that it was nearly impossible to estimate actual damages in cooperative marketing but nevertheless held that the cooperative had to try. Elements of the *Reeves* penalty rule lived on, contrary to Sapiro's public assertion.[46]

The courts' treatment of specific performance yielded equally mixed results. In states where the cooperative movement was particularly strong, courts willingly awarded specific performance as a remedy for breach.[47] Other jurisdictions, rather than award specific performance, enjoined the member from selling to anyone other than the cooperative. The difference was slight but significant. Specific performance affirmatively required the member to deliver to the cooperative, and several courts ruled that this imposed an involuntary servitude that unjustly infringed on the liberty of the individual to deal freely in the market. If the desired result was actual delivery, one judge wrote, there were ways around this objection: "The only adequate remedy is injunction, preventing the member from selling to others and thus forcing the delivery of the wheat to the association." But no injunction could recover crops already sold to outsiders or, as Sapiro found in California, prevent growers from selling their lands and escaping their obligations.[48]

The most damaging constructions of CMA remedies arose in relation to the law's most far-reaching elements: the "conclusive presumption" that tenants were bound to deliver to cooperatives if their landlords had signed contracts and the enforcement powers and criminal penalties against third parties for willfully inducing members to breach their contracts. In Kentucky, the court consistently protected the Bingham Act from sustained assault by tobacco warehouses and private commission brokers.[49] What was essential for Kentucky, however, was overkill for most of the rest of the nation. Despite Sapiro's appearance on behalf of the cooperative, Louisiana struck the entire subsection relating to tenants: "These particular clauses . . . are unconstitutional, null, and void, as being a patent invasion by the Legislature of this state of the constitutional right to liberty of contract, secured to the tenants of defendant under the Fourteenth Amendment to the Constitution of the United States." North Carolina refused to hold members liable for their tenants' failure to deliver. Minnesota rejected the entire section penalizing third parties for inducing breach.[50]

State courts that had approved the statutory policy of cooperative marketing in the early 1920s were finding ways to limit its reach by mid-decade. Five months after upholding the CMA in *Bekkedal* (1923), the Wisconsin court modified its ruling along the lines of the Minnesota decision. The concern in both states was to protect the "ordinary course of a legitimate business" from what could become an oppressive cooperative regime.[51] Most state courts were willing to embrace the CMAs as a permissible relaxation of laws against anti-competitive behavior, but only as to the parties involved and only as long as the cooperative abided by the statutory purpose.

The pressing issue of the day, however, was equal protection, not antitrust. The CMAs presented the same constitutional problem as the earlier state antitrust exemptions, with a crucial difference. These statutes were accompanied by sweeping policy justifications, and the classifications were constructed in terms of organizations, not persons. This time, because there was greater acceptance of economic combination, the belief that efficiencies could be accomplished through associations, information sharing, and other less than freely competitive trade practices, and, especially, a new legislative definition of cooperation, the courts were more inclined to regard these arrangements as reasonable. Once they dropped their objections to the statutes' larger purpose, courts endorsed the means by which the law achieved that purpose—the separate classification of farmers in associations specifically organized to carry out economic policies in the public interest. The Kentucky court concluded, "It is because of basic economic conditions affecting vitally not only the farmers but also the public weal, that the classification based upon agricultural pursuits is reasonable, just, and imperative for the good of the entire nation and every citizen thereof."[52]

The reasonability of the relation between the classification and the purpose of the statute rested in the distinctive nature of agricultural marketing and the state's legitimate aim in stabilizing agricultural markets. The CMA did not permit farmers to manipulate supply, production, or prices. The Wisconsin court ruled that the difference between cooperatives and illegal trusts justified the statute: "That there is an inherent difference . . . between a combination of a strong and powerful few who are able to get together and exercise a complete control over a particular industry or commodity essential to social existence, and a combination of farmers, each weak in his individual strength and power, numbering into the millions and scattered far and wide, is reasonably apparent." The classification was made for a valid purpose; it applied equally to all within the class, and it did not countenance acts otherwise illegal. The state courts could not move the *Connolly* boulder from the road, so they—like the Supreme Court itself—went around it instead. As the Minnesota court noted, "The purposes of our co-operative marketing law are broad and of course include everyone alike. No one engaged in this occupation, no matter how little, is excluded therefrom." Thus *Connolly* could be easily evaded: "The law in question classifies an industry and makes this law apply with absolute equality to all of those who bring themselves within the conditions imposed by the law and we are of the opinion that it in no way violates the Fourteenth Amendment."[53]

The problem with this statutory construction was that to achieve this result, courts presumed that cooperatives did not violate the antitrust laws. Certainly the USDA shared this assumption in its handling of Sun-Maid's affairs. The state courts so overwhelmingly endorsed this view that few commentators examined it critically. Most lawyers accepted the courts' use of the CMA's economic philosophy to ratify the new legal policy.[54] One of the rare critics was the California lawyer (and future state supreme court justice) Mathew O. Tobriner, counsel to several cooperative organizations and coauthor of a treatise on the subject. Echoing Capper-Volstead's opponents, he argued that the CMAs did not enable cooperatives to do anything they were not previously able to do, nor did it permit them to engage in practices prohibited by the antitrust laws. Moreover, the Sapiro organizations in fruit, tobacco, wheat, and dairy engaged in the very activities the courts presumed they would not: fixing prices at the beginning of the selling season and guaranteeing them against later decline. Tobriner asserted that cooperatives would have imposed production controls to deal with chronic surpluses if they could have gotten away with it. These activities, Tobriner asserted, made it illogical, if not unconstitutional, for the courts to evade *Connolly*: "It is apparent that the language of the Connolly case does specifically condemn the cooperative association; and even if the decision itself may be avoided the words can not be."[55]

Tobriner's solution was a conservative legal idea but a radical political concept: return the classification to one of farmers rather than cooperatives. Classification of farmers was reasonable by virtue of their exceptional marketing problems. Classification of cooperatives was not, despite the public supervision provided in the CMAs, because their tendency to restrain trade meant that they were no different from any other illegal combination.[56] This recommendation went nowhere. *Connolly* prohibited separate classification of farmers. As long as that precedent remained good law, legislatures could not classify according to personhood or occupation. The strict equality of *Lochner* jurisprudence presumed that all people stood on level economic ground regardless of their status. It was legally and politically expedient to distinguish by organizations in order to dispense privileges to individuals. The CMAs circumvented the equal protection issue.

The courts, satisfied with the legislatures' declaration of an apparently reasonable policy to meet an economic necessity, looked no further than the constitutional issue in *Connolly* for grounds to uphold the statutes. The courts' failure to inquire more rigorously into the market behavior of cooperatives is surprising, given both the history of cooperatives' treatment under the antitrust laws and the role of substantive due process in shaping the regulatory powers of the state. By the 1920s, however, state legislation and administrative practice across the nation had firmly established the special status of cooperatives. The widespread acceptance of the idea that farmers required distinctive privileges at law to bring order to agricultural markets disinclined the courts from rigorously examining the trustlike behavior of cooperatives.

The courts were forced to press their equal protection analysis further when the CMAs were attacked by persons most likely to be disadvantaged by cooperatives' new status and power: nonfarmers who supplied their main competition. Again the antitrust laws collided with the Fourteenth Amendment. Because state courts were unwilling to reconcile the absolute bans contained in their antitrust laws with the exemption for agricultural associations in the CMAs, some litigants took their claims to federal court. In 1925, four commercial creameries sued to enjoin Colorado from enforcing that state's antitrust law against them. The plaintiffs argued that the antitrust law was unconstitutional because the Colorado CMA permitted farmers to engage in activities for which they were being prosecuted. The federal district court, quoting at length from the equal protection analysis in *Connolly*, agreed and struck down the state's antitrust law. The district court was willing to say what the state courts would not: "Nothing can be plainer than that these combinations authorized through the formation of the associations as provided for in the Act would, in fact, be combinations in restraint of trade and an attempt to lessen competition in the marketing of agricultural products."[57] In other words, the state could not have it both ways; it could either prohibit all combinations that restrained trade, or it could parcel out dispensations as it pleased and send them up against the *Connolly* edifice.

When a case testing the constitutionality of the CMAs reached the U.S. Supreme Court, however, it did not present the same contrast between a state antitrust law and the cooperative marketing law. Sapiro could not have chosen a better case to argue before the Supreme Court than the appeal by the warehouse company in *Liberty Warehouse Co. v. Burley Tobacco Growers' Cooperative Marketing Association* (1928). The Kentucky legislature had the foresight to repeal its antitrust act the year it passed the Bingham Act. Accordingly, the strategy employed in the Colorado case was not available to the attorneys for the warehouse company, and they had to attack the Bingham Act by arguing that it conflicted with the common law. This approach diluted the Warehouse Company's equal protection claim. It also enabled the Supreme Court to avoid deciding whether cooperatives actually violated the antitrust laws.[58]

The Supreme Court unanimously upheld the Bingham Act. Speaking through Justice James McReynolds, hardly a leading light of Progressivism, the

Court confined the Warehouse Company's claims to the question of whether the penalties levied against third parties for inducing breach of contract violated the equal protection clause. The Court ruled that they did not because the statute did not differentiate among potential violators of the provision: "The statute penalizes all who wittingly solicit, persuade, or induce an association member to break his marketing contract. . . . Nobody is permitted to do what is denied to warehousemen." Thus *Connolly* did not apply; it presented the problem in which one firm "was forbidden to do what others could do with impunity. Here the situation is very different. The questioned statute undertakes to protect sanctioned contracts against any interference — no one could lawfully do what the Warehouse Company did." [59]

The antitrust claims were quickly dispatched. Without a state antitrust law on the books, the Warehouse Company could not prove that the combinations authorized by the Bingham Act would otherwise have been unlawful: "Undoubtedly the State had power to authorize formation of corporations by farmers for the purpose of dealing in their own products." The Court found no evidence of any anticompetitive effects on interstate commerce. The constitutionality of the Bingham Act rested, as the state courts had found, on the validity of the statute's purpose and the reasonability of the classification effecting that purpose: "The co-operative marketing statutes promote the common interest. The provisions for protecting the fundamental contracts against interference by outsiders are essential to the plan. . . . The liberty of contract guaranteed by the Constitution is freedom from arbitrary restraint — not immunity from reasonable regulation to safeguard the public interest." [60]

## THE LIMITS OF COOPERATIVES' NEW POWERS

It was at this point, after the legislatures and courts had endorsed cooperative marketing, that the formalism of the *Lochner* era caught up with the cooperative movement. The case of *Frost v. Corporation Commission of Oklahoma*, challenging not the CMA but a statute regulating entry into the cotton ginning business, brought into clearer relief the larger problem of whether cooperatives deserved special privileges at law. By ruling that it was unconstitutional for the state to impose stricter licensing standards on corporate cotton gins than it did on cooperative cotton gins, the Supreme Court limited the degree to which state legislatures could give cooperatives a competitive advantage in the market. [61]

The *Liberty Warehouse* decision apparently had no relevance in *Frost*. Justice George Sutherland ruled that Oklahoma had violated the Fourteenth Amendment by permitting cooperatives to evade the more rigorous regulatory standards. Justice Louis D. Brandeis, in dissent, argued that the purpose of stiffer requirements for private individuals and corporations was to cure the monopoly that existed in the Oklahoma cotton ginning industry. Brandeis agreed with the Oklahoma legislature that cooperatives presented a valid competitive alternative to the existing monopoly.[62]

According to Walton Hamilton, a Yale University constitutional law scholar, *Frost* ended fifteen years of judicial promotion of the cooperative movement.[63] Certainly the decision exposed the complexity and confusion in the law of cooperative marketing while doing little to remedy the disorder, but Hamilton exaggerated *Frost*'s impact. The decision did not disturb the laws of cooperative organization and marketing. The law still granted farmers far stronger and more effective means of economic collectivization than those offered, for example, to workers. Indeed, the decision demonstrated the degree to which cooperatives had entered mainstream economic life. The Court no longer saw cooperatives as in need of special protection. Such a judgment cut both ways. On the one hand, it undermined the rationale for cooperative marketing laws and the declaration that agriculture was affected with a public interest. On the other hand, it also indicated that farmers finally had at their disposal the legal forms and organizational tools to compete with others on the great playing field of the American marketplace as judges envisioned it in the 1920s. Farmers had thus come full circle — no longer passive victims of the unscrupulous middle merchants, no longer in need of the state's overprotective shield, but now capable of competing successfully with the privileges provided by the associationalist state.

The cooperative marketing laws constituted a legislative movement whose major premises were broad enough to command general assent but whose implementation generated specific controversies that called the major premises into question. State and federal statutes declared cooperation to be in the public interest, and they declared that exemptions from antitrust laws and income taxes served this public policy. The courts, however, were uncertain about the permissible scope of these exemptions and even less sure about their relationship to a market defined paradoxically by individual freedom and centralized power. It was tempting to accept Sapiro's redefinition of cooperation as a vehicle of economic justice, parity through self-help, and monopolization in the public interest. But to fulfill the policy goals, courts had to determine whether cooperatives could act like trusts, administer prices, and restrain free exchange within the confines of constitutional assumptions about competition. The judiciary's reluctance to engage in the necessary fact-finding to determine whether cooperatives violated the antitrust laws meant that no one could say conclusively whether cooperation was good for the economy. The result by 1929 was a stand-off: cooperatives were now endowed with extraordinary privileges at law, but no one was sure of the limits of their legitimate behavior.

These changes in the law of cooperative marketing signaled that a reconsideration of "true" cooperation was taking place. The CMAs revised cooperation's legal status by characterizing farmers' relationship to the market in terms of the public benefits of cooperation. The state courts, more sensitive now than before to the immediate political context of agricultural distress, supported the new policy and the means that the legislatures selected to accomplish it. When pressed to reconcile their new findings with earlier rulings such as *Connolly*, state judges responded by saying that the Capper-Volstead Act had changed cooperation into a more recognizable, more stable legal form that justified separate treatment from corporations and protection from other market actors. The federal courts, including the Supreme Court, hesitated to address the broader legal issues involved in the CMAs, confined their rulings to narrow questions of legislative prerogative, and ultimately drew the line when presented with substantive due process challenges to the extraordinary status of cooperatives. Farmers won a legal battle in *Liberty Warehouse* but lost the larger economic war, their legal triumph undermined by chronic economic instability, excess productive capacity, and the inability to control output.

These developments in the legislatures and the courts undoubtedly shaped the outcome of the Sun-Maid case. With the courts backing up the policy declarations of the CMAs, there was little reason to think that a Sherman Act lawsuit filed after Capper-Volstead would have any better chance of success than the one filed before. The USDA had problems with dissatisfied constituents, but it was not about to play the adversarial role that Capper-Volstead's regulatory responsibilities required. Because the secretary was unwilling to make findings of "unduly enhanced" prices and the judiciary was ratifying the CMAs and their broader policy on farmers, the DOJ faced steep obstacles in pursuing Sun-Maid under the consent decree. Small wonder that when these obstacles converged in 1925, the attorney general backed down.

On the eve of the Great Depression, there was little doubt that cooperation occupied an exalted place in American society. It also had become an acceptable tool for defining farmers' place in the industrial market. Despite the contentiousness of national farm politics during and after the post–World War I farm crisis, farm leaders, politicians, and cooperative promoters all recognized that cooperation was essential to the task of stabilizing American agriculture. What they each saw in cooperation, however, was a reflection of their own vision for agriculture, its connection to pastoral visions, and its necessary place in the new economy. Little wonder that, in attempting to control cooperation's meaning, all sides resorted to explicitly religious metaphors. Secretary Wallace "regard[ed] true cooperation as applied Christianity." Herman Steen called cooperation the "Golden Rule" of agriculture.[64] While cooperation was miscast as the savior of American farmers, it continued to shape law and policy on the economic stabilization of agriculture after the Depression hit.

# 10 ～ From Administered Markets to Public Monopoly

*There are over 6 million farms in the United States, each a competitive unit and each*

*conducted for the most part with little regard for the forces affecting the aggregate relation*

*between production and consumption. . . . Agriculture is the sole great basic industry*

*in which there has been no development of centralized control of production policies.*

—*Brief for the Government,* U.S. v. Butler, *1935*

Nothing so neatly embodied the triumph of associationalism as the cooperative marketing acts. They marked the highest recognition yet accorded to collective action as a legitimate form of enterprise in the American economy. Cooperation, of course, did not cure agriculture's ills during the 1920s, nor did it prevent the Great Depression. But because the associationalist state was so influential to New Dealers piloting the nation's recovery, it is worth asking what cooperation contributed to New Deal thought and administrative practice and what the New Deal contributed to the cooperative movement. What would remain of the cooperative tradition when the expanded administrative state not only absorbed cooperation's monopoly over marketing but also superimposed its authority over the price function?

Nearly all historical and legal analysis of the New Deal has focused exclusively on the transformation of federal statutory and administrative authority, the conflict between the Roosevelt administration and its political and intellectual opponents, and the significance of the constitutional revolution of 1937 and the Supreme Court's "switch in time."[1] Most historians now believe that the New Deal did not significantly alter the market inequities resulting from decades of moderately regulated capitalist competition. As a result, the structural advantages enjoyed by corporations and large-scale producers undermined the entire scheme of public control. In short, the New Deal only modestly affected the dynamics of government-business relations, particularly in agriculture.[2] This chapter will revisit that historiographical issue, but in a context almost entirely neglected by historians: the state price and surplus control programs for agriculture.

The New Deal federal agricultural programs supplied the piece missing from the 1920s regulatory regime: they imposed limits on production. The Agricultural Adjustment Act (AAA) of 1933 levied taxes on agricultural processors and used the revenues to fund rental or benefit payments to farmers — in effect, a wage in lieu of producing. The law also empowered the secretary to enter into agreements with groups of producers, associations, processors, and others to provide for orderly marketing of commodities; the secretary also had the power to issue licenses to anyone seeking to market commodities in interstate commerce. The purpose of the act was to reduce commodity surpluses while increasing the purchasing power of farmers to a level equivalent to a base period defined as the prewar years of 1909 to 1914.[3]

The AAA swept broadly; it centralized the marketing of most major stable commodities and it was immediately controversial. Historians have used the AAA as proof of the New Deal's failure to dislodge established economic and political interests. The federal government's strategy was highly risky in view of the Supreme Court's aversion to public control of prices and its longstanding rule that the federal government had no authority to regulate purely local aspects of the economy. It did not help the government that the AAA was a poorly written statute, so much so that its attorneys had trouble finding a decent test case. When the Supreme Court overturned the AAA in 1936, even Justice Louis D. Brandeis, leader of the Court's progressive wing, signed the opinion. Together, the rulings in *Schecter Poultry Co. v. U.S.* and *U.S. v. Butler* made it clear that only states could regulate production.[4]

The states did not do so. In the milk and fruit industries of New York and California, surpluses had been endemic since the 1920s. Yet the legislatures in those states opted for marketing control programs that addressed price without attacking the underlying problem: overproduction. Still wedded to cooperation's promise of private, voluntary adjustment of market relations, these states passed laws that continued to rely largely on voluntary mechanisms, specifically marketing cooperatives, to control production. Cooperatives were no more equipped to limit output after the crash than they had been before. By the end of the decade, New York retreated from even its halfhearted attempt to fix prices, and California avoided explicit price control by largely copying the federal Agricultural Marketing Agreement Act of 1937.[5]

Throughout the 1930s, the economic disarray in these vital industries and the attendant social disorder generated by rural frustration over the law's failures continued to build. The highest irony of the New Deal was not the Supreme Court's capitulation, for whatever reason, to the era of plenary federal regulatory power; rather, it was the failure of the states to use the power the Supreme Court explicitly reserved for them. The states' inability to find ways to limit production fairly and efficiently, paradoxically, followed from their belief that cooperatives could manage the productive capacity of agricultural industries.[6] The state market control programs in milk and raisins relied on this belief but in the end served the majority of growers little better than the cooperative movement had during the previous decade.

### NEBBIA AND PRICE CONTROLS

In the New York milk industry, economic chaos and social disorder walked hand in hand. New York agriculture had been on the decline for more than a century, but through the 1920s, the dairy industry remained the most resilient sector of the state's farm economy.[7] The Depression threatened to reverse that course. Forbidden by existing law to limit production or fix prices, local cooperative associations disintegrated under the pressure of cutthroat competition. The only statewide dairy cooperative, the Dairymen's League, conspired with Borden's Condensed Milk Company to limit prices paid to producers and protect milk dealers from unpredictable swings in retail prices.[8] During the Depression, demand for fluid milk dropped sharply just as an increase in the number of cows pushed supply up. As a result, one legal writer observed, "While distributors earned profits of 9.9 percent in 1932, producers were receiving less than the cost of production, and by 1933 only half that cost."[9]

Under great pressure from farmers, the legislature decided in 1932 to conduct a full inquiry into conditions in the milk market. One farmer wrote a desperate testimonial to a rural newspaper: "Every can of milk we sell leaves us further in debt than we were before we produced it. This robbery must stop soon or reform will be too late to help us." A legislative committee took testimony at hearings and meetings with producers and distributors. It learned that small and independent farmers bore the costs of an inefficient marketing system. The committee's published findings, known as the "Pitcher Report," documented the chaotic conditions accurately without pointing fingers at specific firms or organizations. All the participants in the industry sought to ensure that whatever legislation resulted from the Pitcher Report protected them. Farmers sought some control over the price function. Large distributors wanted regulation to stabilize the retail market in which they were readily undercut by smaller firms that did not divert milk to less valuable dairy products such as cheese and butter.[10]

The New York Milk Control Act, passed on April 10, 1933, expressly intended to treat the milk industry as a public utility for purposes of regulating prices paid to producers and charged to consumers. As "the first legislative attempt to control the economic forces bearing upon the industry," the law broadly extended the state's regulatory power over the marketing and sale of milk. After declaring the production and distribution of milk in New York to be "a business affecting the public health and interest," the law established an administrative agency — the Milk Control Board — to enforce the act and delegated to the board's director the power to set the minimum prices that retail customers would pay for fluid milk.[11]

The law was quickly put to the test. Leo Nebbia, a Rochester shopkeeper, sold two quarts of milk and a loaf of bread for eighteen cents after the state Milk Control Board had fixed the retail price of milk at nine cents. Either Nebbia intended to give away bread, or he sought to circumvent the minimum. The state concluded the latter and convicted him of misdemeanor violations of the statute. Nebbia appealed all the way to the U.S. Supreme Court, where to the shock of contemporary Court observers and the puzzlement of generations of constitutional historians, he lost. The Supreme Court upheld Nebbia's conviction and the Milk Control Act not on the narrow grounds of temporary economic emergency but on the far broader basis that the state's power to regulate economic enterprise was not limited to businesses that were public in nature, as evidenced by the grant of a franchise, the existence of a natural monopoly, or the operation of a public utility.[12]

*Nebbia* raised the very philosophical and economic questions that underlay both twentieth-century agricultural policy and the unfulfilled promise of the cooperative movement. What could the state do to assist farmers when an unregulated market left them operating chronically at a loss? Was it indeed an issue critical to the public interest if agriculture were imperiled by its competitive disadvantage? If the state had the power to rescue a whole class of entrepreneurs from economic catastrophe, did that power embrace regulation of the price function, which, according to the Supreme Court, was the most sacred aspect of market relations?[13]

When Nebbia's case came before the New York and U.S. Supreme Courts, the judges gave lengthy consideration not only to the question of the legitimacy of the regulation but also to the economic conditions of the industry as documented by the Pitcher Report. The state court crisply laid aside the long-standing judicial regard for economic autonomy to protect the industry from economic chaos and the public from threats to "a continuous and adequate supply of pure and wholesome milk": "The policy of non-interference with individual freedom must at times give way to the policy of compulsion for the general welfare." The Supreme Court likewise endorsed the law and restated the state court's test in the form of a specific constitutional rule: "Price control, like any other form of regulation, is unconstitutional only if arbitrary, discriminatory, or demonstrably irrelevant to the policy the legislature is free to adopt."[14]

Thus the state's power to regulate prices for the good of the public was assured. The legislature intended for the law to rescue the dairy industry from economic disaster, and the courts fully accepted both the legislature's expression of intent and the state's argument that the act achieved its stated aim. In fact, the law's chances for helping those who were supposed to be its primary beneficiaries were dim from the start and were evident in the spring of 1933, when the law was still in the drafting process. As initially written, the bill gave the Milk Control Board "authority to investigate the production and distribution of milk in all its phases, to license milk dealers, and, as an emergency measure, to fix the price to be paid farmers for milk for one year." In its final form, however, the law contained no price mandate for farmers, only a maximum that could be charged to consumers. To obtain this change, distributors assured the legislature that they would pass on to the farmers the profits from the stabilized retail prices. The fix was in.[15]

The *Rural New Yorker* reported that the original aim of the legislation had been subverted: "Farmers all over the state protested the change in the original bill." In western New York, violence broke out among farmers determined to stop shipments to New York City. They attacked not only Borden's trucks but also those of the Dairymen's League. The farmers' own cooperative organization had become the target of the farmers' animus: "There were conflicts between the striking farmers and the troopers who were armed with helmets, tear gas, masks, clubs and some with guns. There were bruises and sore heads on both sides. At the Monroe–Wayne County line 400 dairymen obstructed a train of 10 trucks from other counties and had a conflict with the troopers. There were many sore heads, but every quart of milk was dumped." Rather than inducing changes to the bill, the reports of violence and disorder only spurred its swift passage. To pacify farmers, the Milk Control Board promptly held a hearing on producer prices. The result, an increase of about half a cent per quart, was greeted by more violence: "Disillusioned and desperate, dairy farmers rose in a wide area in protest and rebellion. . . . And the Governor sent armed troopers to the disturbed areas." In an effort to quell the unrest, the governor ordered an investigation into the spread between the prices dealers paid and the prices received by producers. His commission reported that many dealers were operating at a loss — a finding that was plainly contradicted by a Federal Trade Commission report that in 1933 Borden netted profits in excess of $7 million.[16]

When the New York high court took judicial notice of the violent milk strikes, it failed to notice that the farmers' anger was targeted at the state as much as at the distributors. Having concluded that the legislature had found a reasonable response to the economic crisis, the court did not stop to inquire how well that response worked for those it was to benefit.[17] As one legal observer noted, it was "a mistake to suppose that having gained a Milk Board the farmers thereby gained full equality of bargaining power with the distributors." Between 1933 and 1937, dealers' profit margin increased by nearly a cent per quart. At the same time, farmers and independent dealers, who constituted two-thirds of the industry, were being grossly underpaid in direct contradiction of the express aims of the law.[18]

The state's efforts to enforce the law were, as Leo Nebbia might have attested, misdirected at the retailers who were caught up in a self-perpetuating cycle of price cutting, created by the law's exemption for cooperatives and small independent dealers. Ironically, the case that tested the constitutionality of the state's power to fix retail prices for ordinary commodities involved a retailer's violation of the minimum price provision rather than the dealers' failure to pay producers their mandated share. As framed by the prosecution, the Nebbia case did not begin to reveal the fraud and deception that occurred under the Milk Control Act as dealers and the Dairymen's League manipulated the law and the market. The state's aim was to obtain a ruling upholding its power to regulate prices; to assure that the issue was presented as clearly and purely as possible, it laid aside all "technical questions" "which might prevent or abort its consideration by the higher courts on the merits."[19] The Milk Control Board's attorney resigned in frustration at being obliged to sue farmers and small dealers for buying and selling below state prices. New York's attempt to stabilize the milk industry had failed. "Dealers did not pay farmers established prices, and practically every retailer in New York City purchased milk at less than the prices fixed."[20]

Because the milk industry was a regional business, the failure of the New York law reverberated throughout the Northeast. Federal and state regulatory control over milk was effectively stymied by opposition within the industry and the obstacles erected by the federal-state system. The inability of individual states to dislodge distributors' control over prices was exacerbated in 1935, when the Supreme Court ruled that states could not regulate prices of milk originating from out of state, whether or not it was in the original package. This decision effectively undercut the state's ability to do what the

Supreme Court said it could do in *Nebbia* because any dealer could easily sub-stitute out-of-state milk, purchased at unregulated prices, for New York milk. In 1933, the federal government issued fifteen AAA agreements to regulate milk marketing, compelling the participation of all entrepreneurs in the milk sheds covered by the agreements. The AAA agreements often increased the spread between the prices received for fluid milk and those for cream, butter, cheese, and other nonfluid products; accordingly, violation of the agreements and licenses was widespread. By early 1934 all the AAA agreements for milk had been terminated, and the secretary proceeded to regulate milk market-ing under an aggressive use of the AAA's licensing provisions. This strategy was undermined by the lack of control over production and resale prices, but most of all by the fact that until 1937, the lower federal courts had "almost uniformly" struck down AAA orders regulating milk, either as an unconstitu-tional intrusion into intrastate commerce or as an attempt to regulate produc-tion. Similarly, a proposal in 1936 for an interstate pact to regulate milk sales and shipments all over New England foundered for lack of support.[21]

The state finally threw in the towel in 1937 and repealed the price-fixing provision of the Milk Control Act.[22] The new law reverted to the more vol-untary model of pre–New Deal cooperation, abandoning the legal ground gained with the attempt to regulate the milk industry as a public utility. Im-plicit in that approach was the assumption that the state had the authority to reorganize the entire market—in effect, to control it as a monopoly that was accountable to the interests of producers and consumers alike. In stripping the Milk Control Board of its authority to fix prices, the state gave up its chance to eliminate wasteful competition and reduce the industry's excess produc-tive capacity. Nationwide, the cooperative movement had never succeeded in acquiring the power to control production. Nevertheless, people continued to believe that, given the proper legal support, cooperatives could indeed fix the economic problems of agricultural marketing. Indeed, the failure of the New York milk control law spurred calls for a renewal of "true" producer co-operation, supported and supervised by government intervention to protect consumers and to make distribution systems operate fairly and efficiently.[23]

Although in theory the New York milk control law aimed to regulate the milk industry as a public utility, in practice the state never came close. For the public utility approach to work, the state not only had to fix prices but also to limit output, either by cutting production or by diverting surplus milk away from the fluid market. Since the distributors exploited the presence of the surplus to manipulate the prices at which they bought and sold milk, they op-posed any efforts by the state to alter supply significantly. In contrast, the AAA directly—and to its opponents, heartlessly—attacked the surplus problem by destroying commodities. To great effect the administration's opponents used the image of the slaughter of thousands of baby pigs by order of Secretary of Agriculture Henry A. Wallace.[24] New York's regulatory plan for the milk industry did not imitate the federal agricultural recovery program. It may well not have succeeded even if it had, given the power of the distributors in New York; but when the state ceased fixing prices in 1937, it conceded that the distributors, not the farmers, controlled the regulatory process.

## THE CALIFORNIA APPROACH TO SURPLUS CONTROLS

In California, the situation and the outcome were different. In 1933, the state enacted its first Agricultural Prorate Act, substantively identical to the first AAA. The purpose of the prorate act was to "provide means for relief from the price lowering effects of increased production" by permitting a majority of growers to compel an entire industry to divert a portion of the crop from the market. The law's authors, University of California professor E. A. Stokdyk and farm leader Edson Abel, intended for the state to accomplish what cooperatives had been unable to do: control agricultural surpluses. A democratic mechanism triggered mandatory participation in all control programs and thus maintained the connection, however tenuous, between the cooperative tradition and the new broker state.[25]

The prorate act offered a solution steeped in New Deal economic philosophy. By directly enlisting the various interests of the industry and economic experts to make regulatory decisions, it employed standard corporativist methods to join business and government. A petition signed by two-thirds of the growers or a majority vote in special elections authorized the Prorate Commission to set up proration zones for each industry. A local program committee, nominated by growers and processors and selected by the commission, ran the marketing program. The costs of administration were distributed evenly to all the producers through the assessment of fees on all crops marketed. Although the law erected elaborate administrative machinery to run the proration programs, it was not until 1935 that the legislature provided substantive regulatory goals and more specific guidelines for instituting and administering prorate programs.[26]

By 1937, California had applied its prorate law to fourteen different commodities in an attempt to alleviate agricultural distress.[27] Unlike its federal counterpart, the prorate law survived judicial scrutiny. Turning away attacks invoking the commerce clause, the due process clause, and limitations on the delegation of legislative and judicial powers, the California Supreme Court upheld the statute in *Agricultural Prorate Commission v. Superior Court* (1936):

"Statutes aiming to establish a standard of social justice, to conform the law to the accepted standards of the community, to stimulate the production of a vital food product by fixing living standards of prices for the producer, are to be interpreted with that degree of liberality which is essential to the attainment of the end in view." [28] This conclusion relied on the enduring belief that a prosperous agriculture was essential to society.

The Prorate Act effectively recast the state's role in agricultural marketing, moving from the New York example of mere price-fixer to a model of a state-conducted marketing monopoly.[29] The state supreme court's 1936 decision, however, was not the vehicle for this contest. Rather, it was the prorate program in the raisin industry that eventually provided the test case for the constitutionality of a state-run monopoly.

## PRORATION IN THE RAISIN INDUSTRY

The California raisin industry flirted with federal and state help during the first half of the New Deal. With Sun-Maid's market control significantly reduced by insolvency, raisin growers and packers could not agree on whether their industry needed public regulation, much less on the form that regulation should take.

Well before the onset of the Depression, Sun-Maid had been relegated to the competitive status of an ordinary commercial packer. From 1932 to 1935 Sun-Maid handled only 35 percent of the California raisin crop. Still weighted down by millions of dollars in debt from marketing crops during the 1920s, Sun-Maid was heavily dependent on loans from the Federal Farm Board and facing a new rival for control of the grape industry in the California Vineyardists' Association. The Farm Board's willingness to continue to bail out Sun-Maid despite tepid grower support enriched the banks that held Sun-Maid's bonds but resulted in no lasting price stabilization or production control programs.[30]

Sun-Maid's lack of leadership was apparent during the early years of the New Deal. Late in 1933, a raisin growers' committee sent a desperate telegram to President Franklin D. Roosevelt, informing him that the packers, including Sun-Maid, were impeding the work of AAA economists and asking him to intercede: "Twelve thousand California raisin growers have implicit faith in you and the New Deal." It took six months to come up with an agreement both packers and growers would accept.[31] In May 1934, the AAA set up a license and marketing agreement for California raisins. The program required packers to pay a minimum price for 85 percent of the raisins they purchased, which they could then sell freely. Only 15 percent of the 1934 crop would be held off the market.[32]

Led by Sun-Maid, the packers pulled out after one year. Their complaint was not that the program was ineffective. Indeed, both sides agreed that the marketing agreement "was responsible for substantial increases in the total returns to raisin growers for the 1934 crop." The problem was that some packers refused to participate, forcing the others to bear the costs of surplus control for the industry. As Sun-Maid president J. M. Leslie explained: "The tremendous loss resulting from failure of enforcement of the 1934 crop control program fell solely upon the processors of raisins, including [Sun-Maid], who had lived up to the obligations imposed upon them by that program." Wallace terminated the license as the 1935 harvest was beginning.[33] So ended federal marketing control for raisins during the New Deal.

Because of the ambiguity of the first state prorate law and Sun-Maid's optimistic crop forecasts for 1935 and 1936, no one attempted to initiate a prorate program for raisins for those years.[34] After Sun-Maid promised and failed to control the surplus in those years, growers' attitudes toward state market controls changed. In the spring of 1936, the Fresno branch of the American Farm Bureau Federation met with Sun-Maid officials to ask them to support a prorate petition, the first step towards state market control. By early 1937, rank-and-file growers had signed the petition in significant numbers, including over half of Sun-Maid's members, but Leslie and Sun-Maid's general manager, William N. Keeler, wanted no part of a prorate program despite mounting evidence of the cooperative's inability to maintain prices and dispose of surpluses.[35] In 1937, Sun-Maid persuaded the industry's prorate committee that its own plan for market stabilization obviated the need for state control. The state's press, including the leading farm journals, vilified Sun-Maid for obstructing the will of the growers; when Sun-Maid's financing for the 1937 crop collapsed and prices dropped, the cooperative was thoroughly discredited as a stabilizing force in the industry.[36]

The following year, the state Prorate Commission obtained $9 million in federal loans to finance the withholding of 20 percent of the crop from the market. When the 1938 crop came in at nearly three hundred thousand tons, the industry program committee scrambled to divert raisins to brandy and stock feed. Growers criticized the committee's disposal of the surplus, convinced it had been directed by the "dominating influence of Sun-Maid on the Program Committee." Opponents of the prorate law funneled the growers' criticism of the program committee into a general condemnation of surplus control. When prices increased under proration, growers scurried to augment their income by selling surplus raisins through normal marketing channels. This entirely defeated the purpose of control. As one analyst noted, "Opposition came not only from the growers who wanted to do away with the prorate entirely, but from others who have never clearly seen the necessity of surplus removal."[37]

The extent to which agricultural interests in California were dissatisfied with both the way proration worked and the underlying purpose of surplus removal was made plain when twelve bills amending the Prorate Act and two more repealing it were introduced in the 1939 session of the California legislature. These bills reflected common complaints with the law: it did not permit growers to withdraw from the programs, proration did not work very well for semi- and nonperishable crops, and the Prorate Commission had no time to supervise local program committees or the programs they instituted. The legislature recognized the administrative lapses in the law but refused to jettison the concept of mandatory participation. Since the California Supreme Court had already ruled on the constitutionality of proration, repeal was not likely.[38] To appease the law's opponents, the legislature abolished the Prorate Commission and established an Agricultural Prorate Advisory Commission that gave growers more voice and deprived packers, processors, and cooperatives of their representatives. The advisory commission, unlike its predecessor, had nothing to do with the implementation of the law. Instead, the state director of agriculture had plenary authority to enforce the act.[39]

The legislature also enumerated the methods that prorate programs could employ to minimize surpluses. Particularly important from the standpoint of the raisin industry was the provision for stabilization pools, from which crops could be sold back into regular trade channels when the program committee administering the prorate determined that such sales would not adversely affect prices. Formerly, under the 1935 amendments, surplus pools were required to divert all surplus to less valuable uses. Now program committees could use two types of pools: stabilization pools, which enhanced prices by reducing the crop available for sale and which then could be sold to take advantage of those higher prices; and surplus pools, which would not compete with crops free to be marketed (free tonnage) or stabilization pool crops but which could be sold for nonhuman consumption. The *Pacific Rural Press* noted that the changes helped to alleviate the political tension between the commission and the growers.[40] As it turned out, the new law furnished a means of controlling the surplus that distributed the burden equally among growers and processors without permanently devaluing a significant portion of the crop.

It took some time before the raisin industry was willing to try the new approach. When members of the industry's program committee met with USDA and Commodity Credit Corporation (CCC) officials in Washington to apply for federal loans for the 1939 crop, they were continually undermined by public opposition. Scores of growers signed a petition asking Secretary Wallace not to provide the program committee with any federal funds. Opponents of proration flooded Washington with telegrams during the negotiations and took to the radio to argue against all "artificial" controls and in favor of permitting "natural events [to] attend to the price." The "show of conflicting messages to federal officials" revealed the disarray within the industry and undermined the program committee's credibility with the federal lending agencies. When the state broke off the talks, the federal government decided not to make funds available for the 1939 crop. Without federal funds, the program committee was forced to cancel the prorate program.[41]

The raisin industry remained uncommitted to federal or state control, but neither the federal or state governments sought to force market control programs on the industry. New administrative solutions continued to emerge at both levels of government. The USDA, with powers augmented by the Agricultural Marketing Agreement Act of 1937, could impose a federal marketing order on the industry. Or, using resources available through the Federal Farm Credit Administration and the CCC, the federal government could induce the industry to accept a state prorate plan over which growers had substantially more control. In 1940, the raisin industry chose the latter. This state-federal cooperation in the regulation of agricultural commodities raised novel constitutional questions just as the nation's involvement in international conflict was beginning to deepen.

In the spring of 1940, the industry continued to flirt with the idea of market controls. Prices remained significantly below 1939 levels, and another large crop was on the vines. Growers were beginning to realize that some market control was better than none. Under the new provisions, no one could market the surplus on the sly and undo the effects of proration. A growers' committee quickly agreed on a program with a 20 percent surplus pool and a 50 percent stabilization pool for 1940. The remainder — a mere 30 percent of the crop, as compared to 80 percent under the 1938 prorate — would be freely marketable. Growers could market no raisins until they delivered to both pools; the program committee issued certificates as proof of compliance. The federal government provided $8 million in crop loans, but Secretary Wallace made the funds contingent "upon the express condition that the program . . . be made effective and maintained in operation." The growers voted their approval in late July, and just before harvest the state director of agriculture officially announced its implementation. Even Sun-Maid, which had independently secured CCC financing to market its share of the crop, supported the prorate:

"In view of world conditions and their effect upon the export market we believe that the program is as good as could reasonably have been expected, and urge its whole-hearted support."[42]

The uncharacteristic dispatch with which the industry settled on such an extensive control program caught Porter L. Brown unaware. Brown had been a raisin grower all his life; in 1939, he established a small packing facility on his modest farm in Fresno County. Fruit packers typically started selling the crop the spring before it was harvested, based on crop forecasts and surplus estimates from the previous year. In early 1940, Brown contracted to sell 762.5 tons of raisins, 300 tons of which he had on hand and the rest he planned to buy on the open market. When the program went into effect on September 7, Brown had not yet purchased all the raisins he needed. He had gambled that there would be no prorate for the 1940 crop, that growers would glut the market, and that he could fill his orders at rock-bottom prices. Although there were plenty of raisins available to packers after the prorate was instituted, they cost $55 per ton, $10 more than the springtime price. Brown's delay in purchasing the raisins he needed cost him an additional $5,625. The prorate held prices firm during the fall, and on December 28, he filed suit in the U.S. District Court at Los Angeles to enjoin its operation.[43]

The district court issued the injunction — the first judicial decision against proration in eight years — and fifteen commodity prorates. Implicitly rejecting the *Nebbia* rationale, the court held that the Prorate Act was not constitutional as an exercise of the state's police power because it did not protect health and safety interests. The court also ruled against the prorate on commerce clause grounds: "It seems clear to us that the program is frankly and simply a means of controlling the supply of raisins into interstate trade channels to meet the market demands. As to this purpose it is not in our province to comment. We think the State has attempted to accomplish this result by a process which impinges upon a grant of power to the Federal Government."[44]

As the state's appeal went forward to the U.S. Supreme Court, the success of the 1940 prorate, in combination with changing world conditions, made the program extraneous. Every ton of raisins sold had commanded prices above the minimum established under the prorate. The *Pacific Rural Press* reported that the success of the 1940 program had "persuaded the [federal] government to put a $72 [per ton] floor under the 1941 crop." With this guarantee and the government's subsequent decision to purchase all remaining surplus raisins for Lend-Lease assistance, the industry decided that proration was no longer needed. The program committee closed its Fresno office for good in April 1942.[45]

The Supreme Court ordered reargument on the relationship between proration and the Sherman Act, the Agricultural Marketing Agreement Act of 1937, the Agricultural Adjustment Act of 1938, the Clayton Act, and the Capper-Volstead Act of 1922. The Court's action suggested that, despite the ruling of the district court, it did not intend to use *Parker v. Brown* to revisit the issues raised in *Nebbia v. New York*. Rather, the California case presented the question of how public administration of markets should proceed in a federal-state system. Was the state program preempted by federal legislation that established the same regulations? Was the prorate law, by setting up a state-run monopoly, invalid under the federal antitrust laws? These questions pointed to the relationship between cooperation in the associationalist state and market control in the broker state. Did room remain for private collective arrangements if the state exercised the authority to regulate markets?

The Supreme Court unanimously upheld the constitutionality of the California prorate law. Writing for the Court, Chief Justice Harlan F. Stone dismissed the notion that proration erected an illegal combination among private persons. In enacting the Prorate Act, the state exercised its sovereign powers, outside the scope of the federal antitrust laws. Stone also adopted the state's argument that the state Prorate Act and the Agricultural Marketing Agreement Act were entirely compatible. It was not the intention of Congress, he wrote, to preempt state efforts to regulate, especially when the secretary of agriculture fully concurred in the cooperative arrangements: "The conditions imposed by the Secretary of Agriculture in the [CCC] loan agreement with the State of California, and the collaboration of federal officials in the drafting of the program, must be taken as an expression of opinion by the Department of Agriculture that the state program thus aided by the loan is consistent with the policies of the Agricultural Adjustment and Agricultural Marketing Agreement Acts." The Court found no interference with interstate commerce because "the regulation is . . . applied to transactions wholly intrastate before the raisins are ready for shipment in interstate commerce."[46] The Court upheld proration because it presented no conflict with federal agricultural policy.

Chief Justice Stone confined his ruling to the antitrust, statutory, and commerce clause issues. One legal scholar has termed *Parker* "a decision rooted in a particular legislative context."[47] The decision also recognized that cooperation affected—and was affected by—the scope and purpose of proration. The state's method of stabilizing the raisin industry borrowed heavily from cooperation's model of market control. By compelling participation, proration extended the model and subdued an unruly band of growers, packers, and cooperative officials—all of whom had contributed to Sun-Maid's inability to maintain control over the market. In contrast to New York, where milk marketing regulations increasingly relied on voluntary participation as the decade progressed, California's prorate act reflected the insight of decades of cooperative experience: the free-rider problem could not be solved unless the state mandated universal participation. Such a mandate was widely perceived as unfair by both growers and processors until the state finally found a way to dispose of the surplus that preserved prices. By allowing growers to sell some of the surplus at state-supported prices, the stabilization pool enabled the industry to recoup some of its losses while spreading the costs of the surplus across all participants. Though the prorate law never limited production, it proved remarkably effective at disposing of surpluses in specialty crops. In view of what the Court had already approved in both *Nebbia* and *Wickard v. Filburn* (1942), if any substantive constitutional or statutory objection remained to publicly administered markets, *Parker* did not raise them. And by then the Court was no longer inclined to trim the New Deal's sails.[48]

The case's implications for the cooperative movement were not so clear. Because the test case arose in the raisin industry, one of the most difficult to organize and one in which cooperation had scored a notorious failure, it is easy to conclude that the public administration of markets as practiced in California and approved by the Supreme Court left no room for cooperatives, particularly those that—like Sun-Maid—aspired to monopolize marketing in private hands. However much Sun-Maid exemplified the associational ideal, it was never capable of stabilizing an entire industry. Sun-Maid's difficulties during the 1920s and 1930s attest to the impossibility of running a monopoly as contemplated by the cooperative marketing laws. But Sun-Maid was always the liminal case. The terms of the coexistence of cooperatives and government marketing programs were forged more easily in other commodities. Proration did not meet nearly as much resistance where there was a strong cooperative. In the lemon and walnut industries, cooperatives retained control over most growers; they used proration and federal marketing orders to maintain that control and to solve their free-rider problems. There the partnerships between the public and private spheres, between the state and the federal governments, were strengthened by industry confidence in cooperative action. *Parker* was not a ruling on the legacy of the cooperative ideal but rather a recognition that a transformation had already taken place in the competitive position of cooperatives. For most California cooperatives, proration made it possible to compete in the market, retain some of the useful privileges cooperatives still held, and avoid some of the pitfalls of cutthroat competition.

State agricultural regulation during the New Deal left a decidedly mixed record. In New York, the state fixed prices but did not attempt to control production. In California, the state focused primarily on surplus management. California's job, if anything, was made easier by the geographic monopolies in horticulture. But both states remained wedded to the agrarian philosophy of ever-expanding production — a philosophy whose somber economic implications were blunted first during World War II and then by the establishment of permanent price control and federal marketing order programs after the cessation of hostilities. That these permanent agricultural regulations were instituted at the federal rather than state level testifies to the continuing reluctance of authorities more closely connected to industry participants to interfere with the farmer's persistent impulse to produce. The New Deal did not bring production and marketing into closer relation. In that respect it prolonged but did not change the pattern established during the heyday of the cooperative movement.

Likewise, the legal ratification of state regulatory authority over agriculture during the 1930s did not disturb the legal privileges with which public policy endowed cooperatives for the benefit of their members and the public. They simply provided new ways for cooperatives to take advantage of those privileges. Whether, as in New York, cooperation was little help to either the state or dairy farmers, or as in California, cooperation was a firmly instituted player in both the market and politics, the cooperative tradition was both embodied and subsumed in the era's administrative innovations. In both states, the laws aimed to use cooperatives to implement market controls by relying on cooperatives' representation of and connection to their members to reinforce the principle of mandatory participation triggered by majority vote. The state laws regulating agriculture during the 1930s revealed that the New Deal relied substantially on the democratic appeal of cooperation, while cooperation survived the decade's economic turbulence by conceding the power of monopoly to the state.

# Conclusion

*Since **Connolly's** case was decided, nearly forty years ago, an impressive legislative*

*movement bears witness to general acceptance of the view that the differences between*

*agriculture and industry call for differentiation in the formulation of public policy.*

—*Justice Felix Frankfurter, 1940*

Farmers emerged from the Great Depression constitutionally set apart from other entrepreneurs in the market and with the cooperative movement essentially immunized under federal antitrust law. The equal protection rule laid out in *Connolly v. Union Sewer Pipe Co.* was overturned in 1940. That year, Justice Joseph McKenna's dissent in *Connolly* became the law of the land in *Tigner v. Texas*, in which the Court ruled that the differences between agriculture and industry were within the state's power to recognize and that these differences provided a constitutionally legitimate basis for preferential regulation. Justice Felix Frankfurter declared for a unanimous Court: "The Constitution does not require things which are different in fact or opinion to be treated as though they were the same. And so we conclude that to write into law the differences between agriculture and other economic pursuits was within the power of the Texas legislature."[1] The Court found that fundamental economic differences between agriculture and industry—the isolation of the individual farmer, the competitive conditions in the national market—justified the different legal treatment for cooperatives.

The changes in the law of cooperative organization that took place between 1890 and 1945 pointed to an inescapable irony: even as cooperatives became special in the eyes of the law because they were a distinct form of organization, farmers were remaking cooperatives in the image of their corporate counterparts.[2]

The Capper-Volstead Act, which embodied the legal transformation of cooperation, remains to this day the most important federal statute regulating agricultural cooperatives. It grants cooperatives substantial freedom to gather producers into combinations, consolidate marketing operations, and exert control over prices short of "unduly enhanc[ing]" them. It legalizes the appropriation of the business corporation by farmers' collective marketing organizations. And although it was enacted principally because of a postwar economic crisis, its endurance testifies to the lasting influence of farmers in American political life. The Capper-Volstead Act has withstood every attempt by processors and corporations to amend or repeal it.[3]

The Capper-Volstead Act, termed "agriculture's Magna Carta" by lawyers and economists, did not confer a blanket of complete immunity on farmers, but its passage enabled the cooperative movement to dodge the bullet of the only serious Sherman Act litigation that an agricultural cooperative faced during the Progressive Era. In light of the ease with which Capper-Volstead was passed, Congress seemed more interested in protecting farmers than in determining whether they monopolized illegally. The statute freed cooperation to carry the banner for agricultural self-sufficiency throughout the 1920s and effectively curtailed enforcement of the antitrust laws against cooperatives throughout that decade and since. When in the late 1930s the federal government reignited its enforcement machinery, it won the ruling that cooperatives were not completely immune to the antitrust laws. Yet the Justice Department did not undertake any reinvigorated antitrust campaigns against farmers' cooperatives.[4]

The Capper-Volstead Act also symbolized a larger cultural ambivalence of the period. On the one hand, it recognized the inherently commercial nature of American agriculture; on the other, it revised the cooperative ideal to protect farmers from industrial competition. Asked to choose between traditional cooperation, which farmers had rejected as unworkable in the national market, and full-blown corporatism, which seemed incompatible with American agriculture, Congress tried for something in between. In the end, the statute represented a compromise among competing priorities: the cultural ideal of farming, farmers' demands for a commercially viable organizational form, and the fear that farmers would hold cities hostage to food monopolies. The debate about the nature of agricultural enterprise that preceded passage of the Capper-Volstead Act provided fleeting moments of national reflection on the agrarian vocation and how industrialization had changed it. Urban society wanted agriculture to survive, but in a form that was compatible with its image of rural life as bucolic, serene, and above all, enduring. As long as individual farmers competed in the market against the large corporations that controlled the distribution system, agriculture's economic stability would remain evanescent. The Capper-Volstead Act marked the beginning of Congress's efforts to address the implications of the changes industrialization had brought to agriculture.

While the law stopped short of granting farmers complete immunity from prosecution, it has resulted in few actual prosecutions of cooperatives. In the seventy-five years since the passage of the act, no secretary of agriculture has ever invoked the statutory powers to declare prices in excess of "fair and reasonable." Over the decades since its enactment, cooperatives have only intermittently tripped over the act's restraints; on the whole, farmers continue to enjoy broadly favorable standing under the antitrust laws. The increasing corporate control of agricultural landowning, production, and processing has made corporate firms such as Archer, Daniels, Midland a far greater threat to competition, fair prices, and consumers' interests than farmers' cooperative associations.[5]

The cooperative movement's larger story is a tale about the conflicts that arose as the legal changes in cooperatives changed cultural perceptions of cooperation and farming. Cooperation's traditional image of an equitable, democratic capitalism had a lasting hold on Americans' hopes for the future. Yet the cooperatives that succeeded during the Progressive Era and survive to the present did not exemplify this ideal; rather, they manipulated it for their own ends. The CARC created and exploited the image of the benevolent trust—a monopoly that served the public interest—even as its leaders sought to blunt the legal implications of the cooperative's unprecedented market power. Aaron Sapiro's image of the cooperative as the prototype of the associational state reinforced attempts to redefine government-business relations during the 1920s. Finally, the example of cooperatives' centralized market control—however incomplete—both tempered and complemented the mandatory control programs of the New Deal broker state. The New Deal regulators' willingness to employ governmental power where their 1920s predecessors relied on the market enabled the New Deal to succeed to the extent that it did; the New Deal's reliance on existing cooperatives to carry out federal market control programs, as in the California prorate program, emphasized the function of cooperatives in bridging the state and the market. The conditions of industrial business changed cooperation, and farmers forced the law to recognize and accommodate those changes.[6]

For California growers, as I have argued, law furnished both a malleable tool and a formal restraint. The raisin industry cooperatives marked the first time that growers took control of marketing on a national scale. In the process, they upended prevailing rules about competition and monopoly and the permissible limits of collective action. In turning a blind eye to the night riders' violence, local law enforcement officials aided and abetted the cooperative's oppression of its own members. The CARC and Sun-Maid first fulfilled, then violated, the cooperative ideal. Indeed, as they borrowed corporate tools, these organizations became as ruthless as any industrial trust. Uninterested from the start in carrying on the Rochdale tradition, California growers unreservedly exploited it. They understood cooperation's appeal well enough, but they had no intention of adhering to the traditional style of cooperation at the expense of the economic growth of their industries. They manipulated urban Americans' perceptions of rural life and agriculture, engaged in illegal market behavior, and got away with it—but only for a time.

The temporary reign of the benevolent trust illustrates the vagaries, the unintended effects and consequences, of recourse to law. Farmers got the law to recognize their special place in the market, but at the expense of the special qualities of traditional cooperation. And what they needed, as it turned out, was the ability to monopolize more effectively than even the changed legal rules permitted. As farmers appropriated the legal powers of corporations, they were tagged as robber barons in overalls. Instead of holding a claim to official sympathy as the victims of monopoly, as they had done effectively though sporadically in the late nineteenth century, farmers became the perpetrators of monopoly. Yet the new laws did not permit cooperative monopolies to eliminate competition or—more essential—to regulate production. Without this power, as students of economics know, the benevolent trust could never act as a true monopoly. In the eyes of the law, however, it was.

Historically, law and markets interact contrapuntally. Sometimes the legal system and the economy parallel each other; other times they move in contrary directions. Courts and legislatures are reactive by nature; they respond to social and political changes after those changes have taken place, and they seldom respond identically or predictably. The cooperative movement was able to secure legislative assent to the far-reaching changes it was making to traditional cooperation. Courts generally were more intent on maintaining formal distinctions between cooperatives and corporations because they saw cooperatives as a threat to free competition and prices that resulted from the interplay of market forces. Farmers saw a threat as well, but in their eyes it came from the middle merchants and could be remedied only if they had greater bargaining power. Once the larger political context shifted to acknowledge industrialization's impact on society, the courts recognized that the nature of competition had changed since the late nineteenth century and permitted the use of law, in the words of J. Willard Hurst, to "keep the market in being."[7]

In view of the far-reaching, if not radical, legal revolution that the cooperative movement won for itself early in this century and the lack of substantial shift in the practice or law of cooperative marketing since then, this book raises the issue of what kind of monopoly modern cooperatives conduct and whether these monopolies last over the long run. Concepts borrowed from economic theory help to frame the answers. Beginning in the 1960s, economists and sociologists applying institutionalist theory to the history of corporate formation and the rise of industrial monopolies argued that the differences between "center" and "peripheral" firms were material to understanding industrial capitalism. Center firms not only grew to large size during the great merger movement ending in 1904 but have remained so over the century. They rapidly integrate technological advances into their production facilities, achieve significant economies of scale, and tend to be vertically integrated—that is, to control all facets of manufacturing from the raw material to retail sales. Businesses such as Standard Oil, American Tobacco, and Carnegie Steel exemplify the model of the center firm. Peripheral firms are, in the words of Thomas McCraw, "everything center firms [are] not." Although they play vital roles in the economy, they do not acquire the market control that center firms do. They tend to compete fiercely at the local level, their attempts to fix prices are usually readily detectable, and their horizontal combinations tend to come and go.[8]

The horticultural combinations formed by the California cooperatives both exemplify and blur the differences between center and peripheral firms. These associations do not meet the definition of center firms: demand for their products is seasonal and fairly elastic; by definition they cannot employ continuous production processes; and their price-fixing policies readily become apparent to consumers and the government. Yet with the exception of Sun-Maid, the major California horticultural cooperatives have maintained their large market shares over the course of the century; they are vertically integrated, achieving the efficiencies of scale that come with vertical integration; and they seem impervious to antitrust enforcement even now. Since the New Deal, Sunkist has reigned as the largest monopolistic cooperative in California; between 1954 and 1982 it averaged a 70 percent share of the market for fresh oranges. Evoking striking parallels to the CARC lawsuit, in 1977 the Federal Trade Commission (FTC) began an investigation into Sunkist's market control, particularly its integration into processing. After agricultural interests brought their influence to bear in Congress, much the same as they had on Sun-Maid's behalf in the early 1920s, the FTC settled the case in 1981 with a consent decree that did as little to reduce Sunkist's monopoly as the 1922 decree accomplished in the Sun-Maid case. Sunkist continues to dominate the market for its goods; its sails have been trimmed only insofar as it has been barred from using the Sunkist brand name on its line of orange soda.[9] In contrast, Sun-Maid's bankruptcies in the 1920s and 1930s enabled other processors to enter the industry. Since the New Deal, Sun-Maid has averaged a 30 percent share of the market. Beginning after World War II, a combination of federal marketing orders and agreements by an industry-run bargaining association have regulated the industry. Since then, Sun-Maid has been one player among many, and Armenian-owned packing firms usually deal the cards.[10]

The history of the cooperative movement shows that short-term economic aims often overwhelm longer-term goals. The aim of the CARC was to maximize prices, but Sun-Maid's leaders learned, to their cost, that focusing on immediate profits undermined their ability to stabilize prices over the long haul. The lack of sustained price stabilization in the raisin industry substantially weakened Sun-Maid and ended its monopoly. Although it possessed significant center-firm advantages, Sun-Maid ultimately was unable to turn them to its profit. Other cooperatives have fared better. Sunkist, Diamond Walnuts, Blue Diamond Almonds, and Ocean Spray Cranberries have significant market shares today, despite — or perhaps because of — ongoing federal and state marketing programs. The difference between these organizations and Sun-Maid is less economic than historical: the night riders left a lasting impression. Memories of the violence remain vivid, and raisin growers today will not support an industry-wide monopoly.[11]

The legacy of the New Deal was that it made price stabilization, not price maximization, the goal of the economic regulation of agriculture. The New Deal's planners understood the difference, but because of political constraints they could only incompletely integrate that insight into the market control programs they wrote into law. The programs to subsidize agricultural prices, limit production, and control the flow of agricultural commodities onto the market have continued for over sixty years, paying farmers not to produce and keeping American food prices low. While rules mandating participation by all producers and processors have theoretically done away with the problem of free riding — one problem cooperatives cannot solve on their own despite the powers of the benevolent trust — U.S. farmers have been perennially dissatisfied with mandated restrictions on marketing and production, perhaps because price stabilization sometimes keeps agricultural profits lower than farmers want. To many Americans the farm program is an anachronism, irrationally protecting inefficiencies that the market would otherwise weed out. The resurgence of political and economic conservatism in the 1980s and 1990s finally led to the slaying of the sacred cow of U.S. farm policy. In 1996 the farm lobby supported the passage of a farm bill that phases out production controls and price subsidies on most commodities by 2003.[12] Yet farmers have by no means ceased to expect that the government should help them compete in the market and get a fair price for their products.[13]

The most lasting irony of all may be that cooperation's legal gains have endured while rural culture has become increasingly marginal to American society. Like the frontier at the end of the nineteenth century, the American farmer is disappearing at the end of the twentieth. The U.S. Census Bureau has announced that it will not conduct a separate count of farm residents in the year 2000 because an increasing number of agricultural producers no longer live on the land they work.[14] Yet agriculture retains its vast political influence, Americans still care about family farms, and law and policy continue to favor agriculture. The nature of this favoritism may be changing, as the phaseout of agricultural subsidies indicates, but even a disappearing farmer, it appears, maintains a hold on the public imagination.

# Appendix
## A Statistical Profile of Small Farms in California, 1850–1940

This appendix showcases the statistical analysis supporting the conclusions about farm size laid out in Chapter 2. Using county-level census data, I separated regions primarily devoted to fruit growing from those primarily dedicated to the production of field crops and ranching. I then tested for the correlation between farm size and type of commodity produced. This analysis permits the small farms that proliferated in fruit growing over time to emerge from hiding.

Data from the published agricultural census vividly illustrate that small farms existed from the beginning in California and grew in number over time.[1] Table A.1 shows that the number of farms 20 to 49 acres more than doubled between 1880 and 1890; those between 10 and 19 acres quadrupled. By 1900, the number of farms from 10 to 19 acres doubled again. During the nineteenth century, the category of farms between 100 and 499 acres contained the lion's share of farms, both in absolute numbers and percentage. When the Census Bureau broke down the 100–499-acre farm size category into three separate farm size ranges beginning in 1900, the smaller farm size ranges contained the most farms. After 1900, most farms fell into the category of 20 to 49 acres. In 1940, the bulk of farms fell into the next smallest category, 10 to 19 acres. In 1900, farms under 50 acres represented 39 percent of all farms in the state; this figure peaked at 63 percent in 1930 and dropped only slightly during the Depression to 59 percent in 1940.[2] Overall, the number of smaller farms (under 100 acres and particularly under 50) grew dramatically after 1900. That year, 36,259 of 72,542 total farms in the state—just under half—were smaller than 100 acres. In 1920, the proportion jumped to about 68.6 percent; by 1940, it was 77.4 percent. Generations of historians overlooked these data, not because the information was unavailable but because their preoccupation with the rise of agribusiness led them to accept contemporary anecdotal assertions about land monopoly and the dominance of large spreads devoted to field crops and stock raising.

Subdivision, colony settlement, and fruit growing together created not one kind of horticultural enterprise but several, based on both farm size and the mix of crops produced: larger, specialized fruit groves; medium-sized, diversified farms growing both fruit and field crops; and small, self-sufficient fruit farms that produced vegetables, fruit, and enough field crops to support a cow

TABLE A.1. Numbers of Farms in California, by Size (in Acres), 1860–1940

| Year | <3 | 3–9[a] | 10–19 | 20–49 | 50–99 | 100–499 | 500–999 | >1000 |
|---|---|---|---|---|---|---|---|---|
| 1860 | | 8,292 | 1,102 | 2,344 | 2,428 | 6,541 | 538 | 262 |
| 1870 | 0 | 2,187 | 1,086 | 3,064 | 3,224 | 12,248 | 1,202 | 713 |
| 1880 | 143 | 1,064 | 1,430 | 3,475 | 3,969 | 20,214 | 3,108 | 2,531 |
| 1890 | n/a | 2,827 | 4,010 | 7,691 | 5,796 | 24,531 | 4,367 | 3,672 |

| | <3 | 3–9 | 10–19 | 20–49 | 50–99 | 100–174 | 175–259 | 260–499 | 500–999 | >1000 |
|---|---|---|---|---|---|---|---|---|---|---|
| 1900 | 1,492 | 5,354 | 8,236 | 13,110 | 8,067 | 13,196 | 4,635 | 8,370 | 5,329 | 4,753 |
| 1910 | 1,269 | 9,324 | 11,932 | 20,614 | 10,680 | 12,015 | 4,689 | 7,862 | 5,119 | 4,693 |
| 1920 | 2,904 | 13,793 | 17,370 | 31,723 | 15,034 | 13,217 | 5,320 | 8,351 | 5,052 | 4,906 |
| 1930 | 8,527 | 20,365 | 22,037 | 34,948 | 16,378 | 11,561 | 4,976 | 6,969 | 4,861 | 5,054 |
| 1940 | 6,476 | 25,070 | 37,826 | 22,295 | 16,641 | 10,669 | 4,725 | 6,498 | 4,551 | 5,265 |

*Source*: ICPSR Data Sets, 1860–1940.

*Note*: n/a = data not available

[a] In 1860, the smallest farm size range was "smaller than 10 acres"; in 1890, the category "smaller than 3 acres" was not used.

or two. Photographs of late nineteenth- and early twentieth-century orchards frequently show alfalfa, which was used to feed livestock, planted between the trees. Even growers with only twenty acres almost always planted market gardens for home use and local selling.[3] Despite this productive diversity, small farms devoted principally to fruit growing had high capitalization costs. Unlike annual crops, fruit trees and grapevines are a long-term investment. Once the land is subdivided and planted into orchards or vineyards, it tends to remain devoted to those uses for years, whereas land planted to annual crops or devoted to grazing can be converted to other uses each year with little loss of investment. Farm size in horticulture is thus likely to be more stable over time than farm size in other agricultural economies. The long-term nature of capital investment in horticulture also promoted stability in ownership and occupancy, though economic downturns caused some people to abandon their farms from time to time.

Differences in climate, soil conditions, and topography reinforced differ-

TABLE A.2. Top Fruit-Producing Counties in California and Value of Crops Produced, in Dollars, 1910–1940

| Rank | 1910 | 1920 | 1930 | 1940 |
|------|------|------|------|------|
| 1 | Los Angeles 6,731,532 | Los Angeles 42,615,691 | Los Angeles 45,405,615 | Los Angeles 14,363,418 |
| 2 | San Bernardino 5,357,098 | Fresno 42,287,283 | Orange 43,102,274 | Tulare 14,220,197 |
| 3 | Fresno 5,279,794 | San Bernardino 23,429,055 | San Bernardino 32,021,883 | Orange 13,694,931 |
| 4 | Santa Clara 4,234,874 | Santa Clara 19,513,693 | Tulare 31,315,542 | Fresno 13,488,879 |
| 5 | Orange 2,497,734 | Tulare 19,416,780 | Fresno 22,685,777 | Ventura 11,287,262 |
| 6 | Riverside 2,393,371 | Orange 19,102,148 | Ventura 18,785,290 | Santa Clara 11,092,705 |
| 7 | Sacramento 2,265,690 | Sonoma 10,029,335 | Riverside 15,637,434 | San Bernardino 10,857,902 |
| 8 | Tulare 2,053,596 | Riverside 9,691,736 | Santa Clara 11,794,020 | San Joaquin 9,016,661 |
| 9 | Sonoma 2,034,805 | San Joaquin 9,432,595 | San Joaquin 6,579,067 | Riverside 5,831,445 |
| 10 | Ventura 1,795,606 | Ventura 8,813,077 | Stanislaus 6,193,575 | Stanislaus 4,943,701 |

*Source*: ICPSR Data Sets. Sorted by PROC SORT function of SAS.

ences among farms. The mountainous areas in the east and north tended to be more suitable for ranching, while the flat lands in the San Joaquin and Sacramento Valleys, the Santa Clara Valley, and southern California were easier to irrigate and improve for horticulture. Some counties lay across geographic regions, and some produced both fruit and field crops (cereals, grains, and hay) in large amounts. Still, county-level data are sensitive enough to permit the fruit-growing regions to be distinguished from field crop, alpine, and unimproved desert areas, bringing into sharp relief the importance of small farms in specific parts of the state.

Let us first examine differences between the fruit and field crop economies.

TABLE A.3. Top Field-Crop-Producing Counties in California and Value of Crops Produced, in Dollars, 1910–1940

| Rank | 1910 | 1920 | 1930 | 1940 |
|------|------|------|------|------|
| 1 | San Joaquin 5,536,030 | San Joaquin 18,312,753 | San Joaquin 10,223,237 | San Joaquin 7,046,080 |
| 2 | Los Angeles 4,455,633 | Stanislaus 12,130,952 | Stanislaus 8,428,676 | Stanislaus 6,679,725 |
| 3 | Ventura 3,991,451 | Colusa 11,851,942 | Imperial 7,832,970 | Imperial 5,460,924 |
| 4 | Merced 3,057,905 | Glenn 11,066,826 | Ventura 7,651,669 | Fresno 5,024,360 |
| 5 | Monterey 2,878,636 | Tulare 10,963,904 | Monterey 6,506,079 | Monterey 4,783,845 |
| 6 | Tulare 2,822,605 | Yolo 10,838,508 | Los Angeles 6,148,584 | Sutter 4,619,957 |
| 7 | Stanislaus 2,756,820 | Merced 10,022,824 | Yolo 5,683,213 | Merced 4,551,909 |
| 8 | Riverside 2,529,906 | Los Angeles 9,062,231 | Merced 5,513,465 | Los Angeles 4,520,217 |
| 9 | Contra Costa 2,326,132 | Fresno 9,014,446 | Tulare 4,920,084 | Tulare 4,488,824 |
| 10 | Santa Barbara 2,226,909 | Ventura 8,919,760 | Sutter 4,834,482 | Kern 3,845,682 |

Source: ICPSR Data Sets. Sorted by PROC SORT function in SAS.

Tables A.2 and A.3 sort and rank the ten top fruit- and field-crop-producing counties for 1910, 1920, 1930, and 1940 according to total value of each kind of crop produced. Among fruit-producing counties, the ranks are fairly constant, with eight counties appearing in both 1910 and 1920 and all the same counties appearing in 1930 and 1940. Counties dominated by citrus tend to rank above Central Valley counties, where vineyards shared space with orchards of all sorts—citrus, nuts, and tree fruits. In contrast, the ten top field-crop-producing counties for the same period exhibit more change over time.[4] Four counties—Los Angeles, Fresno, Tulare, and Ventura—figure prominently in both kinds of crops, yet their farm size demographics follow the pattern expected for fruit growing. Table A.4 shows that all four counties experienced a decline in median farm size between 1900 and 1940; after 1920, median farm size was lower than fifty acres for three of the four, and the fourth was lower than ninety-nine acres. Average farm size in these counties, by contrast, is stuck at well over one hundred acres throughout the same period, except in Los Angeles County. In view of the fact that these four counties all figured prominently in field crop production even as they led the state in fruit production, the presence of so many small farms here is astonishing. Even in areas where field crop production inflated average farm size, the number of small farms increased over time.

The reason for this is deceptively simple. Small farms were ideal for horticulture. Unlike field crops or cattle ranges, fruit farms did not require much land to be profitable. In fruit-growing industries, single families could and did run farms with their own labor most of the year, hiring migrant help only at harvest time. A 1913 pamphlet promoting Fresno agriculture did not exaggerate: "The small raisin grower has advantages which are beyond the reach of the big grower. Where the owner does practically all the work, the profit per acre is higher, and one man can handle as many as forty acres without assistance except in the harvest period."[5] The value per acre of fruit was and still is higher than the value per acre for field crops or for livestock. Although citrus fruit, dried apricots, peaches, prunes, raisins, and nuts were all luxuries to Gilded Age Americans, after 1910 California fruit growers invented new marketing methods and processing technologies to lower production costs and create demand for their products.

An important factor in the economy was the high degree of regionalization. Almonds were grown in the upper Sacramento River Valley; walnuts along the coast near Santa Barbara; table and wine grapes in Yolo, San Bernardino, and Sonoma Counties; prunes, plums, and other stone fruits near the tiny town of San Jose in Santa Clara County; citrus spread across Los Angeles, Orange, and Riverside Counties; and peaches, apricots, and raisin grapes in Fresno. These commodities commanded prices high enough to make small farms profitable. Studies by University of California agricultural economists between 1900 and 1940 repeatedly demonstrated that family farms growing fruit required far less land than field crop and livestock production to make a satisfactory living.[6]

In short, the data show that where fruit was grown, there were more farms overall and more smaller farms. Historians seem to have overlooked a phenomenon that makes perfect sense to an agricultural scientist or a farmer: small farm size is more highly correlated with fruit production than it is with field crop production. The correlation between farm size and type of com-

TABLE A.4. Farm Size in Four California Fruit-Growing Counties, 1900–1940

| Year | Fresno | | Los Angeles | | Tulare | | Ventura | |
|---|---|---|---|---|---|---|---|---|
| | Median[a] | Average | Median | Average | Median | Average | Median | Average |
| 1900 | 50–99 | 136.2 | 20–49 | 479.1 | 100–174 | 390.5 | 100–174 | 435.3 |
| 1910 | 20–49 | 95.7 | 10–19 | 259.9 | 50–99 | 177.2 | 100–174 | 425.5 |
| 1920 | 20–49 | 70.9 | 10–19 | 170.2 | 20–49 | 148.0 | 50–99 | 249.4 |
| 1930 | 20–49 | 41.6 | 3–9 | 166.7 | 20–49 | 144.5 | 50–99 | 268.2 |
| 1940[b] | 20–29 | 39.2 | 10–19 | 202.8 | 50–69 | 136.7 | 30–49 | 228.7 |

Source: ICPSR Data Sets; some averages and all medians computed by hand.

[a] Average farm size given in acres; median farm size given in range reported by census

[b] The 1940 census used different farm size range categories than its predecessors; it broke down the larger categories used in previous decennials, giving better definition to the size ranges. This makes possible a more precise definition of median farm size for that decennial.

modity produced bears out this conclusion.[7] Beginning in 1910, the census recorded county-level data on fruit production, farm size (reported in size ranges), and agricultural production, represented by the dollar value of crops broken down by category. Every California county produced fruit after 1910 (except San Francisco in 1940), and although the counties on the bottom of the fruit-growing ladder were not commercially important, the fact that they reported fruit crop sales suggests that small farms were present there too.

Tables A.5 and A.6 show a clear pattern that remains consistent over time: fruit growing is highly correlated with farms between 3 and 99 acres. Fruit growing is most highly correlated with farms that were between 10 and 19 acres in size. For farms above 100 acres, the correlation is lower and decreases as farm size increases. Field crop production supplies a near mirror image of this pattern; counties with the largest number of farms between 175 and 259 acres tended to produce the most field crops. In general, the correlation between rank for value of field crops and rank for farm size is greatest for larger farms, although the association of field crops with large farms is not as strong as the association between fruit growing and small farms. For fruit growing and farms between 3 and 99 acres, the correlation statistic $r$ was consistently greater than 0.8 ($r \geq 0.8$), a high degree of correlation. For farms greater than 100 acres, $r$ values were around 0.6, a moderately high degree of correlation. For field crops and farm size, $r$ was between 0.4 and 0.7 for farms 3 to 99 acres; no clear pattern emerges among these correlations. For farms 100 acres and larger, $r$ values were more concentrated between 0.6 and 0.8. Field crops were more highly correlated with larger farms than with smaller farms.

The pattern of rise and fall in the correlations across farm size ranges is highly suggestive. The correlation between rank for farm size and rank for fruit production peaks at ten to nineteen acres in all four decennials and then drops off.[8] The number of small farms in a county was an almost perfect indicator of the value of fruit produced in the county, and the reverse also holds true. Nearly all of these correlations are very highly significant, which indicates that the data are not skewed or distorted by the relatively small sample size. Even without production data for individual farms, researchers can still draw conclusive inferences about the relationship between small farms and fruit growing over time.

Separating the top fruit-producing counties from the rest of the state for each decennial shows just how many more small farms there were in the fruit-growing regions. To analyze the statistical differences in farm size between fruit-growing and nonfruit-growing counties, I defined as "fruit growing" the counties that produced 85 percent of the total value of orchard crops in the state and the rest as nonfruit growing.[9] Figures A.1, A.2, A.3, and A.4 show

TABLE A.5. Correlations between Farm Size (in Acres) and Value of Fruit Crop Production in California Counties, 1910–1940

| Year | <3 | 3–9 | 10–19 | 20–49 | 50–99 | 100–174 | 175–259 | 260–499 | 500–999 | >1,000 |
|------|------|------|------|------|------|------|------|------|------|------|
| 1910 | 0.581 | 0.849 | 0.911 | 0.892 | 0.852 | 0.669 | 0.763 | 0.552 | 0.493 | 0.428 |
| 1920 | 0.619 | 0.833 | 0.900 | 0.872 | 0.866 | 0.744 | 0.743 | 0.558 | 0.460** | 0.429** |
| 1930 | 0.749 | 0.860 | 0.916 | 0.898 | 0.859 | 0.786 | 0.734 | 0.568 | 0.375** | 0.328* |
| 1940 | 0.719 | 0.813 | 0.898 | 0.895 | 0.846 | 0.741 | 0.739 | 0.608 | 0.478* | 0.304* |

*Source:* ICPSR Data Sets, 1910–40. Correlations performed using PROC CORR function of SAS.

*Note:* N = 58 (number of observations equals total counties), except in 1940, when N = 48. All Spearman correlation coefficients are significant at 0.001 probability level unless otherwise noted.

** Spearman correlation coefficient significant at 0.01 probability level

* Spearman correlation coefficient significant at 0.05 probability level

TABLE A.6. Correlations between Farm Size (in Acres) and Value of Field Crop Production in California Counties, 1910–1940

| Year | <3 | 3–9 | 10–19 | 20–49 | 50–99 | 100–174 | 175–259 | 260–499 | 500–999 | >1,000 |
|------|------|--------|---------|--------|--------|---------|---------|---------|---------|--------|
| 1910 | 0.488 | 0.705 | 0.742 | 0.766 | 0.748 | 0.639 | 0.678 | 0.702 | 0.725 | 0.698 |
| 1920 | 0.133ˣ | 0.220ˣ | 0.336** | 0.517 | 0.618 | 0.591 | 0.629 | 0.605 | 0.581 | 0.468 |
| 1930 | 0.537 | 0.615 | 0.667 | 0.741 | 0.769 | 0.776 | 0.799 | 0.756 | 0.681 | 0.563 |
| 1940 | 0.396* | 0.477** | 0.615 | 0.656 | 0.767 | 0.722 | 0.799 | 0.760 | 0.666 | 0.550 |

*Source:* ICPSR Data Sets, 1910–40. Correlations performed using PROC CORR function of SAS.

*Note:* N = 58 (number of observations equals total counties), except in 1940, when N = 48. All Spearman correlation coefficients are significant at 0.001 probability level unless otherwise noted.

**Spearman correlation coefficient significant at 0.01 probability level

* Spearman correlation coefficient significant at 0.05 probability level

ˣ Spearman correlation coefficient not significant

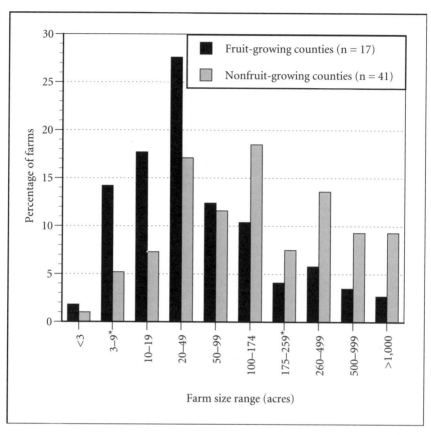

Figure A.1. Farm Size in Fruit-Growing and Nonfruit-Growing Counties in California, 1910
*Source*: ICPSR Data Sets, 1910.
*Differences between fruit-growing and nonfruit-growing counties significantly different at 0.05 level; all others significantly different at 0.001 level

each farm size range as a percentage of the total number of farms for fruit and nonfruit counties between 1910 and 1940.[10] Counties producing most of the state's fruit output had the most small farms. Conversely, counties with more large farms tended to produce field crops (and probably livestock as well, though the Census Bureau's failure to report data for livestock production consistently makes it impossible to include stock ranches in the analysis). In each decennial, the number of farms under 50 acres in fruit counties was significantly lower than the number of small farms in nonfruit counties; likewise, the number of farms over 100 acres was significantly greater in nonfruit counties.[11] After 1910, the only farm size ranges that were not statistically different were farms fewer than 3 and between 50 and 99 acres. There were roughly the

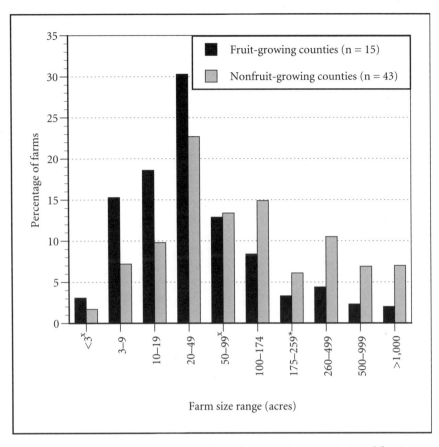

Figure A.2. Farm Size in Fruit-Growing and Nonfruit-Growing Counties in California, 1920
*Source*: ICPSR Data Sets, 1920.
*Differences between fruit-growing and nonfruit-growing counties significantly different
at 0.05 level; ˣ differences not significant; all others significantly different at 0.001 level

same number of farms between 50 and 99 acres in both groups of counties, regardless of the relative commercial importance of fruit or field crops. Above
100 acres, nonfruit counties consistently had more farms. The relationship between farm size and type of crop produced remains almost identical in each
decennial from 1910 to 1940. The number of farms greater than 260 acres in
field crop counties remained more or less constant, at around 10 percent each
decennial.

It follows from these findings that there is a stronger correlation between
the total number of farms and the value of fruit crops than that between total
number of farms and field crops. Table A.7 shows that between 1910 and 1940,
the correlation between total number of farms and fruit growing was always

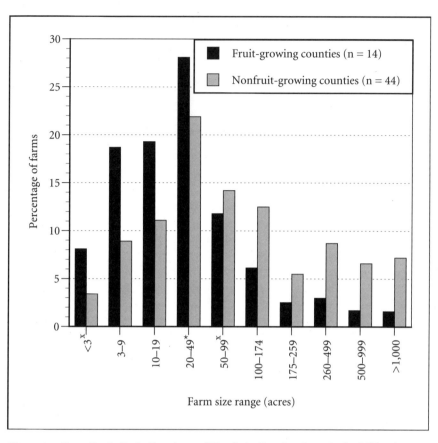

Figure A.3. Farm Size in Fruit-Growing and Nonfruit-Growing Counties in California, 1930
*Source*: ICPSR Data Sets, 1930.
*Differences between fruit-growing and nonfruit-growing counties significantly different at 0.05 level; ˣ differences not significant; all others significantly different at 0.001 level.

higher than the correlation between total number of farms and field crop production. In short, where there was fruit growing, there were more total farms. In 1920, the difference in the two correlations peaked. After the Depression the difference increased once again, suggesting that, contrary to the common assumption, the Depression did not sound the death knell for family farms.

But were these family farms actually owned by families? In fact, they were. Table A.8 shows that in each decennial, the mean numbers of owners, part owners, managers, and tenants in fruit-growing counties were all much higher in fruit than nonfruit counties. Further, the mean number in each category of operators was significantly different between fruit and nonfruit counties. In absolute numbers, horticulture had many more owners; as a result, the

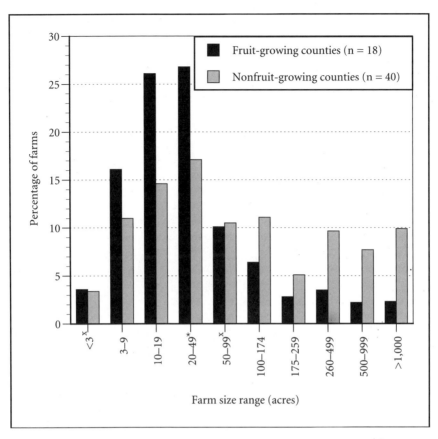

Figure A.4. Farm Size in Fruit-Growing and Nonfruit-Growing Counties in California, 1940
*Source*: ICPSR Data Sets, 1940.
*Differences between fruit-growing and nonfruit-growing counties significantly different at 0.05 level; ˣ differences not significant; all others significantly different at 0.001 level.

proportion of owners to nonowner-operators was much higher in fruit counties than in nonfruit. While there were more tenants in fruit counties, tenants made up a higher percentage of the operators in nonfruit counties.[12] Whatever economic and legal disparities exist between independent owners and dependent operators, they cannot be discerned by looking at horticulture, or any commodity, alone. It is not the case that the small farms in the horticultural industries were disproportionately populated by tenants, managers, or other stand-ins for absentee owners. The prevalence of small farms and individual ownership in the fruit districts was highly significant for the history of the co-operative movement.

TABLE A.7. Correlations between Total Number of Farms and Fruit and Field Crop Production among All California Counties, 1910–1940

| Year | Fruit Crop Production | Field Crop Production |
|------|----------------------|----------------------|
| 1910 | 0.869 | 0.807 |
| 1920 | 0.903 | 0.510 |
| 1930 | 0.890 | 0.755 |
| 1940 | 0.871 | 0.679 |

*Source*: ICPSR Data Sets, 1910–40.

*Note*: N = 58 (number of observations equals total number of counties), except for field crop production in 1940, when N = 48. All Spearman correlation coefficients significant at 0.001 probability level.


TABLE A.8. Operators of Farms in California, 1910–1940

| Year | County Category | N | Owners (Mean, %) | Part Owners (Mean, %) | Managers (Mean, %) | Tenants (Mean, %) |
|------|-----------------|---|------------------|-----------------------|--------------------|--------------------|
| 1910 | Fruit | 17 | 2,394 (78) | n/a | 98 (3) | 586 (19) |
|      | Nonfruit | 41 | 632 (76) | n/a | 10 (1) | 200 (24) |
| 1920 | Fruit | 15 | 2,997 (68) | 379 (9) | 213 (5) | 838 (19) |
|      | Nonfruit | 43 | 719 (60) | 140 (12) | 41 (3) | 293 (25) |
| 1930 | Fruit | 14 | 4,242 (70) | 480 (8) | 372 (6) | 938 (16) |
|      | Nonfruit | 44 | 704 (60) | 146 (13) | 58 (5) | 256 (22) |
| 1940 | Fruit | 18 | 3,544 (71) | 460 (9) | 143 (3) | 862 (17) |
|      | Nonfruit | 40 | 651 (61) | 142 (13) | 20 (2) | 246 (23) |

*Source*: ICPSR Data Sets, 1910–1940.

*Note*: All differences between fruit- and nonfruit-growing counties are significant at 0.001 probability using Student's T-statistics.

# Notes

## Abbreviations

| | |
|---|---|
| AG | *Associated Grower* |
| BAI | Bureau of Antitrust Investigation, Central Classified Subject Files, Antitrust Division, Case No. 60-166-21 (all references, except as noted, in Chapter 8) |
| *Cal. Stats.* | *Statutes of California* |
| CG | *Country Gentleman* |
| *Cong. Rec.* | *Congressional Record* |
| DOJ | U.S. Department of Justice |
| FHS | Fresno City and County Historical Society, Fresno, California |
| *FMR* | *Fresno Morning Republican* |
| FTC | Federal Trade Commission |
| ICPSR | Interuniversity Consortium for Political and Social Research |
| M440 | Microfilm Records, Correspondence of the Secretary of Agriculture, RG 16 |
| *PRP* | *Pacific Rural Press* |
| RG | Record Group, National Archives and Records Administration |
| Secretary's Correspondence | Correspondence of the Secretary of Agriculture, 1906–70 |
| *SMB* | *Sun-Maid Business* |
| *SMH* | *Sun-Maid Herald* |
| Sun-Maid Archives | Sun-Maid Corporate Archives, Kingsburg, California |
| USDA | U.S. Department of Agriculture |

## Introduction

1. On the rise of the national market, see Hurst, *Law and the Conditions of Freedom*; McCurdy, "American Law and the Marketing Structure of the Large Corporation"; Bruchey, *Enterprise*; and Hays, *Response to Industrialism*. Arthur McEvoy's *The Fisherman's Problem*, a study of the commercial exploitation of the California fisheries, argues that law and science were complicit in the exhaustion of natural resources in the capitalist market. A notable exception was the hydraulic mining controversy in California, a rare instance when farmers obtained a permanent injunction against miners, whose industry flooded downstream wheat fields with water and silt. See Kelley, *Gold versus Grain*. The damage inflicted on agriculture by industry and transportation was incompletely compensated, if at all. See Horwitz, *Transformation of American Law, 1780–1860*.

2. For examples of histories of twentieth-century agriculture that look primarily at national political events, see Benedict, *Farm Policies of the U.S.*; Gates, *Farmer's Age*; Fite,

Farmers' Frontier; Shannon, Farmer's Last Frontier; McConnell, Decline of Agrarian Democracy; Perkins, Crisis in Agriculture; Schlebecker, Whereby We Thrive; and Mitchell, Depression Decade. For a recent work that considers agriculture within the administrative state, albeit by focusing more on the administrative state than on farmers' economic activities, see Hamilton, From New Day to New Deal.

3. Hurst, Law and Economic Growth. Hurst did not treat logging as an agricultural enterprise, strictly speaking; he was attracted to the lumber industry as a subject because of its sheer and needless rapaciousness.

4. Beard, Economic Interpretation of the Constitution; Hofstadter, Age of Reform. For one critique of functionalism, see Gordon, "Critical Legal Histories." For the argument that Hurst has been unjustly characterized as a consensus historian, see Soifer, "Willard Hurst, Consensus History, and The Growth of American Law." A recent work that has been criticized for reverting to the functionalist mode is Keller, Regulating a New Economy.

5. See sources cited in notes 7, 8, and 11 below. Criticism also came from neo-Marxists such as Morton Horwitz and William Nelson, who sought to establish the culpability of law in the misdeeds of industrial capitalism. What Willard Hurst characterized as the "release of energy" they saw as a legal system devoted to protecting property interests and promoting technological and territorial expansion, goals whose social costs fell most heavily on the unpropertied. The moral of their story was that law exacerbated inequality and did not promote economic justice (or any other kind) (Horwitz, Transformation of American Law, 1780–1860; Nelson, Americanization of the Common Law). For works that interweave history and Marxist theory, see Thompson, Whigs and Hunters; and Hay, Albion's Fatal Tree.

6. "Belief in the release of private individual and group energies thus furnished one of the working principles which give the coherence of character to our early nineteenth-century public policy" (Hurst, Law and the Conditions of Freedom, 32). Legal historians writing during the 1950s and 1960s — the heyday of Hurst's generation — were primarily interested in the connections between the regulatory traditions laid down during the nineteenth century and the rise of the administrative state in the twentieth. See Levy, Law of the Commonwealth and Chief Justice Shaw; Hartz, Economic Policy and Democratic Thought; Handlin and Handlin, Commonwealth; Scheiber, "Property Law, Expropriation, and Resource Allocation"; Scheiber, "Road to Munn"; Scheiber, "At the Borderland of Law and Economic History"; Scheiber, "Public Rights and the Rule of Law." A recent reinterpretation is Novak, People's Welfare.

7. See, e.g., Edwards, Gendered Strife and Confusion; and Tushnet, The NAACP's Legal Strategy against Segregated Education.

8. Salyer, Laws Harsh as Tigers; Leibhardt, "Law, Environment, and Social Change."

9. My study subtly alters Hurst's conception of the relationship between law and markets. I do not see law and the market solely as supplying "different modes of bargaining among interests"; rather, the market is constituted by law, and as such its operation is contingent on the legal limits that circumscribe the actions of individuals and organizations. The ways in which individuals and corporations resort to law produce mediating effects not only on their competition but also on themselves. See Hurst, Law and Markets in U.S. History.

10. As this discussion makes clear, I draw on the functionalist tradition only in the broadest sense of the term. It is not possible to write sociolegal history without looking at how individuals use the law, but it is an essential aspect of the kind of legal history that I write

that the use of law happens within a specific context and social structure. I do not assume the inevitability of a particular historical outcome as "evolutionary" functionalists do. See Gordon, "Critical Legal Histories," 61–63.

11. Using the labor union as a metaphor between 1890 and 1920 carried no small risk; workers were constantly battling court rulings that applied the antitrust laws to unions. See Tomlins, *State and the Unions*; Forbath, *Law and the Shaping of the American Labor Movement*; Ernst, *Lawyers against Labor*.

12. Thorelli, *Federal Antitrust Policy*; Letwin, *Law and Economic Policy in America*; Sullivan, *Handbook of the Law of Anti-Trust*; see also McCurdy, "*Knight* Sugar Decision"; May, "Political and Economic Theory in Antitrust Analysis"; and Hovenkamp, *Enterprise and American Law*. The history of antitrust has badly slighted farmers. Martin Sklar's revisionist history relied on a social differentiation among "capitalists," by which he meant persons engaged in industrial business. To him, corporate capitalism was no less "natural" a stage of development in economic history than "small-producer, competitive capitalism" that historians have long idealized (*Corporate Reconstruction of American Capitalism*, 2–3). The tendency to construct corporate capitalism as a force displacing nineteenth-century small producers is present in Morton Keller's broad synthesis of economic change during the Progressive Era. Keller's analysis separates "the countryside" from the rest of the economy, suggesting that what was happening in the agricultural sector had little to do with urban industrial society (*Regulating a New Economy*, 148–70). Keller's survey of the rise of cooperatives during the 1910s and 1920s follows the lead of those who fold cooperation into the antitrust movement; as a result, he overlooks the innovative character of the cooperative movement, particularly in California. For example, he repeats the widespread but erroneous view that the attorney Aaron Sapiro organized the first monopolistic California cooperatives whereas those organizations antedated Sapiro's organizing work for the California State Market Commission. (See Chapter 9.) Accordingly, while the works of Sklar and Keller refine the historical context out of which the cooperative movement emerged, both historians treat farmers as homogeneous and their organizational response to industrialism as a variation on the main theme of trusts, mergers, and corporate control of economic relations.

13. On the historical relevance of federalism to Progressive Era regulatory issues, see Scheiber, "Federalism and the American Economic Order."

14. Hurst, *Legitimacy of the Business Corporation*; McCurdy, "*Knight* Sugar Decision"; Grandy, *New Jersey and the Fiscal Origins of Modern American Corporation Law*.

15. The most concise statement of this interpretation is Hofstadter, *Age of Reform*, 113–14.

16. See Tomlins, *State and the Unions*; Forbath, *Law and the Shaping of the American Labor Movement*; Gordon, *Heroes of Their Own Lives*; Kerber, *Women of the Republic*; Leibhardt, "Law, Environment, and Social Change"; Kluger, *Simple Justice*.

17. On the growth of administrative capacities and their relationship to party politics in the late nineteenth and early twentieth centuries, see Evans, Rueschemeyer, and Skocpol, eds., *Bringing the State Back In*; Skowronek, *Building a New Administrative State*; and Skocpol, *Protecting Soldiers and Mothers*. Although the "new institutionalists" tend to slight law in their focus on administrative structures and capacities, this framework, too, sheds light on the relationship between the cooperative movement and the legal system. The historiographical thread concerned with the associationalist mode of government-business relations that flourished under Herbert Hoover's Commerce Department brought to light Hoover's promotion of industrial trade associations and the exchange of market informa-

tion through private channels. Hoover believed that such arrangements encouraged efficiency without endangering free competition. Ellis Hawley's distinguished body of work on Hoover and regulation in the 1920s and 1930s examines agriculture from the perspective of institutional politics. David Hamilton's study of the U.S. Department of Agriculture during the 1920s looks at the limitations on the department's regulatory authority. To apply associationalism to cooperative marketing in agriculture, I examine what the farmers themselves were doing and how state and society reacted to farmers' initiatives in law and policy. On associationalism in the 1920s, see Hawley, "Three Facets of Hooverian Associationalism"; and Hamilton, *From New Day to New Deal*. On the New Deal and agriculture, see Perkins, *Crisis in Agriculture*; and Mitchell, *Depression Decade*.

18. Charles W. McCurdy's work essentially started the field in this direction, arguing that Jacksonian democracy and free labor ideology were central to Justice Stephen Field's conception of law and economic ordering. See McCurdy, "Justice Field and the Jurisprudence of Government-Business Relations"; see also Kalman, *Abe Fortas*. The women's movement used for its own ends the ideology of citizenship and the rights discourse that followed the enactment and early judicial interpretation of the Fourteenth Amendment; see Gordon, *Heroes of Their Own Lives*; Kerber, *Women of the Republic*; Clark, " 'Sacred Rights of the Weak' "; Gordon, " 'Liberty of Self-Degradation.' " On the ideology generated by the labor movement, see Tomlins, *State and the Unions*; Tomlins, *Law, Labor, and Ideology in the Early Republic*; Hattam, *Labor Visions and State Power*; Ernst, *Lawyers against Labor*; Forbath, *Law and the Shaping of the American Labor Movement*. As Lucy Salyer shows in her study of immigration law, due process in the administrative state was not just an institutional practice but an ongoing political process that generated its own ideological and legal norms (*Laws Harsh as Tigers*). Likewise, the temperance movement was threaded through with broader reform movements and ideas of the Progressive Era (Hamm, *Shaping the Eighteenth Amendment*). Robert Stanley sheds new light on the role of capitalist ideology in shaping taxation policy and reform in *Dimensions of Law in the Service of Order*. Herbert Hovenkamp makes ideology the causal agent in his argument that judges based regulatory decisions on economic theory (*Enterprise and American Law*, 67–78).

19. The history of cooperation has been dominated by agricultural economists for decades. See, e.g., Nourse, *Legal Status of Agricultural Co-operation*; Knapp, *Rise of American Cooperative Enterprise*; Knapp, *Advance of American Cooperative Enterprise*; Heflebower, *Cooperatives and Mutuals*; Cobia, ed., *Cooperatives in Agriculture*. Most historians of twentieth-century agriculture have been predominantly concerned with the national farm organizations that have dominated legislation and the regulatory and promotional activities of the USDA. See, e.g., McConnell, *Private Power and American Democracy*; Chambers, *California Farm Organizations*; Shideler, *Farm Crisis*. Social historians have examined the impact of cooperatives on the social structure of local communities. Waldrep, *Night Riders*, and Campbell, *Politics of Despair*, focus on farmers' status and the interaction of race and class, but Campbell's interest lies in the political nature of the growers' organizations, Waldrep's in the social implications of the night riders and the violence they spread within the tobacco communities. Watkins, *Rural Democracy*, has aims similar to mine. She treats organizations of Washington State farmers and understands those organizations as responses not only to hostile national market conditions but also to entrenched local opposition. The movement did not have the same legal or economic impact in Washington as it did in Cali-

fornia, however, and California's agricultural economy was much wealthier. In addition, the dynamics of rural violence in Watkins's study pitted farmers against town merchants, whereas in the raisin industry it was a case of growers attacking other growers.

20. Wylie M. Giffen quoted in Steen, "Story of the California Raisin 'Trust,' " 532.

21. Chandler, *Visible Hand*; Chandler, "Business History as Institutional History"; Chandler, *Scale and Scope*; Galambos, "Emerging Organizational Synthesis," 282. The progressive historians of the New Deal saw clear villains and heroes emerging out of a process, primarily legislative, of rational reform. See, e.g., Higham, Krieger, and Gilbert, *History*. A generation later, historians took direct aim and fired at the Progressive model's emphasis on a rational state staffed by "praiseworthy reformers," arguing instead that the liberal heroes produced ambiguous results: Galambos, "Parsonian Sociology and Post-Progressive History," 26; see Wiebe, *Businessmen and Reform* (arguing that business pursued liberal reforms because corporate leaders perceived such reforms as in their own interests); Hays, *Conservation and the Gospel of Efficiency* (finding that big business supported Progressive conservation policies); Hartz, *Economic Policy and Democratic Thought* (arguing that no single philosophy of government action in the economy existed during the antebellum period); Benson, *Merchants, Farmers, and Railroads* (finding that economic reform consisted of struggles between different groups of business persons); Kolko, *Railroads and Regulation* (arguing that railroads "captured" the regulatory process and corrupted attempts at liberal reform).

22. McWilliams, *Factories in the Field*. For examples of histories of California agriculture that rely on the assumptions flowing from McWilliams's work, see Daniel, *Bitter Harvest*; and Pisani, *From the Family Farm to Agribusiness*.

23. Chan, *This Bittersweet Soil*, 325–26. On the evolution of California agriculture, see Goldschmidt, *As You Sow*; Gates, *Land and Law in California*; Nash, *State Government and Economic Development*; Daniel, *Bitter Harvest*; Pisani, *From the Family Farm to Agribusiness*. For standard assessments of the purpose of California cooperatives and their role in California agribusiness, see Street, "Marketing California Crops at the Turn of the Century"; Blackford, *Politics of Business in California*; Jelenik, *Harvest Empire*; Hutchinson, ed., *California Agriculture*; Erdman, "Development and Significance of California Cooperatives"; Erdman and Larsen, "Development of Agricultural Cooperatives in California"; and Larsen, "A Progressive in Agriculture."

24. On the Chinese, see Chan, *This Bittersweet Soil*; for the new approach to farm-labor relations, see Vaught, "An Orchardist's Point of View."

25. See Olson, *Logic of Collective Action*, for a classic statement of the conflict between self-interest and group action.

26. In 1986, Chan sought to clarify the economic nature of agricultural enterprise with the use of an analytical continuum. On one end is commercial agriculture, "the production of crops for sale." Somewhere in the middle lies industrial agriculture, best understood as "a system of agriculture based on the principles of factory production, including the employment of gangs of agricultural wage laborers." On the other end is agribusiness, the vertical integration of all factors of production, processing, marketing, distribution, and sale of food crops (*This Bittersweet Soil*, 295 n. 55). The idea of an continuum is conceptually helpful but not particularly descriptive of what happens in the story I tell here. Commercial and industrial agriculture and agribusiness are not necessarily exclusive systems, representing

stages in an evolutionary development; I see them as coexisting within agricultural industries where a range of growers' interests weighs against the domination of any single one.

27. Personal interview, Robert Merrill, vice president for member services, Diamond Walnut Growers, Stockton, California, May 24, 1988.

28. The episodes of night riding have remained largely hidden for over seventy years. At the time, local proponents of cooperation avoided discussing the CARC's darker side. Even agricultural newspapers and journals outside California chose to downplay the violence and coercion that made up part of the CARC's recipe for cooperation. Today its few remaining survivors are reluctant to exhume their memories of it. Sun-Maid itself would prefer that this history not come to light, yet Sun-Maid acknowledges that the episodes have not been forgotten. Even now, some growers in the Central Valley refuse to sell to the cooperative because of the crimes committed against their families (personal interview, Gary Marshburn, secretary, Sun-Maid Raisin Growers, Kingsburg, California, July 13, 1993). For Sun-Maid's attitude toward this project, see Kriebel, "Sun-Maid Story."

29. The dynamics involved here differ substantially from those in the world in E. P. Thompson's *Whigs and Hunters*. Instead of a precapitalist society, I deal with a modern society in transition from the frontier to a highly capital-intensive, extractive economy. Instead of an oppressive state, I find an indulgent promotional culture toward agriculture generally and the self-help marketing activities of growers in particular. Instead of the criminalization of petty acts, I note the de facto criminalization of *not* belonging to Sun-Maid and an unofficial but widespread policy of intimidation, injury, and violence. Instead of a simple division between poor and rich, I argue for the presence of divisions based on race, ethnicity, and farm size, all of which defined status, shaped social conflicts, and prolonged their resolution.


*Chapter 1*

1. Taylor and Taylor, *Story of Agricultural Economics*, 117; American Economic Association, *Publications* 8 (1893): 71–72 (paper by rural sociologist F. H. Giddings), quoted ibid., 35; George T. Fairchild, University of Michigan, quoted ibid., 73. For a classic statement of the stereotypical view of the American farmer, see Bryce, *American Commonwealth*, 286, who says that the American farmer "is not free from the usual defects of agriculturalists. He is obstinate, tenacious of his habits, not readily accessible to argument. His way of life is plain and simple, and he prides himself on its simplicity, holding that the class he belongs to to be the mainstay of the country, and regarding city-folk with a mixture of suspicion and jealousy, because he deems them as inferior to himself in virtue as they are superior in adroitness, and likely to outwit him. Sparing rather than stingy in his outlays, and living largely on the produce of his own fields, he has so little ready money that small sums appear large to him."

2. Hoover quoted in Laidler, *Concentration of Control*, 389.

3. Frank Norris, *The Pit*, quoted in Everitt, *Third Power*, 24. A thoughtful treatment of Norris's impact on cultural views of the market is Rothstein, "Frank Norris and Perceptions of the Market." For contemporary statements of the "farm problem," see Carver, *Principles of Rural Economics*, v–vi, 1–13; U.S. Commission on Country Life, *Report*; Croly, *Promise of American Life*. Scholarship on late nineteenth-century agriculture focuses almost entirely

on the economic causes of change in the rural sector. See, e.g., Davis et al., *American Economic Growth*, 369–419; Benedict, *Farm Policies of the U.S.*; Fite, *Farmers' Frontier*; Gates, *History of Public Land Law Development*; Shannon, *Farmer's Last Frontier*; and Cronon, *Nature's Metropolis*.

4. Knapp, *Rise of American Cooperative Enterprise*, 46–47, quoting Jonathan Periam, *A History of the Origin, Aims, and Progress of the Farmers' Movement* (1874). See also Alexander E. Cance, "Farmers' Cooperative Exchanges," *Bulletin of the Extension Service*, Massachusetts Agricultural College, Amherst, 1914, reprinted in Phelan, *Readings in Rural Sociology*, 120–31.

The cast of characters collectively referred to as "middlemen" actually perform distinct functions. A commission broker is an individual or firm that negotiates sales on behalf of buyers or sellers. Such brokers neither physically handle the merchandise nor take title to it. A commission merchant is a receiver who handles more than half his volume on consignment from growers or shippers. A jobber purchases more than half of his volume from wholesale handlers in the local market and sells chiefly to retail outlets. Farmers accused middle merchants of willfully conducting inefficient, sometimes speculative, transactions in order to increase their own profits (Nourse, *Agricultural Economics*, 524–30, 534–35). I have adopted the term "middle merchants" to describe agents of the distribution system working in the range of operations between commercial packers and retail grocers.

In this section of the chapter, I allude to but do not decide the debate over whether the real problem was overproduction or maldistribution; the important issue for my purposes is how farmers defined that problem and tailored their political activism and economic behavior accordingly. For typical treatments of the overproduction/maldistribution problem, see, e.g., Shannon, *Farmer's Last Frontier*; and Benedict, *Farm Policies of the U.S.*

5. As Jeffrey Ostler notes, "A decentralized federal system afforded an agrarian protest movement significant opportunities for state-level political action" (*Prairie Populism*, 7).

6. Smythe, *Constructive Democracy*, 213.

7. The quickening pace of economic change during the eighteenth and nineteenth centuries inspired both workers and farmers to experiment with various forms of collectivism. By the time of the Industrial Revolution in Europe, farmers had pioneered the idea of cooperative rural credit societies and cooperative stores for the purchase of farm implements. Cooperative factories and mills also sprang up after 1815. These organizations were primarily unincorporated pools, dependent on private capitalists for financing and lacking formal legal identity. In England, Denmark, Sweden, and Germany, cooperatives originated in the late Middle Ages as local pools. Most typical were the creamery rings and cheese pools of Switzerland and Denmark. By 1850 cooperation had spread to banking, industry, labor, and consumer purchasing. On the history of cooperation in Europe, see Fay, *Cooperation at Home and Abroad*; Filley, *Cooperation in Agriculture*, 17–21; Hanna, *Law of Cooperative Marketing Associations*, 3–4; Holyoake, *History of Cooperation*; Mears and Tobriner, *Principles and Practices of Cooperative Marketing*, 366–91; Nourse, *Legal Status of Agricultural Co-operation*, 25–35; Steen, *Cooperative Marketing*, 1–4. On rudimentary cooperation in the colonies, see Morgan, *American Slavery, American Freedom*.

8. Barton, "Principles," 24–32, esp. 27; Knapp, *Rise of American Cooperative Enterprise*, 8–15.

9. Holyoake, *History of Cooperation*, quoted in Mears and Tobriner, *Principles and Prac-*

*tices of Cooperative Marketing*, 14-15. On the separation between capital and management in regular corporations, see Berle and Means, *The Modern Corporation and Private Property*, 89; Chandler, *Visible Hand*, 377-483.

10. Knapp, *Rise of American Cooperative Enterprise*, 29-30; Guarneri, *Utopian Alternative*, 32-33, 36-37. Early treatises include Adams, ed., *History of Cooperation*; Adams, *Modern Farmer*; and Coulter, *Cooperation among Farmers*.

11. More than anything else, the Grangers sought to break up the railroads' control over freight rates and routes served; their chief weapon was legislation to regulate rates, which the railroads attacked in court. The Grangers were less successful in obtaining laws to regulate and tax business corporations; as a result, they had little impact on the marketing system. See Buck, *Granger Movement*, 238-78; Shannon, *Farmer's Last Frontier*, 309-11. Buck remains the standard work on the Granger movement. For perspectives on the implications of the legislative changes forced by the movement, see Scheiber, "Road to *Munn*"; Keller, *Affairs of State*; Kolko, *Railroads and Regulation*.

12. Steen, *Cooperative Marketing*, 5; Buck, *Granger Movement*, 69-73.

13. Buck, *Granger Movement*, 258-61, 274-78 (quote on 261); Knapp, *Rise of American Cooperative Enterprise*, 51-55; McKay, *Organization and Development of a Cooperative Citrus-Fruit Marketing Agency*, 9-10.

14. Steen, *Cooperative Marketing*, 4-5; Knapp, *Rise of American Cooperative Enterprise*, 57; Hanna, *Law of Cooperative Marketing Associations*, 5. Buck credits the Granger cooperatives with providing valuable business training to farmers, as well as with spreading the gospel of cooperation throughout the country. He does acknowledge that the 1875 adaptation of the Rochdale principles came too late to save the movement (*Granger Movement*, 258). For a view of Granger cooperatives less celebratory than Buck's, see Cerny, "Cooperation in the Midwest."

15. The Populists have drawn considerably more attention than the Grangers. The first major interpretation was Hicks, *Populist Revolt*, which evaluated Populism as a political crusade. Hicks was revised by Hofstadter, *Age of Reform*, who portrayed the Populists as trying to recapture the romantic agrarianism of the preindustrial past. Lawrence Goodwyn took on both Hicks and Hofstadter in *Democratic Promise*, arguing that the Populists formed a "movement culture" linked to labor radicalism and frontier agrarian discontent. That argument has sparked much debate. See, for example, Brinkley, "Richard Hofstadter's *The Age of Reform*"; Ridge, "Populism Redux"; and Rothstein, "Farmer Movements and Organizations."

Alliance cooperatives were neither as strong as Goodwyn has argued nor as weak as his critics have contended. Some of Goodwyn's critics allege that the Populists did not achieve the successes with cooperatives that he attributes to them. Theodore Saloutos and Gilbert Fite criticized Goodwyn for exaggerating the importance of cooperatives; Robert W. Cherny asserted that Goodwyn wrongly cast Nebraska Populists as a "shadow movement" that detracted from the force of the Populist message in the rest of the nation; and Stanley B. Parsons and three colleagues argued on the basis of a statistical study that the cooperative movement did not play an extensive role in Populism. See Saloutos, "Review of *Democratic Promise*"; Fite, "Review of *Democratic Promise*"; Cherny, "Lawrence Goodwyn and Nebraska Populism"; Ridge, "Populism Redux," 150-51, nn. 24-26; and Parsons, Parsons, Killiae, and Borgers, "The Role of Cooperatives in the Development of the Movement Culture." Works that share Goodwyn's emphasis on ideology and culture but focus on farmers' economic reasons for responding to Populism include Palmer, *"Man Over Money"*; and Hahn, *Roots of Southern Populism*. In *The Promise of the New South*, Edward L. Ayers adds to the critique of the Goodwyn thesis but offers no alternative explanation for the Populists' collapse in the South. None of these works, framed in political and social history, pays any attention to the legal form and economic practices of Farmers' Alliance cooperatives. Rather than deciding whether cooperatives achieved the political aims of the Populist movement, I focus on whether the Alliance changed the marketing practices and legal character of cooperatives.

16. Goodwyn, *Democratic Promise*, 110–53 (quotes on 139 and 143); Ostler, *Prairie Populism*, 111–53 (explaining the lack of Populist strength in Iowa as a result of a committed nonpartisan strategy by the Farmers' Alliance there). Macune was by no means the only farm leader to preach the Rochdale gospel and practice joint stock business enterprise; California cooperative organizers did the same thing. See Chapter 4.

17. McMath, "Sandy Land and Hogs in the Timber," 206.

18. For one expression of the continuing fascination with Rochdale, see "A Conversation with C. Y. Roop, editor of 'The Cooperative Journal,' on the Rochdale Co-operative Movement in California," *Arena* 29 (1903): 541–43; see also Cance, "Farmers' Cooperative Exchanges," 124; and Gill, "Social Effects of Cooperation in Europe."

19. The best statement of the "associational impulse" in California agriculture remains Erdman and Larsen, "Development of Agricultural Cooperatives in California." A substantial network of Rochdale cooperative stores existed in California during the nineteenth century; one account credits the Farmers' Alliance with securing legislation imposing Rochdale standards on equal shareholding and equal voting rights on all cooperative stores organized in the state. See Cross, "Cooperation in Agriculture," 537, n. 12.

20. Plehn, "The State Market Commission of California," 2 (quoting John W. Lloyd, "Cooperation and Other Organized Methods of Marketing California's Horticultural Products" [Ph.D. diss., University of California, 1919]); Cross, "Cooperation in Agriculture," 540. On the impact of commission marketing on California horticulture, see MacCurdy, *History of the California Fruit Growers Exchange*; Steen, *Cooperative Marketing*; Erdman, *California Fruit Growers' Exchange*; Bean and Rawls, *California*, 199–211; and Seftel, "Government Regulation and the Rise of the California Fruit Industry."

21. On the economic beginnings of agriculture after statehood, see Chapter 2; for descriptions of the marketing problems faced by fruit growers, see Hutchinson, ed., *California Agriculture*; Blackford, *Politics of Business in California*, 13–39; Larsen, "A Progressive in Agriculture," 189–90; on the issue of expanding consumption, see Taylor, "Your Inelastic Appetite"; on the early citrus industry, see MacCurdy, *History of the California Fruit Growers Exchange*, 11.

22. Nourse, *Legal Status of Agricultural Co-operation*, 40. The wave of state laws providing for cooperative incorporation in the late nineteenth century, including the leading statutes enacted in California, is treated in Chapter 3. The SCFE was formed to counter the strength and breadth of the national brokerage system that continued to market much of the fruit sold in the United States. It oversaw both the local associations, which handled packing and other processing operations, and the district exchanges, which did the actual selling. The SCFE hired sales agents to represent the growers in eastern markets and coordinated the shipments from locals to eastern commission markets. See Knapp, *Rise of American Cooperative Enterprise*, 86–87; MacCurdy, *History of the California Fruit Growers*

*Exchange*, 22–33; Erdman, *California Fruit Growers' Exchange*, 8–9, 12–15; Steen, *Cooperative Marketing*, 40; McKay, *Federal Research and Educational Work*, 11–12; Powell, "Cooperative Marketing of California Fresh Fruit"; Cross, "Cooperation in Agriculture," 540–42.

23. Steen, *Cooperative Marketing*, 21–50, 272–79. On the Progressives in California, see Mowry, *California Progressives*.

24. Cross, "Cooperation in Agriculture," 542; Erdman, "Development and Significance of California Cooperatives," 181. The CFGE reduced packing costs by about fifteen cents per box between 1895 and 1911 by eliminating commercial packing's profit margins and buying packing supplies at cost for its local associations. The CFGE also successfully bargained with the railroads for lower freight rates and refrigeration charges. See Lloyd, "Cooperative and Other Methods of Marketing," 15–18. See also Meyer, "History of the California Fruit Growers' Exchange"; and Powell, *Organization of a Great Industry*.

25. Steen, *Cooperative Marketing*, 47; Coulter, *Cooperation among Farmers*, 198–214; Filley, *Cooperation in Agriculture*, 186–205. Both the SCFE and the CFGE were organized as cooperatives with capital stock; after the Clayton Act was passed in 1914, exempting non-stock cooperative associations from federal antitrust laws, the CFGE reorganized as a non-stock organization. It raised funds by levying an assessment on each box of fruit shipped. Not all California cooperatives copied this move, however; "centralized" cooperatives such as the CARC remained capital stock organizations, while "federated" organizations such as the CFGE relied on their local associations to raise the capital they needed. See Wickson, *Rural California*, 302; Erdman, "Development and Significance of California Cooperatives," 181. On the Clayton Act, see Chapter 5. For more on Powell's life and work in California, see Powell, *Letters from the Orange Empire*.

26. Cross, "Cooperation in Agriculture," 540; Haskell, "Great Farm Movement," 548.

27. On the walnut industry, see Lloyd, "Cooperative and Other Methods of Marketing," 11, 19; Thorpe, "California Walnuts and Their Co-operative Marketing"; Ordal, "History of the California Walnut Industry," 167–206; California State Commission of Horticulture, "Benefits of Cooperation," 621; Teague, *Fifty Years a Rancher*, 97–111. On the almond industry, see Scholl, "Economic Study of the California Almond Growers Exchange"; and Riley, "History of the Almond Industry in California."

28. Eisen, *Raisin Industry*, 55, 169; Fox, "Co-operation in the Raisin Industry," 1–3; California State Board of Agriculture, "Report of the State Statistician," 135–37; Spence, "Success after Twenty Years," *SMH* 3 (Nov. 1917): 4; Payne, "Cooperation," 14; Cochran, "Results of Co-operation in the Raisin Industry," 7–8; Colby, "California Raisin Industry," 93–94; Crouse, "California Raisin Industry," 5. See Chapter 4.

29. U.S. Tariff Commission, *Grapes, Raisins and Wines*, 142–43; "How Cooperation Developed the Raisin Industry, 1922," typescript (photocopy), Sun-Maid Archives; Meyer, "Development of the Raisin Industry," 79–80; Cochran, "Results of Co-operation in the Raisin Industry," 9–10; Bragg, "History of Cooperative Marketing," 23; Smith, *Garden of the Sun*, 518–19.

30. "Constructive Agitation," *CG* 78 (May 3, 1913): 708.

31. Ross, "Morality of Co-operation," 189; Smythe, *Constructive Democracy*, 217; Croly, *Promise of American Life*, 406; review of Holyoake, *History of Cooperation*, in *Nation* 83 (Aug. 23, 1906): 170–71 (quoting Holyoake, *History of Cooperation*, 2:674). On the political appeal of cooperation during the Progressive Era, see Malin, "Background of the First Bills," 118.

32. Guarneri, *Utopian Alternative*, 1–2, 15–34, 39; see also Thomas, *Alternative America*.

33. Kerr, *Farmers' Union and Federation Advocate Guide*, 8; Carver, *Principles of Rural Economics*, 381; Warbasse, "Cooperative Movement," 565. On the support of the Farmers' Alliance for the Knights of Labor, see Ayers, *Promise of the New South*, 216–22.

34. Lanier, "Cooperation," 196; Cross, "Cooperation in Agriculture," 544.

35. Adams, *Modern Farmer*, 440; Vincent, "Co-operation among Western Farmers," 286. See also Nourse, "Revolution in Farming," 90, 94; Coulter, *Cooperation among Farmers*; Cumberland, *Cooperative Marketing*; Kerr, *Farmers' Union and Federation Advocate Guide*, 10. Other proponents argued that the Rochdale principles, if faithfully followed, would enable farmers to compete as entrepreneurs without sacrificing equity and fairness (Holyoake, *History of Cooperation*; Plunkett, *Rural Life Problem*; Powell, *Cooperation in Agriculture*).

36. Powell, "Fundamental Principles of Co-operation in Agriculture" (1914), 1–4; Powell, "Fundamental Principles of Co-operation in Agriculture" (1920), 1–6.

37. Teague, "Cooperation versus Speculative System of Marketing," unpublished manuscript, n.d. [1920–21], File 1, Box 2, Teague Papers; Herman Steen to Henry C. Wallace, Sept. 8, 1922, Marketing, 1922, Secretary's Correspondence, RG 16. A retrospective on Teague's career is Lillard, "Agricultural Statesman."

38. Weseen, "Co-operative Movement in Nebraska," 478; Knapp, *Rise of American Co-operative Enterprise*, 224–27.

39. On the Grangers' impact on state incorporation laws and the Populists' influence on the California 1895 nonstock law, see Nourse, *Legal Status of Agricultural Co-operation*, 39–43, and Chapter 3.

40. White, "Cooperation among Raisin Growers."


*Chapter 2*

1. Wickson, "California Mission Fruits"; Starr, *Americans and the California Dream*, 197. For a short overview of Bidwell's life, see Hart, *Companion to California*, 46. Bidwell's letter, reprinted in Wickson's article, surveys the condition of the orchards and vineyards at many of the missions. He described what fruits had been grown at which missions, commenting on the variations in quality he observed among them and on the ruined condition in which he found them in 1844 and 1845.

2. See Benedict, "Economic and Social Structure of California Agriculture." In Kentucky and Tennessee, the highly regionalized tobacco economy was also characterized by great disparities in wealth and racial tensions; however, the proportion of renters and tenants was far higher in the tobacco districts. Growers' cooperatives formed there in response to the monopoly held by the American Tobacco Company, the sole buyer for burley tobacco, but these organizations were run by white rural elites whose goal was as much to winnow out black tenants and smallholders as it was to take on their common corporate enemy. See Campbell, *Politics of Despair*; and Waldrep, *Night Riders*.

3. Most historians of California and public lands policy mention the appearance of smaller farms and usually relate this phenomenon to the turn to fruit farming after about 1880, but no one has attributed much meaning to their existence or even bothered to chart the demographics of such farms, though published census data have been available for decades. See Robinson, *Land in California*; and Pisani, "Land Monopoly in Nineteenth-Century California," 19–26.

For studies that focus on the statistic of average farm size to argue for the dominance of large farms in California, see, e.g., McWilliams, *Factories in the Field*, 60–65; Pisani, *From the Family Farm to Agribusiness*, 14–15, 127, 284–85, 450–51; Daniel, *Bitter Harvest*, 17–24; Chan, *This Bittersweet Soil*, 144. More general histories of California that take the large farm as normative include Starr, *Inventing the Dream*; Kelley, *Battling the Inland Sea*; and Jelinek, *Harvest Empire*.

4. Daniel, *Bitter Harvest*, 35; McWilliams, *Factories in the Field*; Pisani, *From the Family Farm to Agribusiness*. For criticisms of agribusiness's emphasis on the commodification of natural resources, see Miller, *Flooding the Courtrooms*; Worster, *Rivers of Empire*; and Stegner, *Beyond the Hundredth Meridian*. The West had many contemporary boosters; see, e.g., Smythe, *Conquest of Arid America*.

The most controversial study of the implications of farm size in a democratic society was Walter Goldschmidt's *Small Business and the Community: A Study in Central Valley of California on Effects of Scale of Farm Operations* (1946), subsequently published as *As You Sow*. Goldschmidt argued that large farms degraded the quality of social life in agricultural communities. This argument has been discredited as based on flawed data, but it left a lasting impression, and its condemnation of California's agricultural economy endures. For criticisms of the Goldschmidt thesis, see Kirkendall, "Social Science in the Central Valley of California"; Taylor, "Walter Goldschmidt's Baptism in Fire," 8; and Hayes and Olmstead, "The Arvin and Dinuba Controversy Reexamined." Among Goldschmidt's defenders is the public lands historian Paul Gates in Gates, "Public Land Disposal in California."

5. My discussion here owes much to Sucheng Chan's trenchant summary of the contributions of Varden Fuller, Carey McWilliams, Paul Taylor, Cletus Daniel, and William E. Smythe, among others, in *This Bittersweet Soil*, 273–301, 324–25. While she is a sensitive reader of the census data analyzed in her study, Chan uses large farms and large landholders as a foil in her argument that the Chinese were unfairly blamed for many of the social evils of agribusiness (319–27). Although I do not dispute her conclusions and agree that the problem of migrant labor ought to remain a central concern for agricultural historians, I view the role and experiences of smallholders as equally important to the task of drawing a multidimensional picture of agricultural society in the Central Valley. As my discussion here and in subsequent chapters will show, not all landowners felt equally empowered by their ownership of land, and smallholders were caught in a bind between their interests as growers and their function as, in effect, hired hands within the larger fruit-growing industry and the cooperative associations to which they belonged.

6. My analysis relies on the published agricultural census schedules in both hard copy and electronic form. Electronic data were retrieved from computer files obtained from the Interuniversity Consortium for Political and Social Research at the University of Michigan (ICPSR). Since the published censuses do not list the size of each individual farm, it is impossible to calculate median farm size with precision. I calculated it by taking the number of farms in each size range, multiplying by the number at the midpoint of each size range, and summing across both the number of farms and average number of acres in each size range. I checked the accuracy of this method against the published figures for total acres in farms in each census and found that my figure for total acres in farms more closely approximated the census's reported figure for *improved* land in farms than for *total acres in farms*, which includes unimproved land. The census's calculation of average farm size used total

acres in farms, meaning that its figures do not reflect how much of the land was actually being used for agricultural purposes and how much was left unimproved.

An important study that is similarly critical of the use of statewide census data is Clarke, "Farmers as Entrepreneurs," 50–51. For example, Clarke argues that geographic factors at the county level were highly significant in determining the rate at which farmers in Corn Belt states purchased tractors.

7. Hendry, "Source Literature of Early Plant Introduction."

8. Wickson, "History of California Fruit Growing"; Blick, "Geography and Mission Agriculture," 38; Wickson, "California Mission Fruits," 501–2. Staples included wheat, barley, corn, beans, and grapes. Grapes, in fact, were widely planted at the missions because the Franciscans made wine for both sacramental use and table drinking. The friars also planted stone fruits, nuts, citrus, and even some tropical crops. The friars recruited Native Americans both to convert to Christianity and to work the mission farms, which soured relations with neighboring Indian villages and thoroughly disrupted native hunting and gathering rhythms. On the transfer of plants and animals from Europe to America, see Crosby, *Columbian Exchange*.

9. Olive trees and grapevines "furnished cuttings for most of the plantations made during the first twenty years or more of American occupation" (Wickson, "California Mission Fruits," 502).

10. Starr, *Inventing the Dream*, 131; Chan, *This Bittersweet Soil*, 79–80; Wickson, "California Mission Fruits," 505. According to Starr, the state's population during the 1850s was around 380,000.

11. Gates, "Adjudication of Spanish-Mexican Land Claims in California." For over sixty years, Gates has argued that land policy is singularly responsible for both the persistence of land monopoly and the rise of agribusiness. After extensively analyzing census data and land records, Gates concluded that legislatures and judges confirmed many dubious Mexican claims on the grounds of equity, while at the same time the federal government delayed in surveying the public lands and opening them to settlement. As a result of the chaos and uncertainty during the critical years of early statehood, many squatters lost not only their claims to the land but also the title to and value of any improvements they made. I argue below and in the Appendix that small farms flourished in California despite the initial pattern of highly concentrated land ownership.

12. I refer to these crops and other field crops not requiring irrigation as "dryland" crops, to distinguish them from fruit, nuts, and other water-intensive field crops such as alfalfa. Cattle ranching and wheat farming were also popular because they required little improvement to property whose legal ownership was still in question. In short, they returned significant profits without requiring substantial investment.

13. Rothstein, "California 'Wheat Kings,' " 1–2; Starr, *Inventing the Dream*, 131; Pisani, *From the Family Farm to Agribusiness*, 10. On Mexican agriculture, see Adams, "Historical Background of California Agriculture," 17–33, and Wickson, *Rural California*, 27–75. For a discussion of agriculture after 1850, see Paul, "Beginnings of Agriculture in California," 19. On the rise of the wheat trade, see ibid., 19–23; Wickson, *Rural California*, 75–83; Rothstein, "West Coast Farmers and the Tyranny of Distance"; on its decline, see Rothstein, "California 'Wheat Kings' "; McWilliams, *Factories in the Field*, 17; Adams, "Historical Background of California Agriculture," 33–35; and Bean and Rawls, *California*, 199–200.

14. John Bigler in California, State Legislature, *Journal of the Fourth Session*, 22, quoted in Nash, *State Government and Economic Development*, 124. Bonanza farming perpetuated land monopoly, created a permanent underclass of dispossessed, rootless farmers and farm laborers, consumed hundreds of square miles of farmland, turned farms into commercial factory-style operations, and made the state dangerously dependent on one crop that was itself subject to the "vagaries of a notoriously unstable international grain market" (Daniel, *Bitter Harvest*, 22). The wheat kings' prosperity put Californians "in a state of ferment over land monopolization and the need for tax reform" (Gates, "The Suscol Principle, Preemption, and California Latifundia," 224).

15. On the shift from wheat to specialized horticulture, see Cleland and Harvey, *March of Industry*; Jelinek, *Harvest Empire*; McWilliams, *Factories in the Field*; Paul, *The Far West and the Great Plains in Transition*; Hutchinson, ed., *California Agriculture*. On California geography, see Bakker, *An Island Called California*; Durrenberger, *California*; and Miller and Myslop, *California*.

16. California State Board of Agriculture, *Fifty-Ninth Annual Report*, 135. The formation of the California State Agricultural Society in 1854 gave farmers an organization within which to exchange information and through which to influence legislature policy. See Adams, "Historical Background of California Agriculture," 39; Nash, *State Government and Economic Development*, 63–65.

17. T. S. Price, "A Twenty-Acre Vineyard," *PRP* 45 (May 27, 1893): 472; Cleland, *Cattle on a Thousand Hills*, 176.

18. On the challenges posed by arid climates to temperate-zone-trained farmers, see Worster, *Rivers of Empire*, 114–25; on litigation over water rights and irrigation districts, see Smith, *Garden of the Sun*, 449–51; Thickens, "Pioneer Agricultural Colonies," 21–24; Miller, *Flooding the Courtrooms*, 39–94; and Pisani, *From the Family Farm to Agribusiness*, 191–249. A good treatment of the complex relationship between law, the state, and market actors battling over natural resources is Walton, *Western Times and Water Wars*.

19. Bioletti, *Phylloxera of the Vine*. To this day phylloxera remains a menace. See, e.g., "Wine-Grape Growers to Best Pest; Phylloxera Epidemic Is Manageable, New Report Says," *San Francisco Examiner*, July 13, 1994, p. Z2; and "New Threat in the Wine Country: Insect Called the Sharpshooter Is Attacking Replanted Vineyards," *San Francisco Chronicle*, Dec. 31, 1993, p. A25.

20. Teague, *Fifty Years a Rancher*, 26–27; Colby, "California Raisin Industry," 89–90; Seftel, "Government Regulation and the Rise of the California Fruit Industry," 377. Seftel argues that growers were more concerned about pests than they were about marketing: "During [the period from 1880 to 1920], the fruit growers' regulatory demands focused primarily on the problem of pest control" (375). Clearly, eastern consumers would not buy decomposed and damaged fruit, but no fruit—healthy or not—would ever reach them unless growers had in place effective marketing mechanisms that protected their economic interests as well as the condition of the fruit in transit.

21. Rothstein, "California 'Wheat Kings,'" 7–8, 10–11; Pisani, *From the Family Farm to Agribusiness*, 18, 107, 120–24; Meyer, "Development of the Raisin Industry," 21–38; Fox, "Cooperation in the Raisin Industry," 2–3. On the agricultural colonies in southern California, see Cleland, *Cattle on a Thousand Hills*, 157–59. On land colonization in California generally, see Hine, *California's Utopian Colonies*.

22. Teilman and Shafer, *Historical Story of Irrigation*, 3–5. The doctrine of prior ap-

propriation, gradually adopted by state courts during the nineteenth century, enabled landowners whose tracts did not lie adjacent to streams to assert claims to the flow and physically remove the water to use for irrigation. See Sax and Abrams, *Water Law*; Worster, *Rivers of Empire*, 88–96; and Pisani, *From the Family Farm to Agribusiness*, 30–53. Studies showing the peculiar development of California's water law include McCurdy, "Stephen J. Field and Public Land Law Development"; and Miller, "Riparian Rights and the Control of Water in California."

23. Thickens, "Pioneer Agricultural Colonies," 21; Rehart and Patterson, *M. Theo Kearney*, 7–8; Elliot, *History of Fresno County*, 111–15; Teilman and Shafer, *Historical Story of Irrigation*, 5–15. According to Wickson, the irrigated communities were larger and more successful than nonirrigated settlements where growers had to stake their own claims to water and build their own canals (*Rural California*, 79–82).

24. Thickens, "Pioneer Agricultural Colonies," 19; Rehart and Patterson, *M. Theo Kearney*, 6. These systems were haphazardly engineered and constructed, resulting in drainage problems, flooding, and alkali accumulation (Colby, "California Raisin Industry," 79–80).

25. W. R. Wood, "Fresno, the Sun-Maid County," *California Cultivator* 65 (Aug. 8, 1925): 122; Pisani, *From the Family Farm to Agribusiness*, 122; Wickson, *Rural California*, 386–87. Church's irrigation company provoked numerous lawsuits during the 1870s and 1880s, particularly from ranchers challenging the diversion of water from their riparian lands by the irrigation company. Riparianism did not give landowners an absolute right of ownership in water; it merely assured the right to use flowing water that abutted the landowner's property. Riparian claimants were able to win judgments and damages against Church and other appropriators only for a time; in California's snarled legal system, riparian rights were undermined by both other riparian rights and the novel doctrine of prior appropriation. Appropriation had its origins in the informal legal norms of mining communities, which gave property rights in fixed quantities of water to miners who diverted streams into ditches and sluice boxes. The legal theory behind prior appropriation was the customary law of the public domain: first in time, first in right. Whether or not they owned riparian lands, first comers had the right not only to divert water but also to retain the use of it, as long as the use was beneficial. The practice spread beyond the mining districts as private irrigation companies began diverting huge quantities of water to colony settlements and individual tracts in the 1870s; the doctrine soon spread to Colorado and Nevada, two arid states, where, as in California, the initial enterprise of choice was mining, a highly water-intensive activity. Even the federal mining code adopted appropriation in 1866. The California courts, however, kept riparianism on the books and so created a distinct, if confused, conception of ownership. The complicated story of water law and policy in the West is beyond the purview of this work, but many historians have taken up the topic. For critical interpretations of the role of appropriation in fostering the development of large-scale, intensive agriculture, see Pisani, *From the Family Farm to Agribusiness*, 440–52; Miller, *Flooding the Courtrooms*, 10–94; and Worster, *Rivers of Empire*, 259–326. A straightforward overview of the law of water rights is Dunbar, *Forging New Rights in Western Waters*, 59–72. On the collision course between hydraulic mining and grain farmers in the Sacramento Valley during the 1860s and 1870s and the legal resolution of this conflict, see Kelley, *Gold versus Grain*.

26. Walker, *Fresno County Blue Book*, 49: "The chief land development [after the advent of irrigation] . . . was the undertaking, toward the close of the century, to cut up the Laguna de Tache Grant into small farms."

27. Circular, n.d. [1887], File 5, Box 1, Kearney Papers, FHS; *Fresno County Facts, Compiled by the Industrial Survey Committee of Fresno County Chamber of Commerce* (Fresno, Calif.: Privately printed, n.d. but probably 1922), 2, Bancroft Library; M. Theo Kearney to Nelson W. Aldrich, June 1, 1897, File 1887–97, Box 1, Kearney Papers, Bancroft Library. See also Thickens, "Pioneer Agricultural Colonies," 24–36, 169–76; Winchell, *History of Fresno County*, 9–11. Chapman and Kearney advertised several of their colonies with pamphlets such as this one. See Rehart and Patterson, *M. Theo Kearney*, 9. The railroad ended dependence on river transportation, brought the last of the miners down from the foothills, carried in people to farm, and, most crucially, gave these farmers a way to send their perishable produce to market (Winchell, *History of Fresno County*, 9; Cleland, *Cattle on a Thousand Hills*, 176). Fresno County's population rose from 4,605 in 1860 to 32,026 in 1890; its total property valuation shot up from $431,403 in 1856 to $8,298,097 in 1882 and $40,000,000 by 1892 (Walker, *Fresno County Blue Book*, 38, 48; Kearney to Aldrich, June 1, 1897, File 1887–97, Box 1, Kearney Papers, Bancroft Library).

28. *Imperial Fresno*, 9; Thickens, "Pioneer Agricultural Colonies," 175, 35–36, 169–77; Meyer, "Development of the Raisin Industry," 52–61; Walker, *Fresno County Blue Book*, 46–48; *Fresno: Where Half the World's Raisins Come From* (Fresno: Fresno County Chamber of Commerce, Sept. 2, 1922), Bancroft Library.

29. Thickens, "Pioneer Agricultural Colonies," 32; Colby, "California Raisin Industry," 58, 60–76; Husmann, "Raisin Industry," 3; Trujillo, *Parlier, the Hub of Raisin America*. The state's first commercial vineyards, in the Yolo County towns of Davisville and Woodland, put raisins on the market in 1873. Between 1890 and 1910, the industry condensed geographically. By 1915, Orange and Riverside Counties stopped producing raisins; Los Angeles was down to "very few." Counties that still had productive raisin grapevines shifted them over to other uses. Yolo County producers sold Sultanas and Thompson's Seedless as table grapes, Riverside and San Bernardino as wine grapes. The wine-producing areas in Napa and Santa Barbara Counties were less well-suited to raisin cultivation; their summers were not as hot or as long as those in the Central Valley. See California State Department of Agriculture, *Fifty-Ninth Annual Report*, 135; California State Department of Agriculture, "Estimated Production of Raisins, 1915, by Counties," in *Statistical Report of State Statistician for 1916*, 137–38.

30. Colby, "California Raisin Industry," 58, 80–85, 87–88; *Imperial Fresno*, 12–13. Initially, Muscat grapes were the dominant variety, despite the seeds and high sugar content that made them sticky and difficult to process. Advances in processing technology produced the seeded Muscat around 1900 and made Muscats more suitable for retail sale. Beginning in 1913, the CARC encouraged new vineyardists to plant Thompson Seedless vines; by the early 1920s, Thompsons had overtaken Muscats for good. Today they are the most widely planted and consumed variety of raisin grape.

31. Colby, "California Raisin Industry," 60, 83–87; *Imperial Fresno*, 14; "California Raisin-Making," *PRP* 45 (June 10, 1893): 505. Californians constantly complained that their labor costs were routinely as much as 50 percent higher than in the East or the Midwest. Not until after 1900 did they make the connection between their state's immigration and exclusion policies and the high cost and scarcity of agricultural labor. By that time, it was politically unfeasible to agitate for Chinese reentry. "[The fruit growers] have been deprived of the most skillful, devoted and trustworthy laborers they ever had, the Chinese. They have had furnished them a large supply of careless, preoccupied, and untrustworthy laborers,

the Japanese. The fruit growers cannot understand what interest of the country is served by forcing them to struggle along with an inferior Oriental; if they must have an Oriental, why not the best for their purposes?" The fruit growers' resolution to this effect at the 1906 annual meeting was branded as "unAmerican and disloyal" ("The Week," *PRP* 72 [Dec. 15, 1906]: 370).

32. *Report of the State Viticultural Commission* (Sacramento: State Superintendent of Printing, 1892), quoted in Fox, "Cooperation in the Raisin Industry," 4; Kearney to Aldrich, June 1, 1897, File 1887-97, Box 1, Kearney Papers, Bancroft Library; M. B. Lerwick, *Fresno County, California* (San Francisco: Sunset Magazine Homeseekers Bureau for the Fresno County Board of Supervisors, n.d. but probably 1913), 21, Bancroft Library; Cleland, *Cattle on a Thousand Hills*, 176.

33. Adams, "Seasonal Labor Requirements for California Crops."

34. The Chinese began to work in the agricultural fields in the 1880s, after they finished building the transcontinental railroad. Exclusion laws reduced their numbers thereafter, and the Japanese replaced them, but only for a time. The Japanese moved quickly from being workers to landowners, an accomplishment that was rewarded by the passage of the Alien Land Laws, which barred Asian landowners from passing their property to their heirs. After the turn of the century, Asians as a class faced stiff discrimination in California; although the rate of immigration slowed almost to zero, they continued to occupy an important place in horticulture, not just as workers but also as growers. In 1912 the *PRP* reported on the formation of the Chinese Agricultural Association of Stockton, an association for Asian fruit growers. The next year, it reported on a method the Japanese were using to evade the restrictions of the Alien Land Laws: forming corporations that purchased their lands; the stock could be transferred to their children (Edward J. Wickson, "The Chinese Take the Hint," *PRP* 84 [Nov. 23, 1912]: 420; "Japanese Farming Companies," *PRP* 86 [Aug. 30, 1913]: 210). White Californians next turned to Mexicans for their seasonal labor needs, a solution that kept migrant Mexicans from becoming permanent residents (Daniel, *Bitter Harvest*; Chan, *This Bittersweet Soil*).

For other studies of the lives and struggles of migrant laborers in California from their own perspective, see Gonzalez, *Labor and Community*; Guerin-Gonzalez, *Mexican Workers and American Dreams*; Iwata, *Planted in Good Soil*; Majka and Majka, *Farm Workers, Agribusiness, and the State*; and Weber, *Dark Sweat, White Gold*. On cannery workers, see Ruiz, *Cannery Women/Cannery Lives*; and Zavella, *Women's Work and Chicano Families*.

35. *Imperial Fresno*, 9; Thickens, "Pioneer Agricultural Colonies," 170-76. The *Oroville Register* recommended "small orchards and small farms with a family on each" to minimize the need for Chinese and Japanese migrant workers and to maintain the white populations in Sacramento Valley towns (quoted in "Gleanings," *PRP* 49 [May 18, 1895]: 328).

36. Thus African Americans and Armenians prospered because they were able to buy land, either on credit or with their own capital. Other institutions—such as the federal Chinese exclusion laws and California's Alien Land Law of 1911—effectively disfranchised Asians economically. On Chinese exclusion, see Salyer, *Laws Harsh as Tigers*, 17-20, 22-23; on the Alien Land Laws, see Higgs, "Landless by Law"; and Ichioka, "Japanese Immigrant Response to the 1920 California Alien Land Law." On ethnicity in the West, see Taylor, "Immigrant Groups in Western Culture."

Examples of the racism against Asian landowners filled the pages of the *PRP* during the 1890s: "Chinese who leased orchards in a wholesale way in the Chico and Fresno districts have fared badly this season. . . . On the whole it is not to be regretted that they have not made it pay, for now it is to be hoped they will be disposed to keep out of the business" ("Gleanings," *PRP* 50 [Nov. 2, 1895]: 276).

37. Mahakian, "History of Armenians," 70. For another historical treatment of ethnic prejudice in California, see Gregory, *American Exodus*.

Several scholars have used the "hidden minority" characterization to describe Armenian Americans. See, e.g., Rollins, *Hidden Minorities*; and Bakalian, *Armenian-Americans*, 4. Interestingly, the U.S. census never classified Armenians in a separate racial category; they were simply folded into those counted as having been born in Turkey.

Not much has been written about the cultural anthropology of Armenians in the United States, but a few recent studies deepen what we know about the Armenian immigrant experience. In addition to the above sources, see Mirak, *Torn between Two Lands*; Okoomian and Schneyer, "Borders of Asian-American Identity"; and Papazian, "Struggle for Personal and Collective Identity." Older valuable sources include Davidian, *Seropians*; and Meyer, "Development of the Raisin Industry" (both give contemporary observations on Armenians in Fresno). LaPiere, "Armenian Colony in Fresno County," contains statistics and interviews documenting white antipathy toward Armenians and the extent of Armenians' presence in farming and the professions in the Central Valley. See also Moradian, "Armenians," 139; Davidian, *Seropians*, 3–14; and Wallis, *Fresno Armenians*, 35–74.

38. Mirak, *Torn between Two Lands*, 22–31, 48–59; Mahakian, "History of Armenians," 8, 10. The Armenian genocide is a contested event in Turkish-Armenian history. The Turkish government contends that no massacre took place. Although an examination of the debate is beyond the scope of this study, I stand with those who believe the genocide occurred. What is relevant to my account is the impact of the genocide on Armenians in America. Armenians' sense of themselves as a people was undoubtedly shaped by their experiences of oppression and murder in Turkey; they responded to discrimination and oppression in the United States by adhering to their culture and religion. I am indebted to Janice Okoomian for pointing out to me the importance of the genocide for scholars and Armenians.

39. In 1909, a federal circuit court ruled that four Armenians—three of whom had been classified as Asiatic—were entitled to apply for naturalization (*In re Halladjian*, 174 F. 834 [C.C.D. Mass. 1909]). Fourteen years later, ignoring stricter laws passed in 1920 and 1924, a federal district court again ruled that Armenians were not Asians (*U.S. v. Cartozian*, 6 F.2d 919 [D.Ore. 1925]). See Lothyan, "A Question of Citizenship"; Okoomian and Schneyer, "Borders of Asian-American Identity"; Bakalian, *Armenian-Americans*, 20; Mirak, *Torn between Two Lands*, 35–44. On the turn of immigration law toward a more exclusionist policy, especially regarding Asians, see Salyer, *Laws Harsh as Tigers*, 121–38.

40. California Commission of Immigration and Housing, *Fresno's Immigration Problem* (Sacramento: State Printer, 1918), 19, cited in Mirak, *Torn between Two Lands*, 113. "Only 22.8 per cent of the 53,004 Armenians who were admitted to the United States with a designated occupation, from the years between 1896 to 1929, were farmers or farm laborers" (Mahakian, "History of Armenians," 20, n. 5). Mirak claims that three thousand Armenians and four thousand Danes lived in Fresno County in 1908. His figures are estimated, however, and he gives no explanation of the method used to arrive at those estimates. As I show below, I think his numbers are high, but probably not significantly so. What matters, as I argue, is not the absolute number of Armenians but the local perception of them as insular and unassimilated and posing an economic threat disproportionate to their numbers.

41. Davidian, *Seropians*, 7; Mahakian, "History of Armenians," 8, 23; Mirak, *Torn between Two Lands*, 113–15. Packinghouse and fieldwork were dirty and difficult. Picking grapes was "stoop work" and was eschewed by European and white immigrants. Most white growers preferred to hire other whites for this work because of "their alleged versatility and skill over the Armenians and Asians" (Mirak, 115), but after about 1885 this became increasingly difficult, and these jobs were reserved for Asians and Mexicans. White men, however, continued to perform the more skilled jobs of "cultivating, hauling, and caring for machinery" (Meyer, "Development of the Raisin Industry," 54).

For evidence of Armenians' capital in hand upon arrival, Mirak relies on U.S. Immigration Commission, *Reports*, Part 25: *Japanese and Other Immigrant Races in the Pacific Coast and Rocky Mountain States*, 573–83, 634, 984–1003. The Immigration Commission selected seventeen Armenian farmers to study and found that they brought an average of $2,781 with them, almost twice as much as the Danes, ranked next at $1,405. Armenians also bought on mortgage far more than other immigrants; the seventeen farmers carried an average debt of $5,702. Japanese farmers were second at $2,982.

For a description of Armenians' success in other agricultural enterprises and in urban professions, see Mirak, *Torn between Two Lands*, 119–22; and LaPiere, "Armenian Colony in Fresno County," 171–87.

42. Meyer, "Development of the Raisin Industry," 60. Mahakian says there were 10,112 Armenians in California in 1920. If his number is correct, then 80 percent of all Armenians in the state lived in Fresno County at the time ("History of Armenians," 10).

43. U.S. Immigration Commission, *Report*, 24:626–27, 634–35 (cited in Mirak, *Torn between Two Lands*, 119); LaPiere, "Armenian Colony in Fresno," 148; Mirak, *Torn between Two Lands*, 282–83; Bakalian, *Armenian-Americans*, 20. On June 16, 1903, the *Fresno Evening Democrat* reported that Armenians produced 25 percent of the raisin crop (File 4, Box 5, Kearney Papers, Bancroft Library). During the 1920s, large numbers of Armenians quit farming and moved to the cities, primarily Los Angeles, because of the depressed prices for raisins after World War I (Mahakian, "History of Armenians," 28).

44. Mahakian, "History of Armenians," 42–43; Bakalian, *Armenian-Americans*, 20; LaPiere, "Armenian Colony in Fresno," 390–415; personal interview, Leo Chooljian, Sanger, California, July 20, 1990. Not surprisingly, the churches were the primary institution retaining the Armenian language in the twentieth century, and the Armenians were not the only group to follow this pattern. Armenians today do not want to complain publicly about their historical experience with residential segregation, in part because they now own homes in the nicest areas of Fresno.

45. William Saroyan, "The Armenian and the American," *Ararat* 25 (Spring 1984): 7, quoted in Bakalian, *Armenian-Americans*, 4; *Fresno Asbarez* quoted in Mirak, *Torn between Two Lands*, 122.

46. Bakalian, *Armenian-Americans*, 20; Personal interview, Dick Markarian, Fowler, California, July 19, 1990.

47. I allude here to the logic of *Plessy v. Ferguson*, 163 U.S. 537 (1896). On racial separation, see, e.g., Kluger, *Simple Justice*; Karst, *Belonging to America*; on the racist denial of access to resources, see Leibhardt, "Allotment Policy in an Incongruous Legal System"; and McEvoy, *Fisherman's Problem*, 41–92.

48. The standard reference is Olson, *Logic of Collective Action*.

Chapter 3

1. Nourse, *Legal Status of Agricultural Co-operation*, 37–38, 43; Hurst, *Legitimacy of the Business Corporation*; Buck, *Granger Movement*, 237–78; McKay, *Federal Research and Educational Work*, 7–10.

2. Nourse, *Legal Status of Agricultural Co-operation*, 39–43; Filley, *Cooperation in Agriculture*, 87; Hanna, *Law of Cooperative Marketing Associations*, 26–39. Aside from sharing the Rochdale fundamental principle of mutual benefit, the laws bore scant resemblance to each other. States singled out different aspects of the Rochdale principles and regulated cooperatives accordingly. For example, the Wisconsin law provided that transactions with nonmembers be conducted in cash; Kansas stipulated only that each member have one vote; the Pennsylvania legislature detailed requirements for voting, patronage dividends, and stock ownership.

3. Meyer, "Law of Co-operative Marketing," 85; Nourse, *Legal Status of Agricultural Co-operation*, 51, 48 n. 27, 53. In regular corporations, capital stock denoted the amount of capital that the organization kept in reserve as collateral; any excess constituted surplus and could be distributed to shareholders. See Elliott, *Law of Corporations*, 328–29; Berle and Means, *Modern Corporation and Private Property*.

4. Nourse, *Legal Status of Agricultural Co-operation*, 54; Cross, "Cooperation in Agriculture." The bill's other influential supporters included David Lubin, founder of the International Institute of Agriculture and an enthusiastic proponent of national marketing systems in Europe and the United States (Nourse, *Legal Status of Agricultural Co-operation*, 94–95).

5. *Cal. Stats.* (1895), chap. 183, p. 222; *Cal. Stats.* (1909), chap. 26, p. 16. The latter statute also gave associations the power to determine policies on voting and shareholding.

6. The nonstock laws were Nevada Laws (1901), chap. 60; Washington Laws (1907), 255; Alabama Act of Special Session (1909), 168; Florida Laws (1909), chap. 5598; Oregon Laws (1909), chap. 190; Idaho Laws (1913), chap. 54; New Mexico Session Laws (1915), chap. 64.

The capital stock laws include Wisconsin Laws (1911), chap. 368; followed by Michigan Public Acts (1913), Act No. 398; South Dakota Laws (1913), chap. 145; Washington Session Laws (1913), chap. 19; Massachusetts Laws (1913), chap. 447; New York Laws (1913), chap. 454; Virginia Acts (1914), chap. 329; Iowa Laws (1915), chap. 145; North Carolina Public Laws (1915), chap. 144; South Carolina Laws (1915), No. 152; Oregon Laws (1915), chap. 226; Rhode Island Acts (1916), chap. 1400; Oklahoma Session Laws (1919), chap. 147. The Nebraska group included Nebraska Laws (1911), chap. 32; followed by Indiana Laws (1913), chap. 164; Colorado Laws (1913), chap. 62; North Dakota Laws (1915), chap. 92; Florida Laws (1917), chap. 7384.

The Wisconsin law limited individual stockholding to $1,000 and permitted no more than one vote per shareholder; it also set the payment of dividends at no more than 6 percent and specified that excess revenues be used for spreading awareness of cooperative ideals. The Nebraska law, passed at the behest of the Nebraska Farmers Cooperative Grain and Livestock Association, defined a cooperative as any enterprise that distributed part or all of its earnings on a patronage basis. It also allowed associations to regulate the ownership and transfer of stock. See Hanna, *Law of Cooperative Marketing Associations*, 31–32, 33 n. 17, 35–38, 113; Nourse, *Legal Status of Agricultural Co-operation*, 45–48; Filley, *Cooperation in Agriculture*, 86–88; Keller, "Pluralist State," 81.

7. These included Oregon (*Laws* 1915, chap. 225), Florida (*Laws* 1909, chap. 5598), Wash-

ington (*Session Laws* 1913, chap. 19), and California (*Cal. Stats.*, 1909, chap. 26). In these states farmers saw to it that the law supplied them with both forms of organization. See Hanna, *Law of Cooperative Marketing Associations*, 33 n. 17, 36 n. 19. Nourse observes that throughout the Granger and Alliance eras farmers relied on existing laws that "were mainly the work of consumer or workman co-operative groups"; laws providing for the incorporation of nonstock associations began to appear well after 1900 (*Legal Status of Agricultural Co-operation*, 43, 59–72).

8. See Thorelli, *Federal Antitrust Policy*; Letwin, *Law and Economic Policy*; McCurdy, "*Knight* Sugar Decision"; Sklar, *Corporate Reconstruction of American Capitalism*, 105–17.

9. On the classical liberal tradition in American law and economic thought, the leading studies are Handlin and Handlin, *Commonwealth*; Hartz, *Economic Policy and Democratic Thought*; and Levy, *Law of the Commonwealth and Chief Justice Shaw*. A useful synthesis is Scheiber, "Government and the Economy." More recently, legal historians intent on recovering the doctrinal origins of late nineteenth-century economic liberalism have repudiated the New Deal generation of scholars represented by the Handlins and Hartz, returning to the myths that generation laid aside. See Hovenkamp, *Enterprise and American Law*, which argues that all nineteenth-century American law was derived from contemporary economic theory, which itself stemmed from "an unreconstructed Adam Smith" (69). According to Horwitz, *Transformation of American Law, 1870–1960*, late nineteenth-century law served as the bastion of conservative protection of property and blocked the redistribution of wealth by the state. An imaginative rebuttal of the revisionism is Novak, "Public Economy and the Well-Ordered Market," 3–10. Federal antitrust law as applied to cooperatives is treated in Chapter 5.

10. For example, the sale of the goodwill associated with a business could be legally protected with an agreement by the seller not to compete against the buyer, although the courts usually imposed time and spatial limitations on covenants not to compete. See U.S. Bureau of Corporations, *Trust Laws and Unfair Competition*, 33–34, 423–27 (state common law cases); 157, 204–5 (state antitrust statutory provisions). But agreements not to compete, made solely to benefit one person or firm or to encourage monopoly, were prohibited by law in several states (ibid., 157 nn. 3–4).

11. An agreement by a seller to refrain from competing against his buyer for a specific period of time, for example, was not invalid at common law, but courts routinely refused to enforce cartel arrangements that attempted to dominate either production of a specialized commodity over a large geographic area or sale of a staple commodity in a locality because both produced anticompetitive effects. See *Central Shade-Roller Co. v. Cushman*, 143 Mass. 353 (1887) (use of a manufacturer's corporation to handle all sales upheld); *Gloucester Isinglass & Clue Co. v. Russia Cement Co.*, 154 Mass. 92 (1891) (agreement among glue manufacturers held valid because glue was not an "article of prime necessity") (*Gloucester*, 94); *Over v. Bryam Foundry Co.*, 37 Ind. App. 452 (1906) (contract for purchase of entire output of firm not void because it did not cover entire supply of commodity); *Jones v. Fell*, 5 Fla. 510 (1854) (pilots' association not per se illegal); *People v. New York Board of Fire Underwriters*, 54 How. Prac. 228 (1875) (association formed to set uniform rates not in restraint of trade). Cases in which the courts identified illegal anticompetitive practices include *Central Ohio Salt Company v. Guthrie*, 35 Ohio St. 666 (1880) (pooling agreement to regulate salt prices and production struck as tending to monopolize); *More v. Bennett*, 140 Ill. 69 (1892) (agreement of association of stenographers not to compete against each other not enforce-

able); *Santa Clara Valley Mill & Lumber Co. v. Hayes et al.,* 76 Cal. 387 (1888) (combination of lumber manufacturers to control prices and limit supply void as against public policy); *Richardson v. Buhl,* 77 Mich. 632 (1889) (Diamond Match Company monopoly struck as trust arrangement in restraint of trade), among others. See Sklar, *Corporate Reconstruction of American Capitalism,* 95–105; and U.S. Bureau of Corporations, *Trust Laws and Unfair Competition,* 36–57.

12. Sklar, *Corporate Reconstruction of American Capitalism,* 105. One critique of Sklar's argument is Carstensen, "Dubious Dichotomies and Blurred Vistas." Carstensen contends that Sklar's analysis flows from an oversimplified approach to legal inconsistency and conflict and that he ignores cases not dealing with corporations as irrelevant even when they helped to shape judicial conceptions of the legality of private control over markets. Sklar may have drawn on a narrow legal history, but his assessment of contemporary judicial and political norms on competition provides the benchmark for my argument about cooperatives and their impact on market competition.

13. U.S. Bureau of Corporations, *Trust Laws and Unfair Competition,* 143–230. For compilations of the state statutes and cases decided thereunder, see U.S. Industrial Commission, *Trusts and Industrial Combinations*; and U.S. Congress, House, Committee on the Judiciary, *Laws on Trusts and Monopolies.* The political and economic origins of state antitrust laws and their effect on business and consolidation deserve more scholarly attention than they have received.

14. Michigan (1889), North Carolina (1889), Mississippi (1890), Louisiana (1890), Wisconsin (1893), Texas (1895), Illinois (1893), Montana (1895), Georgia (1896), Indiana (1897), and Tennessee (1897); U.S. Industrial Commission, *Trusts and Industrial Combinations,* 127–29, 192–96, 141–42, 118–20, 254–56, 230–35, 87–89, 162–63, 79–80, 95–96, 226–27. Illinois, Louisiana, Mississippi, Montana, Texas, and Wisconsin also included laborers and labor unions in the exemption; South Dakota and Nebraska exempted labor but not agriculture. Alabama exempted railroads when the combination was approved by the state railroad commission. Texas and Mississippi codified common law rule that contracts for the sale of goodwill with a business that contained agreements not to compete with the buyer were not void; Missouri exempted insurance combinations in large cities; and North Carolina generously dispensed immunities to charitable institutions, lumber, cotton, wool, fishing, trucking, canning, and merchants "not . . . interested in a trust" (ibid., 195). By 1900, twenty-eight state codes included antitrust statutes; four others had constitutional provisions.

15. Ostler, *Prairie Populism.*

16. Goodwyn, *Democratic Promise*; see also Ayers, *Promise of the New South,* 216–22. Texas and Georgia were the most important sites of Populist political activism, but the exemptions in Louisiana, Mississippi, and North Carolina should not be understood as manifestations of Populist influence. For the argument that southern populism was reactionary, see McMath, *American Populism.*

17. Hanna, "Cooperative Associations and the Public," 163.

18. Act of June 20, 1903, Laws of Illinois (1893), 192, compiled in U.S. Industrial Commission, *Report on Trusts and Industrial Combinations,* 87; *Ford v. Chicago Milk Shippers' Association,* 155 Ill. 166, 178 (1895).

19. *Ford,* 177. Early twentieth-century treatise writers asserted that the case triggered judicial hostility toward agricultural combinations. Edwin G. Nourse and John Hanna

blamed the decision for ruling that farm organizations restrained trade when they attempted to obtain fair prices for producers (Nourse, *Legal Status of Agricultural Co-operation*, 333–34; Hanna, *Law of Cooperative Marketing Associations*, 253; see also Evans and Stokdyk, *Law of Agricultural Co-operative Marketing*, 16, 90, 95). Their analysis, however, removed the case from its larger and more pertinent context: judicial interpretation of restraint of trade under the antitrust statutes. This case was the first to deal with cooperatives' antitrust liability. The parties focused the court's attention on the competitive effects of the exclusive dealing provision, not the legal significance of the organization's cooperative purpose. For a trenchant synthesis of the scholarly debate on late nineteenth-century agricultural distress and protest, see Bruchey, *Enterprise*, 294–307. Bruchey sees a pattern in the late eighteenth and late nineteenth centuries; in both cases, agrarian protests resulted in laws relieving them of various kinds of debts, to which the Supreme Court responded by interpreting the contract clause and the Fourteenth Amendment to defend the property rights of creditors (305).

20. See *People of the State of New York v. The Milk Exchange*, 145 N.Y. Rep. 267 (1895).

21. On the history of "liberty of contract" in nineteenth-century law, see Mensch, "History of Mainstream Legal Thought"; Montgomery, *Fall of the House of Labor*, 46–48; Tomlins, *State and the Unions*, 10–16. Representative judicial expressions can be found in *Ritchie v. People*, 155 Ill. 98 (1895); *In re: Debs*, 158 U.S. 564 (1895); *Holden v. Hardy*, 169 U.S. 366 (1898); *Commonwealth v. Hamilton Manufacturing Co.*, 120 Mass. 383 (1876). On the political and legal tensions generated by labor unrest in Illinois, see Brody, *Steelworkers in America*; Paul, *Conservative Crisis and the Rule of Law*.

22. Except for a notable victory by the Chinese, the Supreme Court recognized no equal protection claims by minorities during the nineteenth century. For cases denying equal protection claims by African Americans, see the *Civil Rights Cases*, 109 U.S. 3 (1883); by women, *Bradwell v. Illinois*, 83 U.S. 130 (1873), and *Minor v. Happersett*, 88 U.S. 162 (1875). The first constitutional case involving the equal protection clause arose not in the areas of racial or gender equality but in economic regulation. New Orleans butchers attacked a state-franchised slaughterhouse as a denial of their right to pursue their trade, but the U.S. Supreme Court refused to find such a right within the equal protection clause (*The Slaughterhouse Cases*, 16 Wall. 36 [1873]). The Chinese victory in *Yick Wo v. Hopkins*, 118 U.S. 356 (1886), followed a significant defeat in *Barbier v. Connolly*, 113 U.S. 27 (1885). In *Barbier*, the Court upheld a San Francisco ordinance requiring all laundries to be housed in brick buildings. Although the law was aimed at Chinese laundries, which were typically housed in wooden buildings, Justice Stephen Field upheld it as a valid health and safety regulation that applied to everyone in the business regardless of race. In *Yick Wo*, the Court struck a municipal ordinance regulating laundries housed in wooden buildings on the grounds that its selective enforcement against Chinese laundry owners denied them the right to equal protection. Together, the decisions held that only legislation that discriminated on its face and whose actual enforcement was directed only against a certain class of persons would be invalidated.

23. *In re Grice*, 79 F. 627 (C.C.S.D. Tex. 1897), 648, 649. The opinion does not identify Grice's occupation but notes that he and his co-defendants were engaged in an "ordinary business partnership" operating in interstate commerce (643). The statute was also overturned on technical grounds. It outlawed all combinations regardless of size or scope; its

failure to "condemn only acts which are oppressive by reason of their magnitude" obscured the distinction between reasonable and unreasonable restraints of trade that was integral to antitrust law adjudication during the 1890s (642).

24. Cited in note 22, above.

25. *Connolly*, 184 U.S. 540 (1902); for the Illinois Antitrust Act of 1893, see note 18, above. On late nineteenth-century substantive due process, see Gillman, *Constitution Besieged*, 45–60.

26. *Connolly*, 560, quoting *Gulf, Colorado and Santa Fe Rwy v. Ellis*, 165 U.S. 150, 159 (1897). *Gulf* invalidated a Texas statute that required railroad companies to pay attorney's fees to livestock owners in successful suits for recovery of the value of animals killed by trains on the grounds that the statute awarded the railroads no similar benefit if they prevailed in such suits. *Gulf*'s contribution to equal protection jurisprudence was the rule that corporations, as legal persons, had rights to equal protection. The danger of railroad operations did not entitle the state to spurn less burdensome alternatives. Cases like *Gulf* indicated that state legislatures had to tread carefully in drawing separate classifications when regulations interfered with market transactions. They stemmed from both principled differences among judges in applying the presumption of constitutional validity to state economic regulations (see McCurdy, "Justice Field and the Jurisprudence of Government-Business Relations") and, as the cases involving agricultural producers and organizations show, differences between judges and legislators in their perceptions of entrepreneurs and their relationships to one another in the market.

27. *Connolly*, 563–64. To illustrate his holding that the classification was arbitrary, Harlan cited *American Sugar Refining Co. v. Louisiana*, 179 U.S. 89 (1900), in which the justices brushed aside equal protection objections to a tax the state levied on persons in the sugar refining business; the tax did not apply to farmers grinding their own sugar. The Court found that the state intended to tax the commercial business of sugar refining, not the incidental use of refined sugar by farmers on their own farms. The tax on sugar refiners did not "impair or destroy rights that are given or secured by the supreme law of the land." The Illinois statute, in contrast, sought to "regulate the enjoyment of rights and the pursuit of callings connected with domestic trade" in a way that promoted the interests of one class of citizens at the expense of another: "It is one thing to exert the power of taxation so as to meet the expenses of government, and at the same time, indirectly, to build up or protect particular interests or industries. It is quite a different thing for the State, under its general police power, to enter the domain of trade or commerce, and discriminate against some by declaring that particular classes within its jurisdiction shall be exempt from the operation of a general statute making it criminal to do certain things connected with domestic trade or commerce" (*Connolly*, 563). The *American Sugar* decision set the power to tax apart from other regulations for purposes of equal protection claims, but the door it opened for treating farmers as a distinct class slammed shut in *Connolly*.

28. *Brown v. Jacobs' Pharmacy Company*, 115 Ga. 429 (1902).

29. *Connolly*, 571. Whereas Harlan had carefully distinguished *American Sugar*, McKenna saw them as identical; in both cases, he said, the states had intended to protect the "growers" of products from statutory burdens placed on "traders" of them. "It would be strange, indeed, if the power of a State is limited and confined by the Constitution of the United States . . . [so] that in one case it may see a difference between manufacturers and planters, and in the other case may not see a difference between traders in commodities

acquired for the purposes of sale and such property when held by farmers by whose labor they were produced" (*Connolly*, 568).

30. I have found only one exception to this general rule. In 1903, the Missouri Supreme Court held that a statutory exemption for insurance companies did not condemn the antitrust law under either the state or federal constitution, under the rationale that insurance was a public utility that the state could regulate (*State ex rel. Crow v. Continental Tobacco Co.*, 177 Mo. 1 [1903]). By 1913 the Missouri legislature had repealed the exemption. See Williams, *Laws on Trusts and Monopolies*, 171–87 (citing Missouri Revised Statutes 1913, chap. 98).

31. *Cal. Stats.* (1907), chap. 530, p. 984; *Cal. Stats.* (1909), chap. 362, pp. 593, 594. On contemporary legal comment, see Nourse, *Legal Status of Agricultural Co-operation*, 362; Arndt, "Law of California Co-operative Marketing Associations," 284–85; and Meyer, "Law of Cooperative Marketing," 95. The California appellate courts upheld the laws in *Anaheim Citrus Fruit Assn. v. Yeoman*, 51 Cal. App. 759 (1921); and *Poultry Producers of Southern California, Inc. v. Barlow*, 189 Cal. 278 (1922). On cooperatives and monopolies in California, see Arndt, "Law of California Co-operative Marketing Associations," 287–88; Plehn, "State Market Commission of California," 8; and Steen, *Cooperative Marketing*, 6–7, 281–82.

32. *Colo. Stats.* (1913), chap. 161, sec. 2, quoted in Williams, *Laws on Trusts and Monopolies*, 64. No reported decision of the Colorado courts construed the state's antitrust statute explicitly to exempt agricultural associations; however, in 1922 the state supreme court refused to overturn the statute just because it exempted labor (*Johnson v. the People*, 72 Colo. 218, 220 [1922]).

33. Nourse, *Legal Status of Agricultural Co-operation*, 159; *Laws of Kentucky*, chap. 117 (1906), in Williams, *Laws on Trusts and Monopolies*, 128.

34. *Owen County Burley Tobacco Society v. Brumback*, 128 Ky. 137 (1908). Because the act did not discriminate on its face and the defendant fell within the classification, the court could duck the equal protection issue: "What effect, if any, the privilege of combination conferred on farmers, to the end that a greater price might be obtained for the product raised by them, would have on other classes of persons if the act in its terms excluded other persons from its operation and effect, or if other persons were complaining, is not now before [the court]" (146–47). The Kentucky Supreme Court distinguished *Connolly* on the grounds that the tobacco industry was too essential to the public interest for the state not to act to protect legitimate interests; see *Commonwealth v. Hodges et al.*, 137 Ky. 233 (1910), and *Commonwealth v. Malone*, 141 Ky. 441 (1911). The only meaningful curb the court placed on the Burley Society was to condemn night riding. The court upheld the acquittal of a farmer indicted for selling pooled tobacco to outside dealers because he had joined only under duress (*Commonwealth v. Reffitt*, 149 Ky. 300 [1912]). On the night riders, see Kroll, *Riders in the Night*; Grantham, "Black Patch War"; Waldrep, *Night Riders*; Campbell, *Politics of Despair*; and Hall, "Breaking Trust."

35. *International Harvester Co. v. Kentucky*, 234 U.S. 216 (1914). In the view of Justice Oliver Wendell Holmes, the state had led farmers into a blind alley. Kentucky authorized tobacco pools to obtain the "real value" of their crops "under fair competition, and under normal market conditions" (*International Harvester Co. of America v. Commonwealth*, 147 Ky. 795, 800 [1912]); *International Harvester Co. of America v. Commonwealth*, 137 Ky. 668, 677 [1910]). But to avail themselves of the privilege, farmers had to know what prices they would have obtained under "normal market conditions." This, Holmes asserted, was

simply absurd: "To compel them to guess on peril of indictment what the community would have given for them if the continually changing conditions were other than they are, to an uncertain extent; to divine prophetically what the reaction of only partially determinate facts would be upon the imaginations and desires of purchasers, is to exact gifts that mankind does not possess" (*International Harvester Co. v. Kentucky*, 223–24).

36. Elsworth, "Statistics of Farmers' Selling and Buying Associations," 41; McKay, *Federal Research and Educational Work*, 14–15; Nourse, *Legal Status of Agricultural Co-operation*, 337.

37. *Reeves v. Decorah Farmers' Cooperative Society*, 160 Iowa 194, 196 (1913). The society never required that it make a profit before it would pay dividends. In two years' time it conducted transactions worth $433,639 but had just $29 in its coffers (Nourse, *Legal Status of Agricultural Co-operation*, 337). The plaintiff was identified by Jensen, "Integrating Economic and Legal Thought," 893.

38. *Reeves*, 198–200.

39. Following legal convention, the court determined the society's legal status by evaluating its commercial practices. Purchasing hogs outright, as the society did, was a policy of corporate firms; the court believed cooperatives should act only as selling agents. Here as elsewhere there was significant overlap between cooperatives and corporations. Rochdale cooperatives often bought "outright from their members" and sold on their own accounts; the California horticultural associations operated on an agency basis but not because state law required them to. According to Nourse, agency operations suited organizations dealing in perishable products or in commodities sold on consignment. Where long-term storage and financing were required, cooperatives socialized the attendant risks by buying outright and in effect guaranteeing members a specific rate of return (*Legal Status of Agricultural Co-operation*, 84–85, 339).

40. *Reeves*, 205. A year later, the Iowa Supreme Court emphatically reaffirmed this decision in *Ludowese v. Farmers' Mutual Cooperative Co.*, 164 Iowa 197 (1914). There the cooperative defended the maintenance clause as liquidated damages rather than a forfeiture, but the court held that *Reeves* was controlling.

41. Jensen, "Integrating Economic and Legal Thought," 893.

42. 208 U.S. 161 (1908) and 236 U.S. 1 (1915). See also Mensch, "History of Mainstream Legal Thought," 18–21.

43. McCurdy, "Roots of 'Liberty of Contract' Reconsidered." See also sources on labor cited in note 21, above.

44. *Georgia Fruit Exchange v. Turnipseed*, 9 Ala. App. 123 (1913); *Burns v. Wray Farmers' Grain Co.*, 65 Colo. 425 (1918). In *Turnipseed*, the legal nature of the cooperative was again determined by corporation law; the cooperative's antitrust liability was confirmed by the fact that the membership agreement was conditioned on the subscription of 60 percent of the 1909 peach crop, thus giving the exchange a "commanding position in fixing the price of peaches in this section of the country" (140).

45. "Cooperative Cooperation," *CG* 78 (May 3, 1913): 708; Guy C. Smith, "How to Get Together: Seventeen States Have Statutes to Aid Cooperative Organizations," *CG* 79 (Sept. 26, 1914): 1595.

46. See, e.g., Hays, *Conservation and the Gospel of Efficiency*.

## Chapter 4

1. The *Pacific Rural Press* reported in 1897 that land in the San Joaquin Valley was available for $20 an acre, compared to the 1887 price of $100; improved fruit land commanded a mere $40, down from $250 in 1890. See "The Field: Present Prices of Agricultural Lands in California," *PRP* 54 (July 17, 1897): 39; see also Winchell, *History of Fresno County*, 150–51; and Walker, *Fresno County Blue Book*, 102–4.

2. "The Field," 39; California State Board of Horticulture, *Annual Report*, 135, 361, quoted in Crouse, "California Raisin Industry," 75; Adams, *Modern Farmer*, 465; "A Home Market vs. the Consignment System," *PRP* 51 (Feb. 1, 1896): 66.

For descriptions of the impact of the depression of 1891–94 on the raisin industry, see California State Department of Agriculture, *Statistical Report of State Statistician for 1912*, 136–37; Fox, "Cooperation in the Raisin Industry," 22, 25–26; Rehart and Patterson, *M. Theo Kearney*, 26; Lloyd, "Cooperative and Other Methods of Marketing," 41; Bragg, "History of Cooperative Marketing," 24; Meyer, "Development of the Raisin Industry," 80; Cross, "Cooperation in Agriculture"; "The Raisin Limit," *PRP* 54 (July 31, 1897): 67; "Note and Comments," *PRP* 54 (Oct. 9, 1897): 227; "The Raisin Problem," *PRP* 45 (Feb. 4, 1893): 94; "Gleanings," *PRP* 51 (Oct. 19, 1895): 243; "Fresno Raisin Industry," *PRP* 52 (Feb. 8, 1896): 82; Edward F. Adams, "The Issue Held to Be between the Californian and the Eastern Merchant," *PRP* 52 (Feb. 15, 1896): 100; "The Raisin Situation in the East," *PRP* 52 (Oct. 17, 1896): 242; and Spence, "Success after Twenty Years," *SMH* 3 (Nov. 1917): 5.

3. "Crush or Co-operate," *PRP* 50 (July 20, 1895): 34; Fox, "Cooperation in the Raisin Industry," 7.

4. *PRP* 43 (July 9, 1892): 18; Fox, "Cooperation in the Raisin Industry," 10–12; Spence, "Success after Twenty Years," *SMH* 3 (Nov. 1917): 5, 16; Fred K. Howard, "History of Raisin Marketing in California," unpublished Ms., n.d. (but 1922), Sun-Maid Archives; "The Raisin Growers at Fresno," *PRP* 45 (Feb. 11, 1893): 126; "The Week," *PRP* 45 (Apr. 1, 1893): 274; "The Week," *PRP* 45 (Apr. 15, 1893): 326. Growers' indebtedness to packers and banks crimped their ability to support a cooperative, and the commission system only made things worse (Spence, "Success after Twenty Years," *SMH* 3 [Nov. 1917]: 4).

5. Fox, "Cooperation in the Raisin Industry," 14–16, 19–24; Spence, "Success after Twenty Years," *SMH* 3 (Dec. 1917): 6; ibid., *SMH* 3 (Jan. 1918): 6; "The Week," *PRP* 45 (Apr. 15, 1893): 326; Adams, *Modern Farmer*, 462–63.

6. "The Situation at Fresno," *PRP* 51 (Mar. 28, 1896): 194.

7. Spence, "Success after Twenty Years," *SMH* 3 (Dec. 1917): 6; Meyer, "Development of the Raisin Industry," 79–81; Ely, "Introduction," 9; Fox, "Cooperation in the Raisin Industry," 24–27. A leading farm reporter for the *Sacramento Bee* and *Fresno Morning Republican*, W. Y. Spence, wrote a celebratory history of the CARC and its predecessors for the *Sun-Maid Herald* with the full backing of his editor, Chester Rowell (File 1, Box 3, Rowell Papers).

8. Rehart and Patterson, *M. Theo Kearney*, 3–21; Kearney to Board of Directors, California State Raisin Growers Association, June 13, 1893, File 1, Box 1, Kearney Papers, Bancroft Library; Walker, *Fresno Blue Book*, 153. Growers declined Kearney's offers to extend time on mortgages and pay their taxes and irrigation rates because the low crop prices made it impossible to recover the costs of production (M. Theo Kearney to Nelson W. Aldrich, U.S. Tariff Commission, June 1, 1897, File 1, Box 1, Kearney Papers, Bancroft Library; Fox, "Results of Cooperation," 30).

9. Payne, "Cooperation," 15; Spence, "Success after Twenty Years," *SMH* 3 (Feb. 1918): 8; "Raisin Growers' Plans," *PRP* 55 (May 21, 1898): 334; Fox, "Cooperation in the Raisin Industry," 31–33; Adams, *Modern Farmer*, 465–67. A similar corporation in the wine industry, Adams observed, "is an effective Trust so long as it controls the output" (ibid., 522).

10. Spence, "Success after Twenty Years," *SMH* 3 (Feb. 1918): 9; "Progress with the Raisin Growers," *PRP* 55 (May 28, 1898): 342; "Progress of the Raisin Pool," *PRP* 56 (Aug. 10, 1898): 117; "The Raisin Combine Solid," *PRP* (Oct. 8, 1898): 234; "Success of the Association," *PRP* (Nov. 5, 1898): 303; Kearney, "The California Raisin Industry," address to the California Fruit Growers' Convention, Nov. 29, 1898, *PRP* 56 (Dec. 10, 1898): 380; "A Significant Success," *PRP* 56 (Dec. 17, 1898): 394; Howard, "History of Raisin Marketing," 23–43; Meyer, "Development of the Raisin Industry," 82; Rehart and Patterson, *M. Theo Kearney*, 30; Adams, *Modern Farmer*, 467–68.

11. "A Significant Success," *PRP* 56 (Dec. 17, 1898): 394. Kearney recognized that newly arrived Italians, German-Russians, and Armenians held the key to the organization's success. He helped the immigrants to acquire land and used his formidable oratorical skills at mass meetings to keep people in line. Still, he had enemies, including Chester H. Rowell, the editor of the *Fresno Morning Republican*, and many of the town's leading bankers and businessmen. See Spence, "Success after Twenty Years," *SMH* 3 (Feb. 1918): 8; Rehart and Patterson, *M. Theo Kearney*, 33; Payne, "Cooperation," 15; Bragg, "History of Cooperative Marketing," 12; Fox, "Results of Cooperation," 31.

12. Howard, "History of Raisin Marketing," 5; Kearney, "Address before the State Fruit Growers' Convention: On the Organization of the Raisin Growers for the Sale of Their Raisins," San Jose, Dec. 14, 1899, File 5, Box 1, Kearney Papers, FHS; Walker, *Fresno Blue Book*, 104; Rehart and Patterson, *M. Theo Kearney*, 30. The other trustees were L. S. Chittenden, owner of a large vineyard in Kings County; Alfred Sayre, a Madera county grower; and full-time raisin growers W. S. Porter and Robert Boot. Kearney included no Armenian growers among the CRGA directors. See Spence, "Success after Twenty Years," *SMH* 3 (Feb. 1918): 8–9; Editorial, *PRP* 55 (June 11, 1898): 370.

13. Adams, *Modern Farmer*, 466; Meyer, "Development of the Raisin Industry," 81. The packers did not object when the CRGA acted to maintain prices by buying the crops of outside growers and selling substandard raisins to the wineries. See "The Raisin Combine Solid," *PRP* 56 (Oct. 8, 1898): 234; Spence, "Success after Twenty Years," *SMH* 3 (Feb. 1918): 9; Fox, "Cooperation in the Raisin Industry," 32–33.

14. Kearney, "The California Raisin Industry," *PRP* 56 (Dec. 19, 1898): 380; Kearney, "Address before the State Fruit Growers' Convention, Dec. 14, 1899," File 5, Box 1, Kearney Papers, FHS.

15. Adams, *Modern Farmer*, 462; Kearney, "The California Raisin Industry," *PRP* 56 (Dec. 19, 1898): 380; R. C. Allen, "Raisin Culture," *PRP* 57 (Jan. 28, 1899): 52; Spence, "Success after Twenty Years," *SMH* 3 (Feb. 1918): 9.

16. Kearney, "Address before the State Fruit Growers' Convention, Dec. 14, 1899," File 5, Box 1, Kearney Papers, FHS; Allen, "Raisin Culture," 52; Spence, "Success after Twenty Years," *SMH* 3 (Feb. 1918): 9–11. The 1899 contract with the packers reduced packing fees by 25 percent and sales commissions by 33 percent, and packers forfeited $4 per ton if they violated any of the terms. See "Fruit Marketing," *PRP* 57 (Jan. 7, 1899): 22; Fox, "Cooperation in the Raisin Industry," 33–34; Adams, *Modern Farmer*, 468.

17. "Fruit Marketing," *PRP* 57 (Jan. 7, 1899): 22; Spence, "Success after Twenty Years,"

SMH 3 (Feb. 1918): 11; Fox, "Cooperation in the Raisin Industry," 34; Smith, *Garden of the Sun*, 521–22; Articles of Incorporation and By-Laws of the California Raisin Growers' Association, Apr. 1, 1899, File 5, Box 1, Kearney Papers, FHS.

18. Fox, "Cooperation in the Raisin Industry," 36.

19. Ibid., 37; "Editorial," *PRP* 59 (Jan. 6, 1900): 2; Spence, "Success after Twenty Years," *SMH* 3 (Mar. 1918): 8; Kearney, *M. Theo Kearney's Fresno County, California*.

20. Kearney to growers, May 22, 1900, File 1, Box 3, Kearney Papers, Bancroft Library; California State Board of Horticulture, "Review of the Fruit Season," 38. See also Fox, "Cooperation in the Raisin Industry," 38–39; Payne, "Cooperation," 44; Rehart and Patterson, *M. Theo Kearney*, 33; and Spence, "Success after Twenty Years," *SMH* 3 (Mar. 1918): 8.

21. Fox, "Cooperation in the Raisin Industry," 38–39; "The Raisin Arrangement Proceeds," *PRP* 59 (May 19, 1900): 309; Spence, "Success after Twenty Years," *SMH* 3 (Mar. 1918): 8.

22. "Stand by the Association," *FMR*, Apr. 2, 1901, p. 1; California State Board of Horticulture, "Review of the Fruit Season," 38; Fox, "Cooperation in the Raisin Industry," 42–43; "Agricultural Review," *PRP* 61 (May 25, 1901): 321; Spence, "Success after Twenty Years," *SMH* 3 (Mar. 1918): 9. The *Republican*'s editor, Chester H. Rowell, a prominent California Progressive, supported cooperation but not Kearney's approach to it.

23. *San Francisco Call*, Apr. 9, 1901, Bancroft Library; "The Raisin Situation," *PRP* 62 (July 20, 1901): 36; *California Cured Fruit Association v. Ainsworth*, 134 Cal. 461 (1901). The court limited the cooperative's damages to 2 percent of the value of the crop withheld by the grower ($14) instead of the full value it sought ($700). The *PRP* reported that industry leaders dismissed rumors of Sherman Act proceedings in 1903 as the scheme of "importers, whose occupation is threatened by the constant advance of the California product" (Editorial, *PRP* 65 [Mar. 3, 1921]: 178).

24. "The Raisin Situation," *PRP* 62 (Aug. 31, 1901): 140; Fox, "Cooperation in the Raisin Industry," 44; "Reorganization of Raisin Association Completed," *FMR*, June 5, 1901, Clippings File 2, Box 5, Kearney Papers, Bancroft Library. I am grateful to Arthur Stinchcombe for suggesting the analogy.

25. "Reorganization of Raisin Association Completed," *FMR*, June 5, 1901, Clippings File 2, Box 5, Kearney Papers, Bancroft Library; "A New Proposition Made to Growers under the Lease," *Fresno Evening Democrat*, Oct. 28, 1901, File 1, Box 3, Kearney Papers, Bancroft Library. On New Jersey incorporation and tax law, see Grandy, *New Jersey and the Fiscal Origins of Modern American Corporation Law*; and McCurdy, "*Knight* Sugar Decision."

26. Kearney, circular letter to growers, Nov. 1, 1901, File 6, Box 1, Kearney Papers, FHS; Editorial, *PRP* 62 (Dec. 14, 1901): 370 (quoting Kearney's speech at 1901 Fruit Growers Convention); Kearney, "The Raisin Industry," *Report of 26th Fruit Growers Convention*, 1901, Box 18, Erdman Papers; remarks of Alexander Gordon, Proceedings of the 30th Fruit-Growers Convention, San Jose, Dec. 6–9, 1904, in California State Commissioner of Horticulture, *First Biennial Report*, 312.

27. Allison, Gordon, and Boot quoted in E. J. Wickson, "The Raisin Situation," *PRP* 62 (Aug. 31, 1901): 140.

28. Wickson, "The Raisin Situation," *PRP* 62 (Aug. 31, 1901): 140; Payne, "Cooperation," 15.

29. *Fresno Evening Democrat*, Mar. 28, 1902, File 6, Box 1, Kearney Papers, FHS; Editorial, *PRP* 62 (Dec. 14, 1901): 370; Editorial, *PRP* 63 (Jan. 4, 1902): 2; Editorial, *PRP* 63 (Jan. 11,

1902): 22; Editorial, *PRP* 63 (Apr. 12, 1902): 246; Spence, "Success after Twenty Years," *SMH* 3 (Mar. 1918): 9. Kearney's departure from the scene did not mean that the CRGA was immune to accusations of restraint of trade; the *PRP* reported in 1903 that rumors abounded of an imminent federal prosecution against the CRGA under the Sherman Act (Editorial, *PRP* 65 [Mar. 21, 1903]: 178).

30. Wallis, *Fresno Armenians*, 35–41; Mirak, *Torn between Two Lands*, 118–19.

31. *Fresno Evening Democrat*, Mar. 28, 1902, File 6, Box 1, Kearney Papers, FHS.

32. Boot quoted in "The Week," *PRP* 65 (Apr. 18, 1903): 242; Gordon quoted in Editorial, *PRP* 65 (May 2, 1903): 274; *CRGA v. Andrew Abbott*, complaint, File 1, Box 3, Kearney Papers, Bancroft Library; *California Raisin Growers' Association v. Abbott*, 160 Cal. 601 (1911). The industry produced a record crop of 106 million pounds, of which the CRGA handled approximately 80 percent. Growers received a price of three and four-fifths cents, lower than 1899 prices. See Fox, "Cooperation in the Raisin Industry," 50–52; Spence, "Success after Twenty Years," *SMH* 3 (Mar. 1918): 9; "The Cooperative Raisin Packers' Associations," *PRP* 64 (Sept. 13, 1902): 166; "California Raisin Growers' Association," *PRP* 64 (Oct. 11, 1902): 239; "Raisin Marketing This Year," *PRP* 64 (Nov. 15, 1902): 310; "Agricultural Review: Fresno," *PRP* 64 (Nov. 29, 1902): 343.

33. "The Raisin Association," *PRP* 66 (July 4, 1903): 15; Smith, *Garden of the Sun*, 523; Fox, "Cooperation in the Raisin Industry," 51; Spence, "Success after Twenty Years," *SMH* 3 (Mar. 1918): 12; Payne, "Cooperation," 44. On Kearney's attempts to retake control after 1902, see Fox, "Cooperation in the Raisin Industry," 64–66; Kearney, letter to *Fresno Evening Democrat*, Mar. 31, 1904, Kearney Papers, File 6, Box 1, FHS; Spence, "Success after Twenty Years," *SMH* 3 (Apr. 1918): 6; Remarks of Alexander Gordon, *Proceedings of the 30th Annual Fruit Growers Convention at San Jose*, in California State Commissioner of Horticulture, *First Biennial Report*, 312; Rehart and Patterson, *M. Theo Kearney*, 37.

34. "The Week," *PRP* 71 (June 2, 1906): 310; Rehart and Patterson, *M. Theo Kearney*, 42. Kearney left behind a revealing epitaph: "Warning — here lies the body of M. Theo Kearney, a visionary who thought he could teach the average farmer, and particularly raisin growers, some of the rudiments of sound business management. For eight years he worked strenuously at the task, and at the end of that time he was no further ahead than at the beginning. The effort killed him" (Rehart and Patterson, *M. Theo Kearney*, 37).

35. Smythe, "Benefactor of the State," 146.

36. Testimony of E. L. Chaddock, *In re California Associated Raisin Company, Investigation under Section 6(e) of the Federal Trade Commission Act*, Transcript of Hearings (typescript), 1 (Nov. 22, 1919), 155, Nutting Papers, Bancroft Library; Smythe, "A Benefactor of the State," 146. A grower with large holdings told Feramorz Fox in 1912, " 'I'd rather lose my vineyard than crawl at the feet of Mr. Kearney as he expected us to do' " (Fox, "Results of Cooperation," 30 n.).

37. Rehart and Patterson, *M. Theo Kearney*, 36–37; Bragg, "History of Cooperative Marketing," 14; see also Smythe, "Benefactor of the State," 146–48.

*Chapter 5*

1. 26 Stat. 209 (July 2, 1890); McCurdy, "*Knight* Sugar Decision," 304–5; Jenks and Clark, *Trust Problem*. For the view that the Sherman Act's drafters far exceeded the common law's antagonism toward monopoly, see Letwin, *Law and Economic Policy*, 52. The inimical bent

of American antitrust law toward cartelization contrasts sharply with British and continental legal attitudes during this period. See Keller, "Pluralist State"; and Freyer, *Regulating Big Business*.

2. Sherman Act, §§ 1 and 2; McCurdy, "*Knight* Sugar Decision," 324; Letwin, *Law and Economic Policy*, 85–95; Thorelli, *Federal Antitrust Policy*, 184–214; Scheiber, "Federalism and the American Economic Order." The shift during the nineteenth century from special charters to general incorporation statutes left intact states' traditional control over corporations' organic structure (Hovenkamp, *Enterprise and American Law*, 241–67).

3. McCurdy, "*Knight* Sugar Decision," 308, 328–30; *U.S. v. E. C. Knight Co.*, 156 U.S. 1 (1895). The strict distinction between production and commerce laid down in *Knight* meant that corporate structure and organization were left to the states to govern, an opportunity on which McCurdy argues the states defaulted after New Jersey's lax corporation laws drew corporations to redomicile, siphoning off corporate tax revenues from other states.

4. *U.S. v. Trans-Missouri Freight Association*, 166 U.S. 290 (1897); *U.S. v. Joint-Traffic Association*, 171 U.S. 505 (1898); and *U.S. v. Addyston Pipe and Steel Co.*, 175 U.S. 211 (1899). In these decisions the Court rejected the common law standard of reasonability and construed the statute strictly, ruling that the statute's ban on "every restraint of trade" condemned each combination. The railroad cartels in *Trans-Missouri* and *Joint-Traffic* fell because the railroads concertedly fixed rates and penalized members who failed to abide by the agreement. The pipe manufacturers' scheme to maintain prices and eliminate outside bidders for coveted urban construction jobs palpably suppressed competition. Only in *Addyston Pipe*, in which the Court limited the injunction against the pipe manufacturers' association to contracts for sales in interstate commerce, was the decision unanimous, even though the Interstate Commerce Act of 1887 regulated the railroads on the basis of their direct involvement in interstate commerce (Thorelli, *Federal Antitrust Policy*, 468). The "stream of commerce" metaphor first materialized in *Swift & Co. v. U.S.*, 196 U.S. 375 (1905), and reached its apogee in *Stafford v. Wallace*, 258 U.S. 495 (1922).

5. *Standard Oil Co. of New Jersey et al. v. United States*, 221 U.S. 1 (1911); *American Tobacco Company v. United States*, 221 U.S. 106 (1911); see Thorelli, *Federal Antitrust Policy*, 37 at n. 113, 466–70, 562, 599; and Sklar, *Corporate Reconstruction of American Capitalism*, 147–48. The "rule of reason" refers to the Court's acceptance of the common law standard of reasonability in determining whether a given combination violated the law. Factors such as size thus did not constitute per se violations of the Sherman Act, particularly in Section 2 cases in which the Court had to consider not only the appearance of monopoly but "the existence of unfair practices as evidencing an attempt to monopolize" (U.S. Bureau of Corporations, *Trust Laws and Unfair Competition*, 17). The Court distinguished its earlier decisions from the reasoning in *Standard Oil* and *American Tobacco*; the acceptance of the rule of reason was not intended to vitiate the Sherman Act as a regulatory tool.

The evolution of antitrust law falls roughly into stages: (1) pre–Sherman Act common law, characterized by a reasonableness standard; (2) early Sherman Act jurisprudence, 1890 to 1897, establishing the production/commerce distinction; (3) *Trans-Missouri* to *Standard Oil* and *American Tobacco*, 1897 to 1911, during which time Justice Harlan's literal reading of the Sherman Act combined with the dual federalism scheme laid out in *E. C. Knight* to produce some notable victories for the government (*Northern Securities v. U.S.*, 193 U.S. 197 [1904], and *Swift v. U.S.* [1905]); finally, (4) the post-1911 "rule of reason" paradigm, under which the Court conceded that insistence on a literal interpretation of the statute endangered the advantage of economic efficiency that accrued through consolidation and combination.

6. Sklar, *Corporate Reconstruction of American Capitalism*, 139–60; Levi, *Introduction to Legal Reasoning*, 72–78. Historiographically, there seems to be little dispute about the impact of the Sherman Act on the common law of restraint of trade: "It is plain . . . that at the end of the first decade of the twentieth century the federal courts since 1897 had consistently required that the Sherman Act be regarded as having superseded the common law with respect to restraints of trade and monopoly" (Sklar, *Corporate Reconstruction of American Capitalism*, 135). The lack of a clear policy rationale behind the Supreme Court's Sherman Act jurisprudence led to a continuation of the established pattern of decision making at common law before 1890.

7. McCurdy, "*Knight* Sugar Decision," 336–42. Roosevelt's reputation as a trust-buster has undergone critical scrutiny of late. See Chambers, *Tyranny of Change*, 146, 151. For an adoring but insightful portrait, see Harbaugh, *Life and Times of Theodore Roosevelt*, 150–64.

8. Lamoreaux, *Great Merger Movement*, 118–86; Hofstadter, *Age of Reform*, 215–56; Chambers, *Tyranny of Change*, 157–58; Horwitz, *Transformation of American Law, 1870–1960*, 65–107.

9. *Steers v. U.S.*, 192 F. 1, 5 (C.C.E.D. Ky. 1911); on Wilson's commutation, see Williams, *Laws on Trusts and Monopolies*, 37. For contemporary accounts of night riding in the Kentucky and Tennessee burley tobacco industries, see Tevis, "A Ku-Klux Klan of Today"; and Youngman, "Tobacco Pools of Kentucky and Tennessee." On Congress's refusal to exempt farmers from the Sherman Act, see Letwin, *Law and Economic Policy*, 93–95.

10. *Loewe v. Lawlor*, 208 U.S. 274, 301–2 (1908). A local of the American Federation of Labor staged a boycott against a hat manufacturer for refusing to discriminate against nonunion labor. The hats made at the Danbury, Connecticut, plant were shipped into interstate commerce, thus paving the way for a Sherman Act suit against the union by the manufacturer. The *Danbury Hatters'* case was only one of a series of legal setbacks for workers. Since the Civil War, labor unions had been continually victimized by the federal courts' willingness to enjoin them from boycotting antiunion employers. The Bureau of Corporations noted in 1915 that the federal courts consistently ruled during the 1890s that labor unions could not restrain trade under the Sherman Act but that "labor combinations generally have not been held unlawful" (*Trust Laws and Unfair Competition*, 93). The leading cases include *United States v. Debs et al.* 64 Fed. 724 (C.C.N.D. Ill. 1894); and *In re Debs*, 158 U.S. 564 (1895); see also *United States v. Workingmen's Amalgamated Council of New Orleans*, 54 Fed. 994 (C.C.E.D. La. 1893); and *Coppage v. Kansas*, 236 U.S. 1 (1915). The subject of labor's lot under the antitrust laws has generated no small amount of scholarship in the past thirty years, most of it documenting the judiciary's role in impeding the labor movement and undermining its political legitimacy. See Brody, *Steelworkers in America*; Paul, *Conservative Crisis and the Rule of Law*; Mensch, "History of Mainstream Legal Thought"; Montgomery, *Fall of the House of Labor*; Tomlins, *State and the Unions*; Forbath, *Law and the Shaping of the American Labor Movement*; Hattam, *Labor Visions and State Power*; and Ernst, *Lawyers against Labor*. For a good review essay, see Fisk, "Still 'Learning Something of Legislation.' "

11. Sklar, *Corporate Reconstruction of American Capitalism*, 147. Rep. John J. Fitzgerald (D-N.Y.) noted in 1913, "The attitude of the last Republican administration was such that there was grave reason to believe that it would have taken much more pleasure in prosecuting labor organizations and farmers' organizations than it would have taken in prose-

cuting the industrial combinations that had been so oppressive to the country" (*Cong. Rec.*, 63d Cong., 1st sess., 303).

12. McConnell, *Decline of Agrarian Democracy*, 36–43 (quote on 38). Gompers reinforced the alliance between labor and farmers in his testimony before the House Judiciary Committee in 1913: "We say that we have been in cooperation with the associated farmers of the country; the Farmers' National Union, the Society of Equity, and others have all been with us and we with them" (U.S. Congress, House Committee on the Judiciary, *Hearings on Trust Legislation*, 12–29, 19). On Gompers and the labor movement, see Tomlins, *State and the Unions*, 60–82, 120–21; on the movement to revise the Sherman Act, see sources in note 13, below.

13. Clayton Antitrust Act, 38 Stat. 730 (Oct. 15, 1914), §6 (emphasis added). The main purpose of the law was to render more explicit the prohibitions implicit in the Sherman Act. For the Clayton Act's enumeration of illegal anticompetitive practices, see Section 2 (prohibition on discriminatory pricing), Section 3 (exclusive dealing and restrictions on resale), Section 7 (mergers and holding companies), and Section 8 (interlocking directorates); U.S. Bureau of Corporations, *Trust Laws and Unfair Competition*, 132–37; Letwin, *Law and Economic Policy*, 273–76; Hofstadter, *Age of Reform*, 113, 251; Sklar, *Corporate Reconstruction of American Capitalism*, 330–31.

14. As the majority report of the House Judiciary Committee commented, the object of the section was to make it clear to the courts that such organizations as described in the act did not "come within the scope and purview of the Sherman anti-trust law" in the same way as "industrial corporations and combinations" (House Rep. 623, *Report on H.R. 15657*, May 13, 1914, in U.S. Senate, Committee on the Judiciary, *Hearings on H.R. 15657*, 36). For labor, the declaration in Section 6 amounted to little more than a "restatement of the existing law" and did little to alter the status of unions (Sklar, *Corporate Reconstruction of American Capitalism*, 331).

15. J. H. Paten, general counsel for the Farmers' Educational and Cooperative Union of America, to Rep. Victor Murdock (D-Kan.), Apr. 13, 1914, in *Cong. Rec.*, 63d Cong., 2d sess., 7146.

16. This interpretation is all the more likely if one accepts James Guth's contention that Senator John Walsh, a Montana Republican who knew little about cooperatives, was the author of Section 6 ("Farmer Monopolies," 73).

17. House Rep. 623, *Report on H.R. 15657*, May 13, 1914, in U.S. Senate, Committee on the Judiciary, *Hearings on H.R. 15657*, 118 (minority report of Reps. John M. Nelson and Andrew J. Volstead).

18. Hobson, "Farmers' Co-operative Associations," 222.

19. Sklar, *Corporate Reconstruction of American Capitalism*, 419–22.

20. *Cong. Rec.*, 63d Cong., 1st sess., 319. The Justice Department's 1912 budget amounted to some $10 million, any of which could be spent on antitrust enforcement. A special allocation for the Antitrust Division, including the rider, was first made in 1912, to address the division's failure to act under President Taft and "to relieve the administration of the excuse that it was hampered by a lack of funds" (ibid., 302 [remarks of John F. Fitzgerald, D-Mass.]). On his last day in office, President Taft vetoed the entire 1912 appropriations bill because he considered the rider unconstitutional (ibid., 1105). See Jones, "Status of Farmers' Co-operative Associations," 601.

21. *Cong. Rec.*, 63d Cong., 1st sess., 1103–12, 1189–97, 1269–92 (remarks of Sens. William

Borah [R-Ida.], Charles Townsend [R-Mich.], and Albert Cummins [R-Iowa]). The bill's supporters were concerned about the *Danbury Hatters'* case and believed that an explicit exemption for labor was necessary to clarify congressional intent. Their opponents focused on the unconstitutionality of class legislation and intrusive governmental regulation. A Washington, D.C., attorney made the point: "Thus indirectly, by a joker in an appropriation act, a vague regulation of the prices of the products of farm organizations is effected and an entirely new principle injected into the regulation of industrial organizations as contrasted with public utilities" (Jones, "Status of Farmers' Co-operative Associations," 602).

22. *Cong. Rec.*, 63d Cong., 1st sess., 1114 (remarks of Sen. Augustus Bacon [D-Ga.]: "There can be no question of the effect that the same logic or system of reasoning which brings the court to the conclusion that the antitrust law does cover agreements among laborers will also be used to rule that agreements among agriculturalists for the purpose of endeavoring to get the best prices for their products will also be unlawful and under the ban of this law"). For the riders, see 38 Stat. 53 (June 23, 1913); 38 Stat. 652 (Aug. 1, 1914); 38 Stat. 866 (Mar. 3, 1915); 39 Stat. 312 (July 1, 1916); 40 Stat. 155 (June 12, 1917); 40 Stat. 681 (July 1, 1918); 41 Stat. 336 (Nov. 4, 1919) 42 Stat. 613 (June 1, 1922); 42 Stat. 1080 (Jan. 1, 1923); 43 Stat. 217 (May 28, 1924); 43 Stat. 1027 (June 30, 1926); 44 Stat. 343 (Apr. 29, 1926); 44 Stat. 1194 (Feb. 24, 1927). There were no riders in the 1920, 1921, and 1925 appropriations bills.

23. Hobson, "Farmers' Co-operative Associations," 222.

24. Jones, "Status of Farmers' Co-operative Associations," 602.

25. Ibid.; *U.S. v. King*, 229 Fed. 275 (D.C.D. Mass. 1915) and 250 Fed. 908 (D.C.D. Mass. 1916). Whether the government could have prevailed with this strategy when the issue was what was a fair and reasonable price is another matter entirely, one on which I found no published decision.

26. Malin, "Background of the First Bills," 117, 112–13; Horner, "U.S. Government Activities"; Ensrud, "History of the Origin and Functions of the Federal Bureau of Markets," 1–6; Taylor and Taylor, *Story of Agricultural Economics*, 483–509.

27. Malin, "Background of the First Bills," 120–29; Taylor and Taylor, *Story of Agricultural Economics*, 542–53. When the first bills were introduced in Congress in 1911 to create a federal agency to study agricultural markets, the USDA's representative indicated that such a regulatory function belonged in the Department of Commerce. The separation of production from marketing for purposes of regulation was also reflected in the assignment of the Bureau of Corporations to the Department of Commerce upon the bureau's creation in 1901. I thank Laura Kalman for pointing this out to me.

28. Malin, "Background of the First Bills," 128; Taylor and Taylor, *Story of Agricultural Economics*, 534, 542–44. Caroline B. Sherman credits Houston with "a decided economic and social viewpoint" ("Legal Basis of the Marketing Work," 294). On the contrast between British and continental economists and philosophers, see Hovenkamp, *Enterprise and American Law*, 67–78; on the impact of these schools of thought on economists, see Furner, *Advocacy and Objectivity*; and Ross, *Origins of American Social Science*. Houston succeeded James Wilson, who served as secretary of agriculture under three Republican presidents (1897–1913), and whose support of cooperation consisted primarily of reporting on its rising popularity among farmers (Knapp, *Rise of American Cooperative Enterprise*, 144–45).

29. Wilson supported government seed loans to farmers to help them recover from severe crop losses, and he agreed that the Cotton Futures Act of 1914 performed an essential function by enabling cotton producers "to secure more equitable prices." See Taylor

and Taylor, *Story of Agricultural Economics*, 616, 921–22; Houston, "Report of the Secretary of Agriculture for 1915," in USDA, *Yearbook for 1915*, 45; Cotton Futures Act, 38 Stat. 693 (Aug. 18, 1914). The act empowered the USDA to prescribe the form of contracts for future purchases; to bar trade in substandard grades of cotton; to settle disputes over grades, standards, and quality of cotton; to regulate spot markets and disseminate information about the spot markets to commercial exchanges; and to collect a "prohibitive tax" on contracts for future deliveries if they failed to conform to the act (Taylor and Taylor, *Story of Agricultural Economics*, 616). Other statutes assigned to the Office of Markets include the Grain Standards and Warehouse Acts, 39 Stat. 482 (Aug. 11, 1916), and the Standard Container Act, 39 Stat. 673 (Aug. 31, 1916); in addition, work in other bureaus related to marketing was transferred to the Office of Markets and Rural Organization in 1915 and 1916 (Sherman, "Legal Basis of the Marketing Work," 298–99).

30. On the Office of Markets' early work, see USDA, *Organization and Conduct of a Market Service in the Department of Agriculture*. For examples of Houston's priorities for the Office of Markets, see David F. Houston to Senator P. J. McCumber, Sept. 9, 1919, Marketing, 1920, Secretary's Correspondence, RG 16; Memorandum to file, David F. Houston, June 14, 1919; Statutes, Marketing, General, 1919–37, Box 59, Solicitor's Records, RG 16.

31. USDA, *Annual Report for 1913*, 25, quoted in Taylor and Taylor, *Story of Agricultural Economics*, 544. In 1915, the Office of Markets and Rural Organization "received daily telegraphic market reports from twenty-one of the principal northern and eastern car-lot markets." It also had representatives in New York, Buffalo, Chicago, St. Louis, Kansas City, and Baltimore; in other cities it collected information from the "most responsible" wholesale traders. See Carl Vrooman, Assistant Secretary of Agriculture, to J. H. Stoltzfus, Oct. 28, 1915, Reel 379, M440, Secretary's Correspondence, RG 16. See also Baker, Rasmussen, Wiser, and Porter, *Century of Service*, 74–80; and Sherman, "Legal Basis of the Marketing Work," 298.

32. Ensrud, "History of the Origin and Functions of the Federal Bureau of Markets," 90–107; Sherman, "Legal Basis of the Marketing Work," 296–97; Brand, "Office of Markets." The Office of Markets was led by a USDA economist who had made his career studying cotton for the USDA; most of its early work was therefore focused on that crop. The Office of Markets and the Office of Rural Organization combined forces on July 1, 1914. See Charles J. Brand, "Work of the Office of Markets and Rural Organization," USDA announcement, n.d. (but probably 1914), reprinted in Nourse, *Agricultural Economics*, 558–61.

33. Ensrud, "History of the Origin and Functions of the Federal Bureau of Markets," 111–12. The change in status from office to bureau signified the nominal importance of marketing in the USDA's promotional work; the change became official on July 1, 1917 (Knapp, *Rise of American Cooperative Enterprise*, 157; Sherman, "Legal Basis of the Marketing Work," 299).

34. Smith-Lever Act, 38 Stat. 372 (May 8, 1914). Houston's annual report to the president for 1919 requested additional funds to hire agents specifically to educate farmers about cooperative marketing, but there is no evidence that he did anything to push for such an appropriation. See Knapp, *Rise of American Cooperative Enterprise*, 159. On the history of cooperative extension, see Rasmussen, *Taking the University to the People*.

35. Knapp, *Rise of American Cooperative Enterprise*, 153. Knapp, like the Taylors, tended to magnify the significance of the USDA's work on cooperation during this period. Although gathering information on the number of cooperatives in existence in the United

States and the type of business they conducted was no doubt a vital first step, neither work recognizes that the USDA's support of cooperation was essentially limited to that kind of information-gathering study. This survey produced the first comprehensive federal statistical account of cooperative purchasing and marketing in the United States (Jesness, "Cooperative Purchasing and Marketing Organizations"; Knapp, *Rise of American Cooperative Enterprise*, 156). The study of cooperative accounting systems was published as Kerr and Nahstoll, "Cooperative Organization Business Methods."

36. Brand, "Work of the Office of Markets and Rural Organization," in Nourse, *Agricultural Economics*, 558.

37. Knapp, *Rise of American Cooperative Enterprise*, 153–54, 157, 160; Charles Frederick Bishop to Carl Vrooman, Dec. 22, 1915, enclosing clipping, "Department of Agriculture Forming New Civilization," *New York Times*, Dec. 12, 1915, both in Cooperation, 1915, Secretary's Correspondence, RG 16. In the *New York Times* article, Vrooman is quoted in a speech as saying, "The Department of Agriculture is laying the foundation of 'a new civilization' based on social co-operation as opposed to the individualism that has distinguished the American of the past." He noted that this work was being accomplished through county agents.

38. A notable exception to this prevailing regulatory style was the railroads, which in the nineteenth century had vividly demonstrated the folly of an unregulated market. Contemporary policy makers, however, regarded railroads as the great exception. Because railroads constituted a natural monopoly and because the cost of transportation affected the cost of virtually everything in a national economy, constitutional limitations on the power of state and federal regulatory agencies to set rate maxima were toppled, sometimes in stunning examples of judicial reversals. See McCraw, *Prophets of Regulation*, 1–56; Skowronek, *Building a New Administrative State*, 121–62. For an economic analysis of the industry, see MacAvoy, *Economic Effects of Regulation*.

39. C. B. Boring, "Suggestions for a Course of Study in Voluntary Co-operation," July 1912; George K. Holmes, Chief, Bureau of Statistics, to Willet M. Hays, Acting Secretary, Sept. 14, 1912, both in Cooperation, 1912–13, Secretary's Correspondence, RG 16. Restricting the survey to incorporated cooperatives indicated that unincorporated organizations were not "true" cooperatives; the presence of laws to permit the incorporation of nonstock, nonprofit cooperatives presumably enabled farmers to adhere to traditional cooperative principles. Interested private observers strongly recommended this approach, including the editor of *Co-operation* ("a monthly magazine of economic progress"). See E. M. Tousley to Hays, Nov. 11, 1912, ibid.

40. Willet M. Hays, Assistant Secretary, to G. D. Bice, July 29, 1912, Cooperation, 1912–13, Secretary's Correspondence, RG 16. There was little support in Congress for legislation to provide for federal incorporation of cooperatives (C. W. Thompson to John H. Gray, Nov. 29, 1912, ibid.). Although the USDA monitored developments in state cooperative laws, at this point it was not ready to elaborate on the specifics of "true" cooperative principles. It encouraged uniformity across state lines, but this approach also had little impact.

41. Francis J. Caffey, Solicitor, to Charles J. Brand, Office of Markets, Feb. 23, 1916; Statutes, Marketing, Cooperative Organization, Miscellaneous; Box 60; Solicitor's Records, RG 16.

42. Brand to Caffey, Feb. 13, 1914; Caffey to Bureau of Markets, June 10, 1914, both in Records of the Office of the Solicitor, Box 60, RG 16. Caffey advised cooperatives to limit

the damages they claimed in breach-of-contract cases. Brand was appointed chief of the Bureau of Markets on May 16, 1913. His emphasis on the use of market information to achieve broader economic reforms was a significant factor in Houston's decision to appoint him. See his report in "Department Conference on the Marketing and Distribution of Farm Products," U.S. Department of Agriculture, Mar. 27, 1913, typescript, U.S. Agricultural Library, reprinted in part in Taylor and Taylor, *Story of Agricultural Economics*, 538–40.

43. Hobson, "Farmers' Co-operative Associations," 222. After its reorganization, the CFGE raised funds by levying an assessment on each box of fruit shipped. See Wickson, *Rural California*, 302; Erdman, "Development and Significance of California Cooperatives," 181.

44. Carl Vrooman, Assistant Secretary of Agriculture, to C. W. Perky, Oct. 8, 1915; C. J. Marvin to J. H. Stoltzfus, Oct. 8, 1915; Carl Vrooman to J. H. Stoltzfus, Oct. 28, 1915; and David F. Houston to S. J. Harrison, Dec. 9, 1915, all in Reel 379, M440, Secretary's Correspondence, RG 16; Houston to W. C. Bromley, Apr. 12, 1916; and Houston to LeRoy Smith, n.d. (but Oct. 21, 1916), Reel 380, M440, Secretary's Correspondence, RG 16. The USDA requested a draft that "would embody all the business and economic features that we consider necessary in such a bill." See Carl Vrooman to George E. Farrand, Nov. 21, 1914; and Farrand to Vrooman, Nov. 6, 1914, Cooperation, 1914, Secretary's Correspondence, RG 16; Nourse, *Legal Status of Agricultural Co-operation*, 79 n. 4.

45. The bill was officially published as USDA Office of Markets and Rural Organization, *Service and Regulatory Announcements*, 20 (Feb. 7, 1917). The quoted provision required members to "contribute to the financial support of the cooperative" (Section 17, "Suggestions for a State Co-operative Law," quoted in Nourse, *Legal Status of Agricultural Co-operation*, 83). The bill also laid out the agency relationship between cooperatives and their members and permitted cooperatives to use exclusive marketing agreements to protect their patronage (ibid., 84–86). The USDA later attempted to distinguish the use of capital stock in its model state law from ordinary corporate practice (Jesness, "Cooperative Purchasing and Marketing Organizations," 64–65).

46. William M. Williams to Bureau of Markets, Sept. 17, 1917, File Statutes, Marketing, General, Box 58, Records of the Office of the Solicitor, RG 16.

47. Powell, "Fundamental Principles of Co-operation in Agriculture" (1920), 2; on the reception to the USDA bill, see Nourse, *Legal Status of Agricultural Co-operation*, 88. For the contrast between the traditionalists and the proponents of the new style, see Powell, "Fundamental Principles of Co-operation in Agriculture" (1914); and Steen, *Cooperative Marketing*.

48. *Burns v. Wray Farmers' Grain Co.*, 65 Colo. 425 (1918). A Colorado cooperative, regularly incorporated with $10,000 in capital stock, traded in grain, hogs, and coal for 230 stockholders. It sued Burns when he refused to pay the maintenance clause penalty for thirty-five hundred bushels of wheat he sold outside the company. On appeal, the Colorado Supreme Court concluded that the maintenance clause increased prices, drove competitors out of business in Wray, and gave the Grain Company a monopoly. Unlike the Decorah Society in *Reeves*, which acted merely as a selling agent and never paid dividends, the Grain Company paid its stockholders dividends every year. The court concluded that the Grain Company was in business to profit itself rather than its members and deemed the maintenance clause an unreasonable restriction on the rights of the stockholders to sell to whom they pleased.

49. Houston to LeRoy Smith, California State Retail Hardware Association, Oct. 23, 1916, Reel 380, M440, Secretary's Correspondence, RG 16; Houston to P. J. McCumber, Sept. 9, 1919, Marketing, 1920, Secretary's Correspondence, RG 16; Memorandum to file, David F. Houston, June 14, 1919, Statutes, Marketing, General, Box 59, Solicitor's Records, RG 16; Carl Vrooman to J. H. Stoltzfus, Oct. 28, 1915, Reel 379, M440, Secretary's Correspondence, RG 16. Houston repeatedly discouraged efforts to pass a bill for the federal incorporation of cooperatives; see Houston to McCumber, Sept. 9, 1919; and sources cited in note 40, above.

50. John H. Gray to C. W. Thompson, Dec. 6, 1912, Cooperation, 1912–13, Secretary's Correspondence, RG 16; A. C. True, Director, States Relations Service, to George Livingston, Chief, Bureau of Markets, Feb. 11, 1920, Cooperation, 1920, Secretary's Correspondence, RG 16. When George Livingston became chief in 1919 upon the resignation of Charles Brand, he took an even more cautious approach to marketing than his predecessor: "When it is apparent that cooperative effort is needed to overcome abuse, remedy inefficiency, or supply a recognized need, [the bureau] does suggest to producers the advisability of considering the formation of a cooperative organization, shows them how such organizations are formed and conducted and explains to them the principles that must be observed in order to be successful. It does not, however, undertake any work of the kind ordinarily done by persons known as 'promoters.' It encourages the use of existing marketing machinery when it seems to be operating efficiently and serving the best interests of a community" (Livingston to Mr. Harrison, Secretary's Office, Feb. 12, 1920, ibid.; see also Taylor and Taylor, *Story of Agricultural Economics*, 546; Sherman, "Legal Basis of the Marketing Work," 301).

51. Nourse, "Revolution in Farming," 97. See also Nourse, "Place of Agriculture in Modern Industrial Society."

*Chapter 6*

1. John W. Preston, opening statement before the Federal Trade Commission, *In Re California Associated Raisin Company, Investigation upon Application of Attorney General under Section 6(e) of the Federal Trade Commission Act*, Transcript of Proceedings, Washington, D.C., 1 (Nov. 24, 1919): 333, typescript, Carton 6, Nutting Papers (hereafter *FTC Transcript*); Walker, *Fresno County Blue Book*, 104, 107; Fred K. Howard, "History of Raisin Marketing in California," photocopy, n.d. (but 1922), 57, 59, Bancroft Library. A series of "million-dollar companies" had been formed by industry committees between 1907 and 1912. Many organizers drew directly on their experiences with the Kearney associations and attempted to remedy his failures by adhering more closely to traditional cooperative principles, particularly by limiting stock ownership to growers (Spence, "Success after Twenty Years," *SMH* 3 [Apr.–June 1918]; Henry E. Erdman, Interview with C. A. Parlier, Feb. 26, 1934, Box 18, Erdman Papers). For a detailed explanation of the cooperative organization efforts between 1906 and 1912, see George E. Popovich, "History of the Sun-Maid Organization and Its Antecedents," *Fresno Bee*, Feb. 16, 1958, Sunday Country Life Section; Spence, "Success after Twenty Years," *SMH* 3 (Apr.–June 1918); and Erdman, "Notes on the *California Fruit Grower*" (1912), Box 18, Erdman Papers.

2. Spence, "Success after Twenty Years," *SMH* 3 (June 1918): 7. The participation of Sutherland and Welsh indicates that local attorneys were actively involved in the CARC's organization and incorporation. See Levy, *History of the Cooperative Raisin Industry of California*, 6.

3. The stock was originally divided into ten thousand shares worth $100 each. Because the shares were so expensive, the CARC was obliged to accept growers' notes in payment (Bragg, "History of Cooperative Marketing," 66; *California Fruit Grower* 46 [Nov. 27, 1912]: 1; see also Woehlke, "Raising the Price of the Raisin," 2039, 2058). Just how many of the stockholders were growers and how many were not cannot be determined from extant records. In later testimony before Congress, CARC president Wylie Giffen put the figures at 80 percent growers, 20 percent nongrowers; attorney John W. Preston, representing the commercial packers, contended that the proportion was two to one, nongrowers to growers (U.S. Congress, Senate, Committee on the Judiciary, *Hearings Authorizing Associations of Producers of Agricultural Products*, 8, 98 [hereafter *1921 Hearings*]). Public statements by the CARC's own officers indicate that many growers signed contracts but never purchased stock; in 1917 CARC secretary C. A. Murdock published figures indicating a membership of over 8,000 growers and 3,569 stockholders ("The Co-operative Production and Sale of Raisins in California," *Grain Growers' Guide* [1917], Box 18, Erdman Papers).

4. Spence, "Success after Twenty Years," *SMH* 3 (June 1918): 7; "Subscription and Voting Trust Agreement Used in Organizing [the] California Associated Raisin Company," n.d. [but 1913], in Howard, *History of the Sun-Maid Raisin Growers*, 19–22. The trust arrangement was to revert to a regular stockholding corporation after seven years (Bragg, "History of Cooperative Marketing," 66; Woehlke, "Raising the Price of the Raisin," 2039).

5. Howard, *History of the Sun-Maid Raisin Growers*, 7–9; CARC, "Original Subscription and Voting Trust Agreement," Nov. 1912, reprinted ibid., 19–22; Bragg, "History of Cooperative Marketing," 66; Spence, "Success after Twenty Years," *SMH* 3 (June 1918): 7; Woehlke, "Raising the Price of the Raisin," 2039; Hanna, *Law of Cooperative Marketing Associations*, 7 n. 11; Steen, *Cooperative Marketing*, 23.

Most of the capital subscribed at the start was in the form of growers' notes (Bragg, "History of Cooperative Marketing," 77).

6. Nourse, *Legal Status of Agricultural Co-operation*, 39–50.

7. Cochran, "Results of Cooperation in the Raisin Industry," 66; Bragg, "History of Cooperative Marketing," 83.

8. Spence, "The California Associated Raisin Company: Organization of the Company," *SMH* 3 (July 1918): 7; Bragg, "History of Cooperative Marketing," 67; Edward J. Wickson, "The Raisin Realization," *PRP* 84 (Nov. 2, 1912): 420; D. J. Whitney, "Sunrise for the Raisin Industry," *PRP* 84 (Nov. 23, 1912): 492; Levy, *History of the Cooperative Raisin Industry*, 5–6; Howard, *History of the Sun-Maid Raisin Growers*, 9; Steen, *Cooperative Marketing*, 23. Giffen was the CARC's second president; he succeeded H. H. Welsh in 1914 (Walker, *Fresno County Blue Book*, 370–71).

9. Bragg, "History of Cooperative Marketing," 24, 55; Winchell, *History of Fresno County*, 109, 279–81; Eaton, *Vintage Fresno*, 59; Elliot, *History of Fresno County*, 212.

10. Levy, *History of the Cooperative Raisin Industry*, 5–6; Jesness, "Cooperative Purchasing and Marketing Organizations," 64–65.

11. Welch quoted in testimony of E. L. Chaddock, *FTC Transcript* 1 (Nov. 20, 1919): 129. On the impact of the Progressives on agriculture in California, see Nash, *State Government and Economic Development*; Erdman, "Development and Significance of California Cooperatives"; Larsen, "A Progressive in Agriculture"; and Spencer, *California's Prodigal Sons*.

12. "First Crop Contract" (covering years 1913, 1914, 1915), reprinted in Howard, *History of the Sun-Maid Raisin Growers*, 23–25; Spence, "The California Associated Raisin Com-

pany: Organization of the Company," *SMH* 3 (July 1918): 6–7, 14. Many growers accepted notes from the CARC in lieu of cash payments for the 1913 crop; the CARC paid 7 percent interest on these notes. Some growers redeemed them immediately, but others were willing to wait until they were due (Bragg, "History of Cooperative Marketing," 77–78).

13. Testimony of E. L. Chaddock, *FTC Transcript* 1 (Nov. 20, 1919): 113–19. The contract stipulated that the packers acknowledged the CARC's ownership of the crop. The packers would process the raisins, sell them at prices set by the CARC, and receive from the CARC $5 per ton as their share. Further, the packers were to assume the expense and risk of holding their portion of the surplus from each preceding crop year, a cost that growers had previously borne. Finally, the packers were bound not to process any raisins other than those of the CARC. The packers felt the deal was a bitter pill forced on them by the CARC's strong bargaining position and the packers' own disunity. On the form and operation of the packers' contract, see Reports of Special Agent Arthur M. Allen, Sept. 2, 1913, Exhibit A (CARC contract with Phoenix Packing Company), and Sept. 25, 1913, File 1, Box 560, BAI, RG 60.

14. Testimony of E. L. Chaddock, *FTC Transcript* 1 (Nov. 20, 1919): 113; Trade Circular, Bonner Packing Company, Mar. 29, 1913, Special Section, 1913–18, Box 560, BAI, RG 60. Outside packers not signing contracts with the CARC in 1913 were J. K. Armsby Co., Guggenhime Co., J. B. Inderrieden & Co., Chaddock & Co., and the American Vineyard Company (Bragg, "History of Cooperative Marketing," 74).

15. T. S. Southgate & Co., Trade Circular, Aug. 14, 1913, File 1, Box 559, BAI, RG 60.

16. W. L. Sheill to Rep. E. R. Bathrick, Feb. 19, 1914, Special Section, 1913–18, Box 560, BAI, RG 60; R. E. Russell to Attorney General, July 21, 1913, File 1, Box 559, BAI, RG 60; Trade Circulars, Bonner Packing Company, Fresno, Mar. 25, 29, 1913, Special Section, 1913–18, Box 560, BAI, RG 60. See also George W. Mueller to Attorney General James McReynolds, Nov. 8, 1913; Daniels, Cornell Co. to Department of Justice, July 30, 1913 ("if the Government does not investigate this Trust they should not touch any of the others"), all in File 1, Box 559, BAI, RG 60.

17. Report of Special Agent Arthur M. Allen, "In Re Raisin Trust, Special Report," May 5, 1913, 7, Special Section, 1913–18; Box 560, BAI, RG 60. CARC officers believed they were not breaking the law but urged the DOJ to inform them if they were. Madison told Allen that the CARC's prices for 1913 were entirely reasonable in view of prices for the preceding six years and that the packers and jobbers were merely upset at the loss of "enormous and unreasonable profit[s]." The purpose of the organization, Madison noted, was to "sav[e] many small growers of raisins from actual bankruptcy" (Madison to Allen, Sept. 3, 1913, copy in Allen Report, Sept. 2, 1913, Special Section, 1913–18, Box 560, BAI, RG 60).

18. Allen Report, "In Re Raisin Trust—Violation of the Anti-Trust Law," May 3, 1913; Fowler memorandum to Bielaski and Allen, July 14, 1913, both in Special Section, 1913–18, Box 560, BAI, RG 60. Two days later, Allen submitted a more extensive report that reviewed the history of the industry; the growers' costs in land, equipment, and labor; the problems of weather, alkali damage to soil, and labor shortages; and the uncertain prices between 1895 and 1913. Allen concluded that if the CARC carried out its threat to do its own packing and marketing by bypassing the private packers, it would certainly be conducting interstate business; however, the CARC's goal appeared to be to reduce prices paid by the consumer, which it intended to accomplish by setting a maximum on the prices charged by the packers (Allen, "In Re Raisin Trust, Special Report," May 5, 1913, ibid.).

19. G. Carroll Todd, Assistant to Attorney General, to Albert Schoonover, U.S. Attorney, Mar. 11, Dec. 9, 1914, June 21, 1915, Jan. 6, 1916; Schoonover to the Attorney General, July 7, 1915; Schoonover to Todd, Jan. 12, 1916 ("This case is one that by reason of its large importance we have been unable to fully investigate"); Todd to Schoonover, Jan. 24, 1916 ("[A] special agent will be instructed to report to you at once for such further investigation to that end as may be necessary"); Memorandum, Todd to Bielaski, Jan. 21, 1916 ("Please send a special agent to Los Angeles under instructions to take these . . . matters up and to devote his time to them exclusively until the investigations shall be completed"), all in File 1, Box 559, BAI, RG 60.

20. Spence, "Associated Established on Firm Foundation," *SMH* 4 (Dec. 1918): 23.

21. " 'Sun-Maid' Trade Mark Born at Fresno, Sept. 11, 1914," *SMH* 5 (Nov. 1919): 8–9. By 1919 the value of the Sun-Maid trademark was "conservatively" estimated at $2.5 million (ibid.). Fifty years later, Sun-Maid gave Loraine Collett Peterson a royalty check in the amount of $2,000. In 1987, the seventy-fifth anniversary of the founding of Sun-Maid, the cooperative gave her bonnet to the Smithsonian Institution. The original portrait is on display at the cooperative's offices in Kingsburg, California.

22. Steen, "Story of the California Raisin 'Trust' "; "New Design Adopted for Sun-Maid Carton," *SMH* 1 (Aug. 1915): 5; "Big Advertising Campaign Started," *SMH* 1 (Aug. 1915), 8–9; "Thousands of Dollars of Free Advertising for Raisins," *SMH* 1 (Aug. 1915): 10, 20; F. P. Webster, Report "In Re California Associated Raisin Co., Alleged Violation of Sherman Act," Apr. 19, 1916, Special Section, 1913–18, Box 560, BAI, RG 60. Levy, *History of the Cooperative Raisin Industry*, 7.

23. "California Boost," *CG* 81 (Oct. 21, 1916): 14.

24. Cross, "Cooperation in Agriculture," 540.

25. Thomas Gallagher and Robert O'Connor to Attorney General, Feb. 11, 1916; G. Carroll Todd to Schoonover, Jan. 24, 1916, and Gallagher and O'Connor to Attorney General, Feb. 11, 1916, 11–13, all in Special Section, 1913–18, Box 560, BAI, RG 60.

26. This is a classic economic problem in cooperation. See Steen, *Cooperative Marketing*, 318–40; and Cobia, ed., *Cooperatives in Agriculture*.

27. Forrester, *Report upon Large Scale Co-operative Marketing*, 141; Spence, "Association Established on Firm Foundation," *SMH* 4 (Dec. 1918): 6; Bragg, "History of Cooperative Marketing," 82, 84–85. The first suits against growers for breach of contract were filed in 1914. See *California Fruit Grower*, Nov. 7, 1914, Box 18, Erdman Papers.

28. Board of Directors, CARC, "Shall We Go On?" *SMH* 1 (Jan. 1916): 2–3.

29. *FMR*, Feb. 6, 1916, quoted in *FTC Transcript* 1 (Nov. 24, 1919): 379–81. The *FMR* was a staunch CARC supporter, as were most Central Valley newspapers.

30. "Thrusts from under the Umbrella," *SMH* 1 (Sept. 1915): 11.

31. "The Dilemma of the Associated Raisin Company," *California Fruit News* 53 (Jan. 29, 1916): 1.

32. On the economics of cooperatives and market control, see Heflebower, *Cooperatives and Mutuals*, 17–23. The lack of records makes it impossible to draw an ethnic profile of CARC members during these years. Anecdotal evidence suggests that no growers were automatically excluded on account of race or ethnicity. It is possible that night riders hoped to drive their victims out of the industry and to find purchasers for the vineyards, preferably white, who would be CARC adherents, but I have found no evidence to support such a theory. The fact that white growers attacked nonmembers regardless of their race reinforces

the explanation of economic necessity and the incoherence of the violence; after all, on what grounds could white CARC members claim racial superiority over whites who chose not to join?

33. George Abe, oral interview, Aug. 25, 1980, quoted in Hasegawa and Boettcher, eds., *Success through Perseverance*, 8. For contemporary Californians' attitudes toward Asians, see California State Board of Control, *California and the Oriental*; Nutting, *Interview with Franklin P. Nutting*, 4-8, 28-29, Bancroft Library; Walker, *Fresno County Blue Book*, 47-48. For a history of the Japanese in California agriculture, see Lukes and Okihiro, *Japanese Legacy*.

34. Spence, "Success after Twenty Years," *SMH* 4 (Aug. 1918): 14. A legal scholar found that most of the Central Valley's major cooperatives had many foreign-born growers among their members: "26 percent of the membership of the Sun Maid Raisin Growers' Association is foreign; 30 percent of the membership of the California Bean Growers' Association [is] foreign; and one entire 'district' association of the California Cherry Growers' Association [is] foreign. . . . Moreover, until the enactment of the Alien Land Law of California, 50 per cent of the membership of the California Berry-Growers' Association was Japanese" (Tobriner, "How Shall Cooperatives Be Organized?," 373).

35. Nutting, *Interview with Franklin P. Nutting*, 38-39, Bancroft Library; Report of Special Agent F. P. Webster, Apr. 22, 1916, Special Section, 1913-18, Box 560, BAI, RG 60.

36. "The Dilemma of the Associated Raisin Company," 1; "Giffen's Reply to Rowley," *SMH* 1 (Mar. 1916): 3-4. The lack of newspaper coverage makes it difficult to determine whether any arrests were made during the 1916 campaign.

37. "Night-Riding," *SMH* 1 (Apr. 1916): 2; Report of Special Agent F. P. Webster, May 9, 1916, Special Section, 1913-18, Box 560, BAI, RG 60; "Associated Raisin Company Continues: A Criticism and A Reply," *California Citrograph* 1 (Mar. 1916): 9-10; testimony of E. L. Chaddock, *FTC Transcript* 1 (Nov. 20, 1919): 138-39.

38. After the campaign ended, Madison instructed the packers not to sell to anyone except "at *such prices and upon such terms* and conditions as may be prescribed from time to time by the Associated Raisin Company." F. P. Webster, Report In Re California Associated Raisin Co., Apr. 19, 1916, Special Section, 1913-18, Box 560, BAI, RG 60; see also Webster reports dated Apr. 22, 24, 25, 28, and May 13, 1916, ibid.

39. Schoonover to Attorney General, Apr. 4, 1917, Special Section, 1913-18, Box 560, BAI, RG 60.

40. Lincoln Clark to G. Carroll Todd, June 6, 1917, Special Section, 1913-18, Box 560, BAI, RG 60; *Connolly v. Union Sewer Pipe Co.*, 184 U.S. 540 (1902).

41. G. Carroll Todd to Schoonover, July 27, 1917, Special Section, 1913-18, Box 560, BAI, RG 60. The "rule of reason" decisions, *Standard Oil Co. of New Jersey v. United States*, 221 U.S. 1 (1911) and *United States v. American Tobacco*, 221 U.S. 106 (1911), are discussed in Chapter 5. Todd's letter speaks to contemporary uncertainty about the meaning of the Supreme Court's judicial amendment of the Sherman Act for federal enforcement decision making.

42. William B. Colver, Chair, FTC, "Memorandum to the Commission in the Matter of the Raisin Situation," Aug. 2, 1918, p. 1, Special Section, 1913-18, Box 560, BAI, RG 60. On the uncertain funding for the Antitrust Division, see Giglio, *H. M. Daugherty and the Politics of Expediency*, 122-24; on the DOJ's efforts to prosecute postwar profiteering, see Coben, *A. Mitchell Palmer*, 157-70.

43. Testimony of E. L. Chaddock, *FTC Transcript* 1 (Nov. 22, 1919): 288-91.

44. Howard, *History of the Sun-Maid Raisin Growers*, 27–28; statement of CARC Board of Directors, quoted in *FTC Transcript* 1 (Nov. 25, 1919): 390–93; Spence, "The California Associated Raisin Co.," *SMH* 5 (Oct. 1919): 8; "Agreement, California Associated Raisin Company, 1918, 1919, 1920," *U.S. v. California Associated Raisin Company*, Exhibit A, Defendant's Reply Points and Authorities, Oct. 25, 1920, Case No. B-67, Folder 2, Box 76, RG 21. Whether the California courts would enforce the liquidated damages clause was by no means certain; in 1901 the prune growers' organization had lost a suit to enforce a contract with a liquidated damages clause. In the mid-1920s, state appeals courts upheld the clause but refused to permit the reorganized Sun-Maid Raisin Growers to collect from growers who had signed contracts with the previous organization, the CARC. See Chapter 8.

45. Giffen, "The Campaign Is On," editorial, *SMH* 3 (Dec. 1917): 14; *FTC Transcript* 1 (Nov. 20, 1919): 393. See also *SMH* 3 (Dec. 1917), 3 (Jan. 1918), and 3 (Feb. 1918).

46. Affidavit of Franklin P. Nutting, *U.S. v. California Associated Raisin Company*, Sept. 1920, File 12, Carton 2, Nutting Papers; testimony of E. L. Chaddock, *FTC Transcript* 1 (Nov. 20, 1919): 136–37, 138–39 (quote on 138). The CARC's attorney subsequently extracted an admission from Chaddock that the CARC offered a $1,000 reward for the arrest and conviction of "anybody committing any act of depredation" (ibid., 228).

47. Testimony of Charles Bonner, *FTC Transcript* 2 (Nov. 28, 1919): 936–39, 942. In his testimony Bonner paraphrased Giffen's statements; Giffen did not testify at the hearings about this conversation.

48. "Don't Lose Your Temper," *FMR*, Jan. 24, 1918, quoted in *FTC Transcript* 2 (Nov. 20, 1919): 225–26; testimony of Charles Bonner, *FTC Transcript* 2 (Nov. 28, 1919): 944–46. Bonner's testimony about this victim was hearsay and would have been inadmissible in court.

49. Testimony of Charles Bonner, *FTC Transcript* 2 (Nov. 28, 1919): 946; testimony of E. L. Chaddock, *FTC Transcript* 1 (Nov. 20, 1919): 137.

50. Giffen, "The Final Results," *SMH* 3 (Mar. 1918): 1.

51. Testimony of George H. McCandless, *FTC Transcript* 1 (Nov. 24, 1919): 503–5; testimony of E. L. Chaddock, *FTC Transcript* 1 (Nov. 20, 1919): 126; *1921 Hearings*, 110–11 (testimony of Wylie M. Giffen); 20 (testimony of John W. Preston); "Associated Raisin Co.," notes from *Western Canner and Packer*, Dec. 1918, Box 18, Erdman Papers.

*Chapter 7*

1. On the short crops in 1916 see USDA, "Report of the Secretary of Agriculture," 9–48; Food Production Act, 40 Stat. 216 (Aug. 10, 1917); Food Price Control Act, 40 Stat. 276 (Aug. 10, 1917). The purpose of the Food Price Control Act was to maintain an adequate food supply and prevent monopolization of and speculation in food commodities; the president was solely responsible for carrying out the duties specified in the statute. By declining to repose the authority for enforcing the statute in existing departments, Congress intended to emphasize the emergency (and therefore temporary) nature of the powers conferred by the statute.

2. Bailey Millard, "Big Business and the Farmer," *San Francisco Bulletin*, Feb. 10, 1920, Raisin Case Clippings File, Carton 4, Nutting Papers.

3. The Virginia farmer is quoted in Gabriel, "Farmer in the Commonwealth," 577; Shideler, *Farm Crisis*, 1, 19.

4. *Ex Parte Baldwin County Producers' Corporation*, 203 Ala. 345, 346 (1919). The Alabama court ignored its own hostile precedent of *Georgia Fruit Exchange v. Turnipseed*, 9

Ala. App. 123 (1913), as well as *Reeves v. Decorah Farmers' Cooperative Society*, 160 Iowa 194 (1913), the leading state case in the law on maintenance clauses. This trend had far more to do with the greater visibility of agricultural economic distress after World War I than with a breakthrough in the legal conception of cooperation, a conclusion that is reinforced by similar decisions in states that lacked either a separate incorporation statute or, after 1922, special cooperative marketing laws: *Washington Cranberry Growers' Association v. Moore*, 117 Wash. 430 (1921); *Anaheim Citrus Fruit Association v. Yeoman*, 51 Cal. App. 759 (1921). See Chapter 9.

5. *Bullville Milk Producers' Association. v. Armstrong*, 178 N.Y.S. 612, 613 (1919); *Castorland Milk and Cheese Association. v. Shantz*, 179 N.Y.S. 131 (1919). Both plaintiffs were regularly incorporated with small amounts of capital stock; in *Armstrong* farmers declined to organize under a new cooperative organization law passed in 1913. The marketing agreements in both cases specified liquidated damages of $2 and $10 per cow, respectively, in case of failure to deliver to the associations. In *Castorland*, the court upheld the marketing agreement on common law grounds, completely contradicting *Reeves*: "It is apparent, therefore, [that] this agreement possessed all the elements of a valid and binding contract. . . . I cannot see how the agreement in any respect offends the law, as being an illegal monopoly and an unreasonable restraint of trade" (134).

6. *Bullville*, 613.

7. John W. Preston to Russell J. Poole, Nov. 2, 1921, File 32, Carton 3, Nutting Papers; see also Alex Goldstein to Breed, Abbott & Morgan, Oct. 1, 1919, Raisin Committee Correspondence File, Carton 7, Nutting Papers.

8. Nutting to George H. McCandless, June 30, 1919, File 13, Carton 5, Nutting Papers. A lawyer and the son of CARC founder W. R. Nutting, Franklin Nutting owned a large vineyard, a packing business, and a fairly valuable brand for the Thompson's seedless raisins in which his firm specialized. He was one of the "inside" packers who agreed to pack raisins for the CARC in 1913.

9. Louis C. Drapeau to Franklin P. Nutting, Apr. 3, 1919, Raisin Committee Correspondence File, Carton 7, Nutting Papers. Drapeau told Nutting that "a number of growers . . . would very much like to break the contract," but that same month a jury had upheld the California Prune and Apricot Growers' contractual right to liquidated damages.

10. In his suit for specific performance, Nutting alleged that the CARC failed to deliver the full quantity agreed upon, failed to grant discounts as specified, and delivered substandard raisins (Nutting to Hartley F. Peart, June 21, 1918, File 2; Fred A. Seymour to CARC, Nov. 4, 1914, File 1; Nutting to James Madison, July 5, 1916, File 1, all in Carton 3, Nutting Papers).

11. G. Carroll Todd, Assistant to Attorney General, to Albert Schoonover, U.S. Attorney, July 27, 1917, File 1, Box 559, BAI, RG 60; on Nutting's background and relationship to his father, see Nutting, *Interview with Franklin P. Nutting*, 9–20. For the complaints of Inderrieden and others, see Nutting to ASRC, July 10, Aug. 2, 1918; Nutting to Alex Goldstein, July 18, 20, 1918; Nutting to Simon W. Hathaway, July 22, 1918, all in File 2, Carton 3, Nutting Papers. The FTC opened its files to DOJ attorneys, who were "glad to examine [the FTC's] data" on the CARC (John Walsh, Chief Counsel, FTC, to T. W. Gregory, U.S. Attorney General, May 25, 1918; G. Carroll Todd to Walsh, May 28, 1918, both in File 1, Box 559, BAI, RG 60).

12. Nutting to ASRC, July 20, 1918, File 2, Carton 3, Nutting Papers; Nutting to ASRC,

July 25, 1918, File 9, Carton 5, Nutting Papers. Nutting reported to his Fresno office: "The officials of the Commission seem very much impressed with our case, so much so that we would not be able to withdraw it now if we wanted to" (July 20, 1918). Giffen had also begun marshaling forces; at his behest, *FMR* editor Chester H. Rowell wrote to FTC commissioner Victor Murdock to help him understand the "Fresno atmosphere" from the growers' point of view (Giffen to Rowell, Aug. 8, 21, 1918, both in Box 14, Rowell Papers).

13. William B. Colver, Chair, FTC, "Memorandum to the Commission in the Matter of the Raisin Situation," Aug. 2, 1918, pp. 1, 3–5, File 1, Box 559, BAI, RG 60. The memorandum recommended that public officials be appointed as trustees and that every phase of the operations of the association be conducted under the supervision of the state and federal governments.

14. Nutting to George H. McCandless, Sept. 4, 1918, File 13, Carton 5, Nutting Papers. On the FTC's first years and the political issues raised by its investigations, see McCraw, *Prophets of Regulation*, 143–52; and Sklar, *Corporate Reconstruction of American Capitalism*, 328–32.

15. Nutting to ASRC, Aug. 2, 1918, File 2, Carton 3, Nutting Papers; Nutting to ASRC, Aug. 21, 1918, File 13, Carton 5, Nutting Papers: "Though it may be very difficult to get the Department of Justice to take action I have arranged to take up the matter through the White House if necessary to get action."

16. Nutting to ASRC, Aug. 17, 1918, File 9, Carton 5, Nutting Papers; Bonner to Nutting, Aug. 27, 1918, File 6, Carton 3, Nutting Papers.

17. Bonner to Nutting, Aug. 27, 1918, File 6, Carton 3, Nutting Papers; Nutting to ASRC, Sept. 15, 1918, File 9, Carton 5, Nutting Papers. Nutting worried that Giffen was trying to get the Food Administration to intervene with the FTC and forestall the packers' complaint: "I have some fear from the attitude of the Food Administration that Giffen might possibly get them to indicate to the Trade Commission that the Food Administration would prefer that there be no interference with the Raisin Association" (Nutting to Alex Goldstein, July 20, 1918, File 2, Carton 3, Nutting Papers).

18. Guth, "Farmer Monopolies," 69. On the Justice Department's investigations of the milk producers and the milk industry between 1917 and 1933, see File 60-139-7, Boxes 488, 490, 2805 (*U.S. v. Borden*); 493, 493A, 494 (Chicago milk shed); 489 (Southern Dairies Inc., of Miami); 489C (Milwaukee milk shed), Department of Justice Central Files, Classified Subject Files, RG 60; and Milk, 1936, Box 2383, Milk 2 (Hearings), 1938, Box 2826; Milk, 1938, Box 2827 (Boston milk case), Secretary's Correspondence, RG 16. The *Borden* case is discussed in the Conclusion. For the history of the lawsuits against milk producers and dairy associations and an assessment of the political movements in Congress for new legislation on cooperatives, see Guth, "Farmer Monopolies," 76; and U.S. Congress, Senate, Committee on the Judiciary, *Hearings Authorizing Associations of Producers of Agricultural Products*, 143–46 (testimony of Milo Campbell, president of the National Milk Producers' Federation) (hereafter *1921 Hearings*).

19. Rep. J. A. Elston (Alameda) to A. C. Wyckoff, Sept. 20, 1919, Raisin Committee Correspondence File, Carton 7, Nutting Papers.

20. Howard, *History of the Sun-Maid Raisin Growers*, 13. Others disputed the impact of Prohibition on the raisin industry: "Home brewing is said to have been only an unimportant factor in the price of raisins in 1919" (Camp, "Organization of Agriculture," 306 n. 1).

21. Testimony of George H. McCandless, in Federal Trade Commission, *In Re California*

*Associated Raisin Company, Investigation upon Application of Attorney General under Section 6(e) of the Federal Trade Commission Act*, Transcript of Proceedings, typescript, Washington, D.C., 1 (Nov. 24, 1919): 540 (hereafter *FTC Transcript*); statements introduced in evidence by John W. Preston, attorney for packers, *FTC Transcript* 1 (Nov. 25, 1919): 569–70; F. A. Seymour, "Concerning 1918 Crop Raisin Prices," *SMH* 4 (Aug. 1918): 1; H. G. Johnson, "New Raisin Prices Assure Fair Profit," *SMH* 4 (Aug. 1918): 2; "Good Season Ahead," *SMH* 4 (Aug. 1918): 8; "Concerning Rain Damage," *SMH* 4 (Nov. 1918): 5.

22. Giffen, "Raisin Prices for 1919," *SMH* 5 (Sept. 1919): 1–2; Nutting, *Interview with Franklin P. Nutting*, 41, Bancroft Library.

23. Steen, "California Raisin Trust," 2356; Giffen, "The Price of 1919 Raisins," *SMH* 4 (July 1919): 1–2; *FTC Transcript* 1 (Nov. 25, 1919): 577–83; Howard, *History of the Sun-Maid Raisin Growers*, 13; "Raisin Prices for 1919," *SMH* 5 (Sept. 1919): 1; Bragg, "History of Cooperative Marketing," 184; William A. Sutherland to Henry S. Mitchell, Assistant Attorney General, Sept. 24, 1919, File 2, Box 559, BAI, RG 60; Camp, "Organization of Agriculture," 305, 307.

24. George H. McCandless to Nutting, Sept. 12, 1919; Nutting to McCandless, Oct. 23, 1919, File 13, Carton 5, Nutting Papers; "Raisin Trust May Have Competition," *New York Journal of Commerce and Commercial Journal*, Aug. 20, 1920; Alfred W. McCann, "What Is Sauce for the Goose Isn't for Gander," *Guide and Commercial Advertiser*, Feb. 13, 1920; Fred R. Drake to A. Mitchell Palmer, Aug. 23, 1919, File 1, Box 559, BAI, RG 60; various letters from jobbers, wholesalers, and retailers to the DOJ, August and September 1919, Boxes 559 and 560, BAI, RG 60.

25. Wylie M. Giffen to Henry S. Mitchell, Sept. 30, 1919, File 2, Box 559, BAI, RG 60; see also W. A. Sutherland to Henry S. Mitchell, Sept. 30, 1919, ibid.

26. Breed, Abbott & Morgan to Alex Goldstein, Sept. 22, 1919, Raisin Committee Correspondence File, Carton 7, Nutting Papers. On the retention of counsel, see Nutting to Breed, Abbott & Morgan, Jan. 15, 1920, File 4, Carton 3, Nutting Papers.

27. Goldstein to Breed, Abbott & Morgan, Sept. 23, 1919, Raisin Committee Correspondence File, Carton 7, Nutting Papers; A. Mitchell Palmer to William B. Colver, Sept. 30, 1919, File 2, Box 559, RG 60; Coben, *A. Mitchell Palmer*, 156–70; remarks by Commissioner Murdock, *FTC Transcript* 1 (Nov. 20, 1919): 6. Apparently Giffen, William Sutherland, and Frank H. Wilson, president of the California Peach Growers' Association, were all present in the House gallery when an amendment to strip the appropriations bill of the rider was defeated on Sept. 20 ("Trusts" Scrapbook, Carton 6, Nutting Papers).

28. Guth, "Farmer Monopolies," 72; Alex Goldstein to Breed, Abbott & Morgan, Oct. 28, 1919; Rep. J. A. Elston to A. C. Wyckoff, Sept. 20, 1919, both in Raisin Committee Correspondence File, Carton 7, Nutting Papers; Act of Nov. 4, 41 Stat. 336, chap. 93 (1919).

29. C. B. Ames, Assistant to Attorney General, to President, CARC, Sept. 2, 1919, File 2, Box 559, BAI, RG 60; statement of John W. Preston, *FTC Transcript* 1 (Nov. 25, 1919): 562; Nutting to Breed, Abbott & Morgan, Jan. 15, 1920, File 4, Carton 3, Nutting Papers; Sutherland to Henry S. Mitchell, Sept. 24, 30, 1919; Sutherland to Victor Murdock, Oct. 1, 1919, all in File 20-1-4, Box 31, RG 122. A former San Francisco district attorney, Preston had with Albert H. Elliot and Hartley F. Peart represented a group of milk distributors in a *quo warranto* proceeding against milk associations during the war. See *San Francisco Herald*, Sept. 19, 1919, "Trusts" Scrapbook, Carton 6, Nutting Papers; *1921 Hearings*, 7 (testimony of John W. Preston).

30. Nutting to McCandless, June 12, 1919; Nutting to ASRC, Oct. 23, 1919; McCandless to Nutting, Oct. 30, 1919, all in File 13, Carton 5, Nutting Papers.

31. McCandless to Nutting, Oct. 30, 1919, File 13, Carton 5, Nutting Papers; Guggenheim and Co., Rosenberg Brothers and Co., Bonner Packing Co., and Chaddock and Co., to James W. Ackerly, Breed, Abbott & Morgan, Sept. 19, 1919, File 2, Box 559, BAI, RG 60.

32. Bragg, "History of Cooperative Marketing," 133. In 1918, Giffen said the CARC would have gone out of business if it were not for the commercial packers (*FTC Transcript* 1 [Nov. 24, 1919]: 370).

33. Statement of John W. Preston, *FTC Transcript* 1 (Nov. 24, 1919): 11–12, 111.

34. Ibid., 81, 82, 87, 17, 19, 61. Sutherland responded that the CARC's profits came only from the one-quarter-cent fund; the association distributed all proceeds from raisin sales to the growers. He claimed the CARC's prices were not designed to build up the surplus (although it in fact had accumulated larger surpluses every year since 1914; see Table 6.2).

35. Testimony of Charles Bonner, *FTC Transcript* 2 (Nov. 28, 1919): 878–87.

36. *FTC Transcript* 1 (Nov. 25, 1919): 653, 661. To explain the CARC's 1919 price, Sutherland enumerated and placed a value on all the tasks of a season's work in a vineyard. Sutherland argued that the growers were entitled to compensation for the work they performed themselves, not just reimbursement for the labor they hired. And when one crop was unprofitable because of circumstances beyond the growers' control, as had occurred in 1918, their organization was entitled to make up the shortfall the following year (*FTC Transcript* 2 [Nov. 28, 1919]: 967–1019).

37. William A. Sutherland, opening statement, *FTC Transcript* 1 (Nov. 24, 1919): 331; ibid. 1 (Nov. 25, 1919): 723; see also Sutherland to Henry S. Mitchell, Assistant Attorney General, Sept. 24, 1919, File 2, Box 559, BAI, RG 60.

38. Statements of Preston, *FTC Transcript* 1 (Nov. 24, 1918): 12, 545; ibid. 2 (Dec. 3, 1919): 2059. The government's own attorneys said little during the hearings. Assistant Attorney General Henry S. Mitchell occasionally asked witnesses questions, but he did not formally examine anyone called by either side and throughout gave no indication of the DOJ's position.

39. *Report of the FTC to the Attorney General*, June 8, 1920, p. 8, File 3, Box 559, BAI, RG 60.

40. Ibid., 23–24; Sklar, *Corporate Reconstruction of American Capitalism*, 328.

41. William B. Colver, Chair, FTC, "Memorandum to the Commission in the Matter of the Raisin Situation," Aug. 2, 1918, p. 1, File 1, Box 559, BAI, RG 60.

42. Mitchell to Sutherland, July 7, 1920, File 3, Box 559, BAI, RG 60. The government decided on the 60 percent figure on the basis of federal court decisions dealing with monopolies in corn oil, cash registers, meatpacking, tobacco, and anthracite coal; see *Patterson v. U.S.*, 222 F. 599 (1915); *U.S. v. American Tobacco Co.*, 164 F. 700 (1908); *U.S. v. Corn Products Refining Co.*, 234 F. 964 (1916); *Swift and Co. v. U.S.*, 196 U.S. 375 (1905); *U.S. v. Reading Co.*, 226 U.S. 324 (1912); Plaintiff's Points and Authorities, *U.S. v. CARC*, Sept. 15, 1920, Folder 2, Box 77, RG 21.

43. Giffen to Mitchell, July 16, 26, 1920, Sutherland to Henry S. Mitchell, July 6, 9, 1920; Mitchell to Sutherland, July 7, 1920; all in File 3, Box 559, BAI, RG 60; on Sutherland's interpretation of the agricultural antitrust exemption, see Sutherland to Mitchell, Sept. 30, 1919, File 3, Box 559, BAI, RG 60.

44. Howard, *History of the Sun-Maid Raisin Growers*, 13–14; Giffen to Victor Murdock, Feb. 18, 1920, File 20-1-3, Box 30, RG 122.

45. "Raisin Trust May Have Competition," *New York Journal of Commerce and Commercial Bulletin*, Aug. 27, 1920; Giffen to Mitchell, July 16, 1920, both in File 3, Box 559, BAI, RG 60; *New York Journal of Commerce and Commercial Bulletin*, Aug. 7, 1920, "Trusts" Scrapbook, Carton 6, Nutting Papers. See also "What Is Sauce for the Goose Isn't for the Gander," *Guide and Commercial Advertiser*, Feb. 13, 1920; John Preston to Attorney General, Aug. 2, 1920, both in File 3, Box 559, BAI, RG 60; J. B. Inderrieden Co. to John Preston, Aug. 3, 1920, File 20, Carton 3, Nutting Papers. Preston also suggested that the government interview former CARC general manager and treasurer James Madison (Preston to Mitchell, Sept. 4, 8, 1920; Mitchell to Preston, Sept. 9, 1920, all in File 3, Box 559, BAI, RG 60). F. Y. Foley, an independent packer, purchased almost all of the raisins sold at the auction at eight cents higher than 1919 prices (Giffen to C. B. Ames, Aug. 6, 1920, File 3, Box 559, BAI, RG 60). On Palmer's antitrust enforcement decisions, see Coben, *A. Mitchell Palmer*, 188–95.

46. Palmer to O'Connor, Sept. 5, 8, 1920; O'Connor to Palmer, Sept. 7, 1920; all in File 3, Box 559, BAI, RG 60; Bill of Complaint and Petition, *U.S. v. California Associated Raisin Company, et al.*, Sept. 8, 1920, Case No. B-67 in equity, U.S. District Court, Southern District of California, Northern Division, File 3, Box 559, BAI, RG 60; case file, *U.S. v. CARC*, Folder 2, Box 77, RG 21.

47. Preston to Mitchell, Sept. 24, 1920; "Raisin Industry Must Be Fostered, Not Hampered," Hearst Papers editorial, clipping, n.d.; see also Chester H. Rowell, "Fight Raisin Case Through," *FMR*, clipping, n.d. but probably September 1920, all in File 3, Box 559, BAI, RG 60. John Walsh replaced Sutherland as CARC counsel in August 1920 (Sutherland closed his practice when he went into banking) and was in turn replaced by Carl Lindsay and David Ewing; by October 1920, Ewing had been replaced by his partner O. L. Everts (Giffen to Ames, Aug. 10, 1920; John H. Atwood to Henry S. Mitchell, Sept. 23, 1920, Stipulation, Oct. 15, 1920, *U.S. v. CARC*, all in File 3, Box 559, BAI, RG 60).

48. Steen, "California Raisin Trust," 2356.

49. John H. Atwood to Henry S. Mitchell, Sept. 17, 23, 1920; James D. Phelan to A. Mitchell Palmer, Sept. 11, 1920, all in File 3, Box 559, BAI, RG 60. Phelan also registered a protest with President Wilson: "Confidentially, what [Palmer] has already done jeopardizes any chance Cox or myself had for the California vote and is a gross injustice not only to the producer but to his party candidates" (Phelan to J. P. Tumulty, Secretary to the President, Sept. 11, 1920, File 3, Box 559, BAI, RG 60). The Democrats—including Phelan—were swept from office in the November elections.

In response to the campaign to have growers sign affidavits of noncoercion, the U.S. attorney told the Fresno newspapers that proving the night riding was not crucial to the government's case. Yet, when the case was settled in 1922, among the consent decree's minimal prohibitions was one against the taking of contracts under duress (*FMR*, Sept. 16, 1920, "Trusts" Scrapbook, Carton 6, Nutting Papers; Consent Decree, *U.S. v. CARC*, Jan. 18, 1922, File 12, Carton 2, Nutting Papers). Armenians were probably willing to sign affidavits because they resented the implication that all Armenians were disloyal. In fact, not all Armenians had been coerced into joining the CARC, but many Armenians found it shameful to be treated as a racial minority (*FMR*, Sept. 10, 1920, in "Trusts" Scrapbook, Carton 6, Nutting Papers).

50. Ruling on Motion to Discharge Order to Show Cause, *U.S. v. California Associated Raisin Company*, Sept. 22, 1920, File 3, Box 559, BAI, RG 60.

51. Bledsoe to Robert O'Connor, U.S. Attorney, Nov. 24, 1920, File 3, Box 559, BAI, RG 60; Transcript of Oral Argument, *U.S. v. CARC*, Oct. 15–16, 1920, U.S. District Court, Southern District of California, Northern Division, Documents Library, University of California at Berkeley; Nutting to Breed, Abbott & Morgan, Nov. 30, 1920, File 4, Carton 3, Nutting Papers.

52. Dana T. Ackerly to Nutting, Nov. 23, 1920, File 7, Carton 3, Nutting Papers; *U.S. v. E. C. Knight*, 156 U.S. 1 (1895); *Swift and Co. v. United States*, 196 U.S. 375 (1905).

53. O'Connor to Attorney General, Dec. 2, 1920; Atwood to O'Connor, Dec. 3, 1920; Atwood to Palmer, Jan. 13, 1921, all in File 3, Box 559, BAI, RG 60; Atwood to Franklin K. Nebeker, Special Assistant to the Attorney General, Jan. 4, 1921; Order, *U.S. v. CARC*, Jan. 3, 1921, both in File 4, Box 560, BAI, RG 60.

54. Preston to Nutting, Jan. 6, 1921, File 32, Carton 3, Nutting Papers; John H. Atwood to Palmer, Jan. 15, 1921, File 3, Box 559, BAI, RG 60; Preston to Robert Duncan, Jan. 22, 1921, File 32, Carton 3, Nutting Papers. A settlement conference on January 14, 1921, produced no accord. The packers were theoretically free to pursue contracts with growers, but they took a dim view of their prospects of wresting away much of the CARC's control (Charles Bonner to Nutting, Jan. 15, 1921, File 6, Carton 3, Nutting Papers). The government had prior experience with the judge; Bledsoe ruled against the government in a case to recover California lands sold to the Southern Pacific Railroad that turned out to have substantial oil reserves, after deciding in the government's favor in an earlier case (Coben, *A. Mitchell Palmer*, 192–93).

55. Charles G. Bonner to John W. Preston, Feb. 19, 1921, File 6, Carton 3, Nutting Papers; *1921 Hearings*, 18. Bonner reported that growers were unhappy with the long term of the contract, its lien on the land, and the requirement that growers sign up all their land; in addition, growers had received only ten cents per pound on the 1920 crop, after the CARC had promised fifteen. Bonner predicted that the CARC would be lucky to sign up 75 percent (Bonner to Preston, Feb. 17, 1921, File 6, Carton 3, Nutting Papers).

56. Preston to Harry M. Daugherty, May 23, Aug. 11, 18, 1921; Guy D. Goff, Assistant to the Attorney General, to Preston, Aug. 22, 1921, all in File 4, Box 560, BAI, RG 60.

57. *U.S. v. U.S. Steel*, 251 U.S. 417 (1920). *U.S. Steel* all but overruled the jurisdictional barriers erected in *Knight* and established criteria upon which the government's amended brief heavily relied: whether the company controlled a preponderance of the product, arbitrarily fixed and maintained prices, engaged in unfair competition (preferential dealing and coercion), restricted production, and used secret rebates to undercut competitors.

58. Plaintiff's Points and Authorities, Sept. 15, 1920; Defendant's Reply Points and Authorities, Oct. 25, 1920, both in *U.S. v. CARC*, Folder 2, Box 77, RG 21. U.S. Attorney O'Connor, Assistant Attorney General Mitchell, and their staffs were all replaced by the Harding administration (Preston telegram, June 10, 1921, File 32, Carton 3, Nutting Papers). On the decline in antitrust enforcement, see Keller, *Regulating a New Economy*, 33–37.

59. Text of H.R. 13931, *Cong. Rec.*, 66th Cong., 2d sess., 7851. The bill was named for Senator Arthur Capper of Kansas and California representative Hugh S. Hersman; the following year, Minnesota representative Andrew Volstead reintroduced the legislation in the House. Capital stock organizations were included apparently because of the influence of western

cooperatives, particularly the California fruit organizations (Remarks of Senator Smith and Senator Walsh, *Cong. Rec.*, 66th Cong., 3d sess., 364–65, 316–18, 360–76; Guth, "Farmer Monopolies," 72–73, 78). The House gave enforcement to the secretary of agriculture; the Senate, concerned that the USDA would not know how to determine what constituted "unduly enhanced prices" under federal law, transferred this responsibility to the FTC.

60. Compare drafts of H.R. 13931, *Cong. Rec.*, 66th Cong., 2d sess., 7851 (Capper-Hersman), with H.R. 2373 (Capper-Volstead), in *1921 Hearings*, 139; Guth, "Farmer Monopolies," 73. The Farm Bloc was formed by a group of senators from agricultural states in May 1921 to formulate "legislation beneficial to agriculture" (Lowitt, *George W. Norris*, 172–73). Lowitt contends that most of the bloc's members, including Nebraska Republican Norris, deserted the group when it became apparent that the Farm Bloc was "being used for partisan political purposes" by the incumbent Republican administration (ibid., 173–74). The only biographer of Arthur Capper implies that the Farm Bloc thwarted as much as facilitated the president's agricultural relief program (Socolofsky, *Arthur Capper*). Capper's memoir, *The Agricultural Bloc*, is not useful on this point.

61. *1921 Hearings*, 153.

62. The farm sector was never satisfied with the Clayton Act. Immediately after its passage in 1914, CFGE attorney George Farrand commented on the law's ambiguity, noting, "If its purpose was to exempt labor unions and farmer organizations . . . from the effect of all anti-trust laws, more apt language could have been used to convey that idea" (Farrand, "The Federal Trade Commission Act and the Clayton Bill, Remarks to the Board of Directors of the California Fruit Growers Exchange," Nov. 11, 1914, Cooperation 1914, Secretary's Correspondence, RG 16).

63. Jensen, "Integrating Economic and Legal Thought on Cooperatives," 898. With one exception, the courts had provided little guidance as to the limits of the Clayton Act exemption. In *U.S. v. King*, 229 F. 275 (D.C.D. Mass. 1915) and 250 F. 908 (D.C.D. Mass. 1916), a district court held that farmers could not engage in tactics prohibited by the Sherman Act, specifically a secondary boycott enforced against nonmembers through blacklists and threats. The U.S. attorney ignored the Clayton Act, sued association members individually, and won a ruling that the Aroostook Potato Shippers' Association did not come within the Clayton exemption. Farmers believed that this ruling effectively nullified the Clayton Act. As Franklin D. Jones, one of the attorneys for the raisin packers in *U.S. v. CARC*, wrote, "The section was so drawn that such an organization [a nonstock cooperative] or its members committing acts of a character violating the anti-trust acts remained liable to the penalties of these laws" ("Status of Farmers' Co-operative Associations," 596). The Clayton Act permitted farmers to engage only in activities that were not "otherwise illegal," and courts had ruled in labor cases that the Sherman Act prohibited secondary boycotts. The farm groups misconstrued *King* as antagonistic to farmers when it was hostile to illegal interference with interstate commerce.

64. Keller, *Regulating a New Economy*, 36. For example, when the Supreme Court held that labor unions could not engage in secondary boycotts despite the Clayton Act exemption, the farm groups claimed that the exemption might also be judicially circumscribed as to agriculture. See *Duplex Printing Press Co. v. Deering*, 254 U.S. 443 (1921).

65. Bonner to Nutting, May 18, 1920, File 6, Carton 3, Nutting Papers; Dana T. Ackerly to Preston, May 19, 1920, File 7, Carton 3, Nutting Papers.

66. *Cong. Rec.*, 67th Cong., 1st sess., 1033–46; Nutting to Breed, Abbott & Morgan, Nov. 30, 1920, File 4, Carton 3, Nutting Papers.

67. *1921 Hearings*, 139 (statement of Senator Walsh), 38–39 (testimony of Carl E. Lindsay), 146 (testimony of Milo D. Campbell, National Milk Producers' Federation).

68. Henry C. Wallace to George W. Norris, June 4, 1921, reprinted in *1921 Hearings*, 133.

69. *1921 Hearings*, 139.

70. *Cong. Rec.*, 67th Cong., 2d sess., 2050, 2157, 2123; Guth, "Farmer Monopolies," 73. In the hearings, Walsh implied that charges of DOJ harassment were greatly exaggerated. The farm groups played their role as victims of a vengeful attorney general to the hilt and professed innocence at all charges of possible monopoly. The Senate amendments compelled the bill's proponents to carry the farmers-as-victims characterization still further. Senator Thomas Sterling (R-S.D.) argued, "Ambitious U.S. district attorneys or persons envious of or feeling that their business might possibly be injured by a company or an association of farmers would be quick to seize upon section 2 of the Sherman Antitrust Act for the purpose of instituting a prosecution" (*Cong. Rec.*, 67th Cong., 2d sess., 2218; Shideler, *Farm Crisis*, 113).

71. *Cong. Rec.*, 67th Cong., 2d sess., 2058.

72. The political advantage in championing the cause of farmers meant, of course, that the cooperative marketing bill became a political football, lobbed by president and Congress, Republicans and Democrats, agricultural representatives and urbanites alike. See Shideler, *Farm Crisis*, 95–123, 217–42, and Socolofsky, *Arthur Capper*, 145–65.

73. Wallace to Senator George Norris, Aug. 2, 1921, Bills, 1921, Secretary's Correspondence, RG 16; *Cong. Rec.*, 67th Cong., 2d sess., 2270, 2282, 2582. The House quickly agreed to the Senate changes, and President Harding signed the bill on February 18. See *Cong. Rec.*, 67th Cong., 2d sess., 2453–55, 2523, 2582, 2715, 3172; Capper-Volstead Act, 42 Stat. 388 (Feb. 18, 1922); Guth, "Farmer Monopolies," 80–82; Shideler, *Farm Crisis*, 113–14, 169–70.

74. Nutting to Bonner, May 17, 1920; Nutting to Preston, May 10, 1920, both in File 4, Carton 3, Nutting Papers; Rep. Clarence F. Lea to Preston, May 19, 1920; Preston to Alex Goldstein, Dec. 7, 1920; Preston telegram to office, June 3, 1921, all in File 32, Carton 3, Nutting Papers.

75. Consent Decree, *U.S. v. CARC*, Jan. 18, 1922, Folder 1, Box 76, RG 21; Breed, Abbott & Morgan to Preston, Dec. 12, 1921, File 32, Carton 3, Nutting Papers. As originally drafted, the proposed decree would have canceled all existing crop contracts and the voting trust agreement, required the CARC to redeem all par stock that growers wished to surrender, and "enjoin[ed] defendants from in any manner acquiring so large a part of trade and commerce in raisins produced in California and shipped throughout the country as to enable defendants to arbitrarily control and fix prices" (Proposed Decree in Raisin Case, n.d., File 5, Special Correspondence Section, Box 560, BAI, RG 60). The final decree said nothing about invalidating the voting trust agreement or redeeming par stock (Consent Decree, *U.S. v. CARC*, Jan. 18, 1922, File 12, Carton 2, Nutting Papers).

76. Preston to Senator William H. King (D-Utah), Dec. 17, 1921, File 32, Carton 3, Nutting Papers. Preston pleaded with King to intercede with the DOJ; King reported that he urged "that any decree entered be fair and fully protect consuming public" (King to Preston, Dec. 19, 1921, File 32, Carton 3, Nutting Papers).

77. Steen, "Story of the California Raisin 'Trust,'" 532.

78. Hence the consent decree's prohibition on the CARC's use of "firm at opening price" contracts or any device the CARC used to maintain prices against subsequent decreases (Consent Decree, *U.S. v. CARC*, 2).

*Chapter 8*

1. Jensen, "Integrating Economic and Legal Thought on Cooperatives," 899. See also Jesness, "Bill of Rights of U.S. Cooperative Agriculture."

2. Wylie M. Giffen, "Government and Raisin Growers Reach Agreement," *AG* 3 (Feb. 1922): 6.

3. *In re Application of California Associated Raisin Company for Change of Name*, Case No. 28348, Feb. 17, 1922, Civil Docket, Fresno County Superior Court; Al C. Joy, "Sun-Maid Raisin Growers," *AG* 3 (Mar. 1922): 5.

4. Robert Welles Ritchie, "Rescuing a Raisin Maid: Heroic Efforts Won the Great Cooperative's Fight," *CG* 88 (July 7, 1923): 3–4; F. A. Seymour, "The Sun-Maid Raisin Record for 1922," *AG* 5 (Jan. 1923): 5, 38; Wylie M. Giffen, "Analysis of 1921 Raisin Crop Returns," *AG* 5 (Jan. 1923): 11. By 1922, raisins were falling out of favor as a raw material for bootleg liquor (Ritchie, "Rescuing a Raisin Maid," 3–4). In 1921 and 1922, Sun-Maid spent $2.4 million on advertising and invested $3.5 million in building and leasing packing plants; when prices dropped in 1922 Sun-Maid's cash reserves were depleted (Franklin Bell, "Raisin Advertising Appropriation," *AG* 4 [Aug. 1922]: 5; Wylie M. Giffen, "The Raisin Option and the Packers," *AG* 4 [Aug. 1922]: 7; Giffen, "The 1922 Raisin Prices," *AG* 4 [Sept. 1922]: 5; Bragg, "History of Cooperative Marketing," 214). Sun-Maid billed the growers for overpayments on two highly specialized varieties in 1920. Although growers still netted far more than the guaranteed advance on standard varieties, bitterness over the billback was exacerbated by the apparently widespread impression that the Sun-Maid office was becoming "a sort of 'orphans' home' for relatives and friends of officers of the association" (Report of Special Agent Ralph H. Colvin, June 1, 1923, p. 3, File 60-166-29, Box 561, BAI, RG 60 [hereafter Colvin Report]; Giffen, "1922 Raisin Prices," 12).

5. Ralph P. Merritt to Henry C. Wallace, May 31, 1923; Associations, 1923; Secretary's Correspondence, RG 16; Merritt, "Sun-Maid Program," *AG* 5 (Mar. 1923): 5 ("The contract calls for payment to our Growers at an average price of $85.00 per ton, yet the banks would only loan us $70.00 and $80.00 per ton"); Ritchie, "Rescuing a Raisin Maid," 4. Sun-Maid borrowed $8 million to pay the growers' advance in 1922, using warehouse receipts for the crop as collateral. DOJ agent Ralph H. Colvin later reported this figure at $9 million, $8 million of which had come from a bank with close connections to Sun-Maid. After W. A. Sutherland left his practice during the CARC litigation, he became president of the Pacific-Southwest Trust and Savings Bank, the Fresno branch of which he co-owned with Wylie Giffen (Colvin Report, 3).

6. P. Malloch, "Report on the Raisin Industry in California (1929)" typescript [photocopy], 1, Giannini Foundation of Agricultural Economics Library, University of California, Berkeley. The high prices growers received during the early 1920s drew four thousand growers and speculators into the industry. Raisin acreage rose from 322,000 to 664,000 by 1922. See "Raisins: California Leads the World in a World-Wide Industry," *Western Canner and Packer*, July 1934, 10; C. F. Tucker, "Sun-Maid Side-Lights," *California Cultivator* 62

(Feb. 16, 1924): 187, 194; Ritchie, "Rescuing a Raisin Maid," 3; S. Q. Grady, "How to Help Move 20,000 Tons of Thompsons in 5 Cent Packages," *AG* 2 (Aug. 1921): 6.

7. Bragg, "History of Cooperative Marketing," 115; Forrester, *Report upon Large-Scale Cooperative Marketing*, 48. The DOJ noted that Sun-Maid officers and directors were "personally liable" for the bank loans (Colvin Report, 3). The raisin-growing community respected Giffen as an honest and selfless person—he never took a salary from the CARC in twelve years as officer and trustee. When the financial crisis in the industry cost him his own vineyards in 1923, he quietly resigned as president. Of Giffen, local attorney Louis C. Levy wrote, "From shirt-sleeves to millionaire and bank president, and back to shirt-sleeves usually requires three generations. It is said Mr. Giffen made the course in fourteen years." Giffen paid his debts and avoided bankruptcy; he emerged to lead the Grape Stabilization Program during the early 1930s. He died of complications of diabetes in 1936 (Levy quoted in Bragg, "History of Cooperative Marketing," 215 n.; "Ex-President Giffen," *AG* 5 [Nov. 1923]: 12; Jean Giffen Wiley, "Three Generations of Giffen Ranchers in California's San Joaquin Valley," 32A, unpublished manuscript, 1989 (photocopy in author's possession); Al C. Joy, "Raisin Association Saved from Crash," *San Francisco Journal*, Feb. 1, 1924).

8. Merritt, *"After Me Cometh a Builder,"* 3–4, 112–17, Bancroft Library; Ben R. Walker, "Merritt, Man of Accomplishments," *FMR*, Apr. 2, 1923, p. 11. For one luminous account of Merritt's "rescue" of the rice growers and its importance to Sun-Maid's directors, see Joy, "Raisin Association Saved from Crash."

9. Ritchie, "Rescuing a Raisin Maid," 4; Tucker, "Sun-Maid Side-Lights," 194; Ralph P. Merritt, "Sun-Maid Program," *AG* 5 (Mar. 1923): 1. Merritt had no experience in raisin production or marketing, and the packer Franklin Nutting believed he was appointed president of Sun-Maid to pacify Fresno bankers (Nutting, *Interview with Franklin P. Nutting*, 53–59; see also George E. Popovich, "History of the Sun-Maid Organization and Its Antecedents, 1958," typescript [photocopy], 18–19, Sun-Maid Archives).

10. Merritt to Wallace, Mar. 1, 1923, Associations, 1923, Secretary's Correspondence, RG 16. The concept of separate incorporation for the warehousing/packing facilities and marketing functions was not new; Wylie Giffen had done the same thing briefly in 1914, and at this time attorney Aaron Sapiro was recommending the idea to tobacco, cotton, and wheat farmers in the rest of the nation (Ellis, "Robert Worth Bingham and the Crisis of Cooperative Marketing").

11. The subsidiary was organized because under California law all stock had to carry voting privileges and thus nonproducers buying stock in the cooperative would have the right to vote (Merritt to Wallace, Mar. 1, 1923, Associations, 1923, Secretary's Correspondence, RG 16).

12. Ritchie, "Rescuing a Raisin Maid," 4; Merritt to Wallace, Mar. 1, 1923, Associations, 1923, Secretary's Correspondence, RG 16; "Valley Bankers to Aid Growers," *FMR*, Mar. 27, 1923, p. 1. Merritt planned to sell $2 million in bonds on the plants and equipment and another $2.5 million in new stock subscriptions. The bonds would be gradually retired through the use of a sinking fund, funded by a $4 per ton charge levied on all members. Growers would receive the face amount in preferred stock of the Delaware company, and, once "sufficient working capital ha[d] been accumulated," the money would be used to pay off the bonds (Merritt to Wallace, Mar. 1, 1923; Henry E. Erdman, interview with Mr. Wrightson, Sun-Maid Raisin Association, Dec. 6, 1924, Box 18, Erdman Papers).

Wrightson noted that all the common stock in the Delaware company was controlled by the board of directors of the Sun-Maid Raisin Growers of California.

William A. Sutherland, through the Pacific Southwest Trust and Savings Bank, helped to secure temporary financing for Sun-Maid during the reorganization; he informed growers that the demand notes would be acceptable as collateral for bank loans, and he urged them to seek loans rather than demand payment from Sun-Maid before the reorganization was completed ("Valley Bankers to Aid Growers," 2).

13. Henry E. Erdman, "Impressions of Sun-Maid Raisin Growers," Nov. 12, 1924, typescript [photocopy], Box 18, Erdman Papers; Erdman, interview with F. A. Stewart, Growers' Service Division of Sun-Maid, Apr. 24, 1925, Box 18, Erdman Papers; F. A. Stewart, "The Work of the Contract Department," *AG* 5 (Mar. 1923): 19. On the banks' support for Merritt's reorganization plan, see "Bankers Endorse Sun-Maid Plan," *AG* 3 (Mar. 1923): 8.

14. See Henry E. Erdman, interview with Harry M. Creech, Sun-Maid Raisin Growers, Nov. 12, 1924, Box 18, Erdman Papers. Merritt claimed to be a raisin grower by occupation, a claim that stemmed from his part-ownership of his friend Herbert Hoover's ranch near Wasco, which had "some" acres of raisin vines (Levy, *History of the Cooperative Raisin Industry*, 15). Whether this connection gave his attorney the right to claim on his behalf that he had "personally suffered along with others in the industry by reason of the stress thru [sic] which we have gone," however, is another matter. See Harry M. Creech, Sun-Maid Raisin Growers, to L. M. McLeod, Dec. 8, 1924, "Miscellaneous Information" File, Packing Plant Vault, Sun-Maid Archives.

15. 1923 Crop Contract, Associations, 1923, Secretary's Correspondence, RG 16; also published in *AG* 5 (Apr. 1923): 8, 28; Merritt to Wallace, May 31, 1923, Associations, 1923, Secretary's Correspondence, RG 16; "Sound Contract Is Basis of Association," *AG* 5 (Apr. 1923): 8. The withdrawal clause ran as follows: "The Seller is hereby given the right to file with the Buyer, between Dec. 20 and Dec. 31 of either of the years 1924, 1926, 1928, 1930, 1932, 1934, and 1936 a written notice of his desire to withdraw from this contract as to any or all of the varieties of raisins covered hereby, and thereupon said contract shall be canceled as to the designated variety or varieties of the crops of the succeeding years of the contract period."

16. Colvin Report, 21; "Fresno Banks Back Raisin Growers," *FMR*, Apr. 3, 1923, p. 10. The ad's sponsors were the Bank of Italy, the Growers' National Bank, the Pacific Southwest Trust and Savings Bank, United Bank and Trust Co., and Valley Bank. See also "Bankers Endorse Sun-Maid Plan," *AG* 5 (Mar. 1923): 8.

17. Ritchie, "Rescuing a Raisin Maid," 4; Memorandum, Lloyd S. Tenny to Wallace, Mar. 31, 1923, Sun-Maid Raisin Growers, General Subject Files, RG 83; "Campaign Extended Ten Days," *FMR*, Apr. 26, 1923, 1. Wallace wrote on his copy of the Tenny memo: "Mr Tenny: Had a fine meeting at Fresno 4-9-23." Merritt called upon the "financial powers" of San Francisco and Los Angeles for assistance, and "bank heads and industrial captains" in those cities contributed $250,000 and $150,000, respectively (Ritchie, "Rescuing a Raisin Maid," 4).

18. "Captains in Sun-Maid Drive Are Announced," *FMR*, Mar. 27, 1923, p. 13; Colvin Report, 5–6, 18–19; "Growers Call Mass Meeting," *FMR*, Apr. 25, 1923, p. 1.

19. Tucker, "Sun-Maid Side-Lights," 194; Giffen quoted in Ritchie, "Rescuing a Raisin Maid," 3; "Raisin Acreage Drive Extended through May 5th," *Fresno Bee*, Apr. 26, 1923, p. 1.

20. "Campaign Extended Ten Days," *FMR*, Apr. 26, 1923, pp. 1, 5. The paper reported

that a few incidents of night riding had taken place before Raisin Day but that no one had reported injuries or property damage.

21. "Woman Admits Shooting Man at Monmouth," *FMR*, Apr. 29, 1923, pp. 1, 6; "Jury Will Shift Ranch Shooting," *Los Angeles Times*, May 1, 1923, p. 9; Saroyan, " 'Turn Back the Universe and Give Me Yesterday,' " 54. Saroyan's account differs from the newspapers' in several key respects; he put the size of the mob at "two dozen men," and he said that Nazaret Torossian [*sic*] "stood on his front porch and in broken English said, 'Stop, I shoot' " (ibid.). In 1990 a leading grower recalled that Osterhaudt was employed by Sun-Maid (personal interview, Dick Markarian, Fowler, California, July 19, 1990).

22. Ritchie, "Rescuing a Raisin Maid," 4; "Bankruptcy for Sun-Maid Raisin Growers Monday If Acreage Campaign Fails," *Fresno Bee*, May 3, 1923, p. 1; "Raisin Drive at Standstill," *FMR*, May 4, 1923, p. 1.

23. "Grand Jury Anticipates Raisin Riots," *FMR*, May 4, 1923, p. 13; "Grand Jury to Investigate Raisin Frays," *FMR*, May 3, 1923, p. 1; "Fresno Celebrates Signing of Sun-Maid Contracts by 14,000 Growers of Raisins," *San Francisco Chronicle*, May 7, 1923, p. 4. At the town of Yettem, near Visalia, grower K. Khpehigian and Constable I. D. Sayre were wounded in separate incidents; in Biola, Philip Folmer was charged with assault with a deadly weapon in the shooting of Albin Johnson, who had been soliciting a signature from Folmer's Japanese tenant farmer. Another mob tied a rope around the neck of a defiant grower and tossed him into the San Joaquin River three times before the man "shook the water from his eyes and hair like a water spaniel and 'voluntarily' signed up." The paper reported that the district attorney's office was preparing to "press charges of riot against growers who engage in the destruction of property" ("Grand Jury Anticipates Raisin Riots," 13).

24. Sun-Maid Raisin Growers, "Report to the Secretary of Agriculture upon the Completion of Sun-Maid Raisin Growers Reorganization and Refinancing," Feb. 1–June 1, 1923, Associations, 1923, Secretary's Correspondence, RG 16; "Victory Crowns Drive," *FMR*, May 6, 1923, p. 1; Giffen, "New Era Is Here," *FMR*, May 6, 1923, p. 1.

25. "Shooting Is Investigated by Grand Jury," *FMR*, May 6, 1923, p. 1A.

26. "Notice to Any Growers," *FMR*, May 8 and 9, 1923; "Contracts Signed under Coercion to Be Relinquished," *FMR*, May 23, 1923, p. 1; Colvin Report, 7–8. The Justice Department counted fifty-five growers asking for cancellation. On the disposition of these contracts see text at note 50.

27. "Grand Jury Stops Probe of Raisin Riots," *FMR*, May 9, 1923, p. 9; Robert M. Walsh, "Raisin Industry Growth Marked by Violence," *Fresno Bee*, Aug. 19, 1984, p. G4; Colvin Report, 23–24; "Fresno Celebrates Signing of Sun-Maid Contracts," *San Francisco Chronicle*, May 7, 1923, p. 4. See Lovejoy to Gerald Thomas, Sun-Maid Raisin Growers, May 31, 1923, Associations, 1923, Secretary's Correspondence, RG 16 ("in no case was their [*sic*] sufficient evidence in any alleged case of coercion to justify the returning of an indictment").

28. "Advisory Council Hears of Continued Progress," *AG* 5 (Oct. 1923): 6; "The Raisin Crop Advance," *AG* 5 (Oct. 1923): 10; S. W. Shear, "Cooperative Marketing and the Raisin Industry," typescript, Oct. 1922, 3, Box 18, Erdman Papers.

29. Merris Belford Harris to Chester H. Rowell, Feb. 21, 1924, Box 15, Rowell Papers; "Full Payment of Old Association's Debts Assured by Adoption of Merritt's Plans," *AG* 6 (July 1924), 8; Milo L. Rowell, Trustee, to Members, Sun-Maid Raisin Growers, Sept. 29, 1924, File 60-166-29, Box 561, BAI, RG 60. The California Packing Corporation "snapped up"

many of the bonds issued by the Delaware Sun-Maid in 1923 (Henry E. Erdman, Interview with Wrightson, Sun-Maid Raisin Association, Dec. 6, 1924, Box 18, Erdman Papers).

30. J. T. Turner to Merritt, Sept. 18, 1924; A. M. Thomas to Merritt, Nov. 5, 1924; Jay L. Reed, Sun-Maid Comptroller, to Gentlemen [Fresno bankers], June 25, 1924, all in "Miscellaneous Information re Company" File, Packing Plant Vault, Sun-Maid Archives; "Full Payment of Old Association's Debts Assured by Adoption of Merritt's Plans," *AG* 6 (July 1924): 8; Rowell to Sun-Maid Raisin Growers, Sept. 29, 1924, File 60-166-29, Box 561, BAI, RG 60; "Liquidating Old Association," *California Cultivator* 62 (Feb. 16, 1924): 195; "How about Next Year?" *AG* 6 (Sept.–Oct. 1924): 12. The old Sun-Maid filed an involuntary petition in bankruptcy on December 22, 1923, and Milo L. Rowell, a director of the new organization, was appointed receiver to oversee the transfer of all assets to the California cooperative. See C. W. Hughes, N. H. Castle, and W. H. Treadwell, "California Associated Raisin Company, Sun-Maid Raisin Growers, Sun-Maid Raisin Growers of California, Sun-Maid Raisin Growers Association, Violation of Decree, Jan. 18, 1922, Joint Summary Report," 13–14, 41, Sun-Maid Raisin Growers, General Subject Files, RG 83, (hereafter "Joint Summary Report"). On grower discontent, see V. N. Cooley to Attorney General, Oct. 17, 1924, and A. Simi to Henry A. Guiler, U.S. Attorney, Oct. 15, 1924, both in File 60-166-29, Box 561, BAI, RG 60; A. B. Bevans to Merritt, Nov. 6, 1924; and John C. Rorden to Merritt, Sept. 24, 1924, "Miscellaneous Information" File, Packing Plant Vault, Sun-Maid Archives. For Sun-Maid's intent to hold members to the 1921 contract, see H. C. Goodwin to Members of the Sun-Maid Raisin Growers who have not signed the New Contract of 1923, May 29, 1923, File 60-166-29, Box 561, BAI, RG 60. To avoid paying the full advance guaranteed by the 1921 contract, Sun-Maid paid two cents in cash and gave growers notes for the remainder. The notes were not negotiable, as V. N. Cooley discovered; "all banks [are] refusing to cash them" (Cooley to Attorney General, Oct. 17, 1924, File 60-166-29, Box 561, BAI, RG 60).

31. The sales network opened markets in Europe and China; domestically, Sun-Maid pushed for tariff protection and spent millions more on advertising. To increase industrial use of raisins, Sun-Maid spent $2 million to build a raisin syrup manufacturing plant, but it never went into operation. Merritt also eliminated specialized and unprofitable packs of raisins. Finally, Sun-Maid streamlined its packing operations, staff, and overhead expenses. "What greater demonstration," Merritt asked rhetorically, "of the finer qualities of cooperative marketing could one ask for?" (Merritt, "California Cooperatives Set an Example," 149–58; F. A. Stewart, "Old Sun-Maid Entirely Replaced by New," *AG* 6 [Jan. 1924]: 5; "Sun-Maid's Gross Expenses Reduced Sharply," *SMB*, Dec. 15, 1924, 5; Shear, "Cooperative Marketing and the Raisin Industry," 10).

32. To the Members of Sun-Maid, Demand for Delivery of 1924 Crop, Oct. 13, 1924; "Sun-Maid Moves to Compel 100% Delivery," *SMB*, Sept. 15, 1924, 11; Henry E. Erdman, interview with H. M. Creech, Nov. 12, 1924; To All Members of Sun-Maid, Oct. 13, 1924, all in Box 18, Erdman Papers. Growers interpreted these moves as an attempt by Sun-Maid to gather evidence against them (Erdman, "Impressions of Sun-Maid Raisin Growers," Nov. 12, 1924, Box 18, Erdman Papers).

33. Hans Dahl to Merritt, Nov. 2, 1924; J. T. Turner to Merritt, Sept. 18, 1924, both in "Miscellaneous Information" File, Packing Plant Vault, Sun-Maid Archives; Erdman, interview with E. T. Cunningham, Manager, and Carl Wagner, Assistant Manager, Bank of Italy, Sept. 1924, Box 18, Erdman Papers.

34. "Joint Summary Report," 27, 2. See also "Merritt Will Make Report in 22 Speeches,"

*SMB*, Oct. 15, 1924, pp. 4–5; "Withdrawal Period Ends for Sun-Maid Growers' Members," *FMR*, Jan. 1, 1925, p. 15 ("the total number is much less than was generally anticipated"); "Small Percentage of Raisin Growers Leave Association," *FMR*, Jan. 4, 1925, p. 1 ("Large numbers of the foreign element . . . who had apparently been favoring withdrawal up til the time of the annual report made by Ralph P. Merritt . . . have openly come out in support of the association"). According to grower George F. Hodges, who worked at Sun-Maid from August 20, 1917, to August 31, 1925, "Banks friendly to the Association were supplied with waivers. . . . To give the effect that community interests were behind the desire to eliminate the withdrawal clause of the contract, and . . . Mr. Merritt met with the bankers regularly before the withdrawal period of 1924" ("Joint Summary Report," 63–64).

35. "Members and Community Have Their Responsibility," *SMB*, Jan. 15, 1925, 5; Henry E. Erdman, interview with H. A. Wrightson, Sun-Maid Raisin Growers, Aug. 2, 1924, Box 18, Erdman Papers; "Joint Summary Report," 62; "Small Percentage of Raisin Growers Leave Association," *FMR*, Jan. 4, 1925, p. 1: "Growers of the foreign element who have shown themselves unable to cooperate along American lines . . . will be asked to leave the association." It was clear which "foreign element" the paper had in mind; of the twenty-one withdrawing growers named, ten were Armenian.

36. "Sun-Maid Raisin Withdrawals in County Total 234," *FMR*, Jan. 5, 1925, 1. The county recorders supplied the paper with the names. Only four of these growers filed affidavits with the Justice Department ("Joint Summary Report," 28–61).

37. "Highlights in Merritt Speech at Dedication," *FMR*, Feb. 9, 1925, p. 1.

38. "Chamber Raps Growers Who Quit Sun-Maid," *FMR*, Jan. 7, 1925, p. 6; "Business Men to Interview Large Growers," *FMR*, Feb. 8, 1925, p. 1A; "Prominent Business Men on Committee to Sign Big Growers," *FMR*, Feb. 10, 1925, p. 9; "Growers Vote Ostracism of Non-Members," *FMR*, Feb. 11, 1925, p. 2; "Joint Summary Report," 50, 55–58. The Progress Committee, anxious to dispel the "lethargy" displayed by "business men who are vitally affected by the condition of the raisin industry," promptly secured the contracts of several large growers and two banks that owned vineyard property. The signed growers were Jens Kjar, H. B. Neilsen, R. J. Venn, Hugh Sparkman, and J. W. Meux; the banks included the Pacific Southwest Trust and Savings Bank and the First National Bank of Fresno. The other members of the business committee were Edward Hughson, a Fresno insurance broker; Milo Rowell, trustee for the old Sun-Maid organization; and A. Emory Wishon, vice-president of the San Joaquin Light and Power Corporation and a former CARC director ("Business Men to Interview Large Growers").

39. "Time of Night Rider Past, Madera Raisin Men Told by Collins," *FMR*, Jan. 7, 1925, p. 7.

40. On February 9, attorney Tom Okawara complained to the Fresno sheriff that his clients K. Yaji and S. Miyamoto were assaulted by a mob and forced to sign Sun-Maid contracts. Members of the mob were apparently angered at the news that the two Japanese growers had placed their acreage under contract to independent packers ("Violence Is Alleged by 2," *FMR*, Feb. 11, 1925, p. 1). Krikor Arakelian, one of the area's most prominent Armenian growers, had been a Sun-Maid recruiter in 1923. By 1925 he had become disenchanted, breached his contract, and refused to pay liquidated damages. Sun-Maid's suit against him for the damages was still pending when the night riders raided his farm and destroyed four hundred acres of vines. Another victim, Godfrey Jensen, reported that a caravan consisting of about twenty vehicles cut down "four or five acres of vineyard" and about

one hundred of his peach trees ("Mob Destroys Vineyard, It Is Reported," *FMR*, Feb. 12, 1925, pp. 1, 2). H. E. Thiele reported to the newspaper that a mob destroyed six acres of vines on his ranch in Oleander ("Rancher in Oleander District Says Vines on Ranch Cut Down," *FMR*, n.d. (but probably Feb. 1925), File 60-166-29, Box 561, Anti-Trust Section, RG 60).

41. John E. Pickett, "The Raisin Renewal Fight," *PRP* 109 (Feb. 21, 1925): 226–27, Box 18, Erdman Papers; "Joint Summary Report," 33, 35, 61–62.

42. "Joint Summary Report," 35, 38, 45, 52, 50; "Mob Destroys Vineyard, It Is Reported," 1, 2. The victims, some of whom "wore bandages and had indications of rough treatment," identified their attackers before the grand jury. Ten persons were arrested on these complaints and charged with battery and malicious mischief; all pleaded not guilty and were released without bail. The prosecutor later dropped the charges for lack of evidence. All the grand jury did was to recommend that the sheriff be given more assistance in suppressing the riots ("Alleged Acts of Violence Being Probed," *FMR*, Feb. 14, 1925, 11, 14; "Court Actions Dismissed," *FMR*, Feb. 25, 1925, File 60-166-29, Box 561, BAI, RG 60).

43. "Contracts Obtained through Coercion Will Be Returned," *FMR*, Feb. 12, 1925, 9; "Merritt Warns against Coercion by Workers in Soliciting of Contracts," *FMR*, Feb. 13, 1925, 1; "Joint Summary Report," 37. Fred A. Stewart, head of Sun-Maid's Growers' Service Division, "*confidentially*" disclosed to Henry Erdman that most of the "unsatisfactory element" was in Biola, "where there is a large Russian-German settlement." Sun-Maid's officers thus knew that the Armenians were not the single most "disloyal" growers, nor were they the only growers withdrawing. But because of the persuasive force of equating "outsiders" with "aliens" and the ease with which Euro-Americans saw Armenians as both, Sun-Maid saw no reason to discontinue scapegoating Armenians (Henry E. Erdman, "Memorandum on Growers Service Department in Sun-Maid Raisin Growers, Interview with F. A. Stewart, in charge, Feb. 2, 1926," Box 18, Erdman Papers). On the 1923 USDA-DOJ agreement, see text below at note 53.

44. "Joint Summary Report," 28, 30, 31, 33, 35, 38, 45, 46, 58. Attorney H. J. Wildgrube gave DOJ agents the names of thirteen growers who filed affidavits with Sun-Maid pertaining to coerced 1923 contracts, but the agents were unable to interview any of them because their investigation was suspended ("Joint Summary Report," 72). On the suspension of the DOJ investigation, see text at note 61.

45. *Papazian v. Sun-Maid*, 74 Cal. App. 231 (1925), quote from "Joint Summary Report," 67; *Regier v. Sun-Maid Raisin Growers of California*, trial court decision described in "Joint Summary Report," 67–68; *Reitz and Reitz v. Sun-Maid*, complaint summarized in "Joint Summary Report," 68; *Entz v. Sun-Maid*, judgment summarized in "Joint Summary Report," 70. In cases in which Sun-Maid sued for delivery of raisins purportedly under contract, growers pled coercion as an affirmative defense. The DOJ agents were unable to ascertain how many growers prevailed using this defense ("Joint Summary Report," 70). In *Sun-Maid v. W. F. Jones, Sheriff*, Sun-Maid won a trial judgment entitling it to "replevin the crops of growers who refused to deliver to the Association." See William N. Keeler to R. H. Ellsworth, July 24, 1924; Keeler to L. S. Hulbert, Apr. 26, 1929; both in Sun-Maid Raisin Growers, General Correspondence, RG 83.

The model cooperative marketing statutes that some states had begun to enact in the early 1920s provided for both liquidated damages and specific performance. Before the legislature's enactment of the model law, California courts refused to grant specific performance; see *Anaheim Citrus Fruit Association v. Yeoman*, 51 Cal. App. 759 (1921), and *Poultry*

*Producers of Southern California v. Barlow*, 189 Cal. 278 (1922). After the statute was passed, the state court still refused to grant specific performance (*California Prune and Apricot Growers' Assn. v. Pomeroy Orchard Company*, 195 Cal. 264 [1925]. See Chapter 9.

46. H. E. Armstrong, *Clovis Tribune*, to M. H. Castle, Dec. 22, 1925, File 60-166-29, Box 561, BAI, RG 60; A. W. Sunderman to Secretary Wallace, May 3, 1923; Wallace to Sunderman, May 5, 1923, Fruits, 1923, Secretary's Correspondence, RG 16.

47. Merritt to Wallace, Mar. 1, 1923, Associations, 1923, Secretary's Correspondence, RG 16; Wallace to Merritt, Mar. 10, 1923, Sun-Maid Raisin Growers, General Correspondence, RG 83, also in Associations, 1923, and Reel 166, M440, Secretary's Correspondence, RG 16. This exchange followed a less congenial correspondence between the USDA and the DOJ in February, in which an assistant attorney general told Secretary Wallace that Capper-Volstead did not cover organizations that included nonproducers. At this point, the DOJ was taking an active role in interpreting the Capper-Volstead Act (Assistant Attorney General to Wallace, Feb. 20, 1923, File 60-166-29, Box 561, BAI, RG 60). A USDA press release made it clear that the USDA considered Sun-Maid a Capper-Volstead cooperative: "Under the new system devised by Managing Director Merritt with governmental approval, the requirements of the Capper-Volstead Act have been met and it is expected that the new set up will be accepted as the standard pattern for other cooperative marketing organizations throughout the United States" (D. M. Reynolds, Press Release, Apr. 25, 1923, Sun-Maid Raisin Growers, General Correspondence, RG 83).

48. Colvin Report, 1. Sun-Maid promptly issued a denial: " 'Reports of illegal methods used during the drive were greatly exaggerated and the investigator will find that repeatedly the association urged with every effort through public utterance that the workers desist from any coercive methods' " ("Echo of Raisin Drive Is Heard," *FMR*, May 29, 1923, 11). The Rosenberg and Guggenheim firms immediately denied that they were responsible for the presence of the federal agent, who informed the paper that no complaints had been received by the firms. He did not reveal how the DOJ had learned about the coercion, however ("Deny Local Firms Asked Probe," *FMR*, May 30, 1923, 4).

49. Colvin Report, 5, 8, 15, 12. In addition to the Der Torosian case and others in the newspapers, Colvin described an incident in which one grower who warned the mob away with a shotgun "which he did not use" was nevertheless arrested on a charge of "assault with a deadly weapon with intent to commit murder." The grower was apparently "so disgusted with the present state of 'American freedom' " that he sold his farm at a great loss. Elmer Erickson, one grower who was willing to give Colvin his name, told the agent that the contract he was forced to sign "became smeared with blood from cuts [he] received in the struggle." Knowing that they could not submit the bloodied paper to Sun-Maid, the night riders tried to get him to sign another one, but he refused. Augustus Jewett claimed to have senatorial connections—he told Colvin that both Hiram Johnson and Samuel Shortridge were lifelong friends—yet he was unwilling to commit details to writing or to permit Colvin to disclose his name, for fear that "the Raisin K. Klux Clan" [*sic*] would seek revenge (ibid., 8, 15, 12–13).

50. Ibid., 10, 12, 24–25. Sun-Maid admitted that most of the night riding was committed by "growers who had always been associated with the Sun-Maid" but depicted night riding as an unavoidable hazing ritual. Colvin did not believe that the "practical reign of terror" established by the campaign was owing solely to zealous extralegal behavior; he held Sun-Maid responsible for the acts of its members. Indeed, he saw Sun-Maid's published

statement that it would return all contracts obtained under coercion—and the thirty-four contracts actually returned to growers—as an admission of complicity, if not guilt. Other growers wanted their contracts back too, but they "fear[ed] to make demand, in dread of still further violence and destruction of property" (ibid., 22, 6, 10).

51. "Is the Government to Persecute the Farmers?" *Fresno Bee*, July 18, 1923; "Energy Needed for Work, Not Lawsuits," *Fresno Bee*, July 19, 1923; "Growers under U.S. Probe on Reports of Drive Violence," *Fresno Bee*, July 19, 1923, all in "Sun-Maid Raisin Growers" File, Historical Files, Office of *Fresno Bee*.

52. "Merritt States Action Brought at Own Request," *Fresno Bee*, July 19, 1923; "Alter Raisin Crop Contract Edict of U.S.," *Fresno Bee*, Aug. 4, 1923; Special Assistant Henry A. Guiler to Attorney General Harry M. Daugherty, Sept. 20, 1923, File 60-166-29, Box 561, BAI, RG 60. The ultimatum required Sun-Maid to eliminate the liquidated damages clause, the lien on the land, and the cause of action in equity for nonperformance from the crop contract, or face dissolution. John T. Williams, U.S. Attorney, to Daugherty, Oct. 15, 1925, ibid.: "I believe that with the cooperation and collaboration of my many powerful friends here in California, who have absolute confidence in my integrity of purpose as your subordinate here, that we could bring about beneficial results for you . . . without the necessity of instituting legal proceedings."

53. "Sun-Maid Raisin Controversy Is Closed by U.S.," *Fresno Bee*, Sept. 13, 1923, 1; Guiler to Daugherty, Sept. 20, 1923, "Sun-Maid O.K., Declares Aid to Daugherty," *San Francisco Chronicle*, Sept. 16, 1923, both in File 60-166-29, Box 561, BAI, RG 60. Daugherty wanted to position himself as a presidential candidate after Harding's death in August 1923. He saw an opportunity to gain politically by proceeding with restraint against Sun-Maid and other West Coast commercial interests accused of antitrust violations. His subordinates saw to it that the settlement in the raisin case did not hurt his purposes: "It is now generally understood and known in the San Joaquin Valley that the Attorney General of the United States not only caused certain ill practices in the Raisin Growers' Association to be corrected, but, because of the constructive manner in which he acted the Raisin Growers' Association was saved and that its membership and their friends are most grateful to you" (Williams to Daugherty, Oct. 15, 1925, File 60-166-29, Box 561, BAI, RG 60). Merritt claimed in his oral memoirs that his timely assistance to President Harding, who lay dying in San Francisco during the height of the Sun-Maid/DOJ negotiations, had produced the DOJ's withdrawal in the case, but there is no evidence to support this claim. See Merritt, *"After Me Cometh a Builder,"* 125–26, Bancroft Library; "Sun-Maid Raisin Controversy Is Closed by U.S.," *Fresno Bee*, Sept. 13, 1923, Historical Files, Office of *Fresno Bee*; Schruben, "An Even Stranger Death of President Harding."

54. "Advisory Council Hears of Continued Progress," *AG* 5 (Oct. 1923): 6; "Joint Summary Report," 27.

55. Merritt to Wallace, June 9, 1924; Draft of June 30 letter, Wallace to Merritt, both in Sun-Maid Raisin Growers, General Correspondence, RG 83; Wallace to Merritt, June 30, 1924, Reel 168, M440, Secretary's Correspondence, RG 16.

56. Jensen to Department of Justice, Mar. 29, 1925; Seymour, Memorandum for the Attorney General, Feb. 25, 1925; Atkinson to Attorney General, Nov. 14, 1925, all in File 60-166-29, Box 561, BAI, RG 60.

57. "Joint Summary Report," 41, 59, 46–47. Harry Daugherty's immediate successor as

attorney general, Harlan F. Stone, was swiftly appointed to the Supreme Court and took no action on the raisin case while in office.

58. Compiled from "Joint Summary Report," 27–63. The professionals included a newspaper editor, a postmaster, a high school principal, and a veterinarian.

59. "3 Operatives Investigating Sun-Maid Here," *FMR*, Dec. 22, 1925, 11, 18; Sargent to Atkinson, Dec. 23, 28, 1925; William J. Donovan to William M. Jardine, Dec. 24, 1925; Atkinson to Attorney General, Dec. 24, 1925, all in File 60-166-29, Box 561, BAI, RG 60. On the tug of war between Agriculture and Commerce during the 1920s, see Hamilton, *From New Day to New Deal*, 33–34.

60. Merritt to Jardine, Dec. 22, 1925; Russell to Merritt, Dec. 23, 1925; Merritt to Jardine, Dec. 23, 1925; Associations, 1925, Secretary's Correspondence, RG 16; Tenny to Merritt, Dec. 29, 1925, Sun-Maid Raisin Growers, General Correspondence, RG 83. Merritt also asked the USDA to inform Commerce Secretary Hoover, in case his help would be required; the USDA replied that the secretary would resolve the situation with the attorney general (Merritt to Frank M. Russell, Dec. 22, 1925; Russell to Merritt, Dec. 23, 24, 1925, all in Associations, 1925, Secretary's Correspondence, RG 16.

61. "Press Release," Dec. 29, 1925, File 60-166-29, Box 561, BAI, RG 60; "Statement for the Press," Dec. 29, 1925, Associations, 1925, Secretary's Correspondence, RG 16; Sargent to Atkinson, Dec. 28, 1925, File 60-166-29, Box 561, BAI, RG 60; Sargent to Jardine, Jan. 6, 1926, Sun-Maid Raisin Growers, General Correspondence, RG 83.

62. L. S. Hulbert, "Statement Relative to Investigation of the Sun-Maid Raisin Growers of California by Representatives of the Department of Justice," Dec. 28, 1925, pp. 2, 7, File 60-166-29, Box 561, BAI, RG 60; Tenny to Russell, Feb. 3, 1926; Jardine to Attorney General, Feb. 4, 1926, Sun-Maid Raisin Growers, General Correspondence, RG 83; "Department of Justice Investigation Is Satisfactorily Settled," *SMB*, Jan. 15, 1926, p. 5. A year before, Merritt had acknowledged that the Capper-Volstead enforcement structure created a logical absurdity. On the one hand, the USDA was responsible for both enforcing the law and promoting the broader policy behind it, even though it lacked the expertise to analyze the legal effects of cooperatives' market behavior. On the other, the DOJ had no sympathy for cooperation or for farmers and wanted to enforce the antitrust laws but was barred from doing so under Capper-Volstead. See Henry E. Erdman, "Interview with Mr. Ralph Merritt, Sun-Maid Growers Association," Feb. 5, 1925, Box 18, Erdman Papers.

63. H. B. Teegarden, Assistant Attorney General, Memorandum for Mr. Myers, DOJ, Feb. 3, 1926, Sun-Maid Raisin Growers, General Correspondence, RG 83. Sun-Maid officials openly celebrated the outcome of the DOJ's aborted investigation and the protected position the cooperative now confidently enjoyed: "The prompt termination of this affair is indicative of the value of the close relationship and understanding that has been established between the Management of Sun-Maid and high governmental authority" (Minutes of Sun-Maid Advisory Council Meeting, Jan. 20, 1926, pp. 9–10, Box 18, Erdman Papers).

64. Docket Books, Fresno County Superior Court, 1920–30, Fresno County Courthouse, Fresno, California.

65. *Sun-Maid Raisin Growers of California v. Paul A. Mosesian and Son, Inc.*, 90 Cal. App. 1, 4 (1928), and *Sun-Maid Raisin Growers of California v. K. Arakelian, Inc.*, 90 Cal. App. 10 (1928). In California, the courts had ruled that liquidated damages were an appropriate remedy for a cooperative association in seeking redress for breach by members because it

had no way to estimate actual damages (*Anaheim Citrus Fruit Association v. Yeoman*, 51 Cal. App. 759 [1924]). This rule did not apply to regular commercial corporations, which did not stand to suffer the loss of patronage arising from member breach. Chapter 9 discusses marketing contracts and liquidated damages under new state laws passed during the 1920s.

66. *Mosesian*, 7.

67. Ralph P. Merritt, "Something Ought to Be Done about It," *SMB*, Dec. 15, 1927, p. 3. In late 1926, Henry E. Erdman noted that "banks have gone as far as they dared in refusing credit to those who withdrew at last withdrawal period" (Erdman, "Interview with Walter Shoemaker, Cashier, Pacific Seedless Raisin Co.," Dec. 17, 1926, Box 18, Erdman Papers).

68. "1923 Crop Final Settlements Rendered and Checks Mailed," *SMB*, Feb. 15, 1926, p. 4; "Analysis by Receiving Classification and Pool 1923 Crop Raisins," *SMB*, Feb. 15, 1926, p. 9; "Progress Payment Authorized on 1924 Pools," *SMB*, July 15, 1925, p. 5; Merritt, "Who Killed Cock Robin?" *SMB*, Apr. 15, 1926, p. 3; Popovich, "History of the Sun-Maid Organization," 20; "Sunland Methods Proving Successful in English Raisin Markets," *SMB*, Apr. 15, 1926, p. 4; Malloch, "Report on the Raisin Industry," 1; "Members' Acreage," Oct. 31, 1927, Document 270, Sun-Maid Archives. The decline in membership between 1925 and 1928 resulted from the alienation of many white members—through depression, falling prices, and violence—and the movement of many Armenian growers to the cities, particularly Los Angeles (Mahakian, "History of Armenians in California," 46–51).

69. Merritt, "Who Threw the Brick," *SMB*, Nov. 15, 1927, 3; Popovich, "History of the Sun-Maid Organization," 18–21; Malloch, "Report on the Raisin Industry," 2; Merritt, *"After Me Cometh a Builder,"* 120–36, Bancroft Library; U.S. Tariff Commission, *Grapes, Raisins, and Wines*, 155–56; Watson, "Analysis of Raisin Marketing Controls," 16. In 1927, Sun-Maid secured loans of $5 million from the Wall Street firm of Dillon, Read & Company; in 1928, Merritt borrowed a total of $15.5 million (B. C. Forbes, "Wall Street Makes History by Financing Raisin Growers," *San Francisco Examiner*, Feb. 4, 1927).

70. Popovich, "History of the Sun-Maid Organization," 21; Merritt, "Now It Can Be Told," *SMB*, June 15, 1928, p. 3; Memorandum relating to the conversation, n.d., Document 291; Merritt to Board of Directors, Sun-Maid Raisin Growers of California, n.d. [but Aug. 30, 1928], Document 303; "Strictly Confidential Information for Members of Sun-Maid Raisin Growers of California," Dec. 1928, Document 318, all in Numbered Documents File, Sun-Maid Archives.

71. Malloch, "Report on the Raisin Industry," 2.

72. Merritt, "California Cooperatives Set an Example," 155.


*Chapter 9*

1. Hawley, "Three Facets of Hooverian Associationalism"; and Hamilton, "Building the Associative State."

2. Seitz, "The Kind of Cooperation That Will Afford Farm Relief," 668.

3. Hamilton, *From New Day to New Deal*, 22.

4. Jensen, "Integrating Economic and Legal Thought," 899; Hobson, "Farmers' Cooperative Associations," 225; Capper-Volstead Act, 42 Stat. 388 (Feb. 18, 1922) (enabling clause).

5. *Cong. Rec.*, 67th Cong., 2d sess., 2262 (remarks of Sen. Gilbert M. Hitchcock, R-Neb.),

2172 (remarks of Sen. Frank B. Brandegee, R-Conn.); Nourse, *Legal Status of Agricultural Co-operation*, 256.

6. Wallace to A. S. Ennis, July 17, 1922, Reel 165, M440, BAE Correspondence, RG 16; "Statement Relative to and Text of Capper-Volstead Act," Mar. 7, 1922, Bills (H.R. 2373), Secretary's Correspondence, RG 16. Examples of USDA advice to cooperatives regarding prices and contracts for purchase and sale of commodities can be found in Wallace to Ennis, July 17, 1922, Reel 165, M440, BAE Correspondence, RG 16; Will L. Barter to USDA, Feb. 5, 1923; R. W. Williams to Barter, Feb. 26, 1923, Marketing, Cooperative, Capper-Volstead Act, Miscellaneous, Box 59, Solicitor's Records, RG 16.

7. 49 Stat. 449 (1935).

8. Erdman, "Thought on Cooperation," 915.

9. Larsen and Erdman, "Aaron Sapiro," 251. Sapiro was born in San Francisco in 1884 and orphaned at the age of nine. Sent to Cincinnati to study for the rabbinate, he changed his mind, returned to California, and took his law degree from Hastings Law College of the University of California in 1911. There is no full-length biography. See "Aaron Sapiro, 75, Lawyer, Is Dead," *New York Times*, Nov. 25, 1959, p. 29. Scholars continue to credit Sapiro with the legal innovations of the major California horticultural cooperatives although he had nothing to do with their organization. See, e.g., Hamilton, *From New Day to New Deal*, 15.

10. The intent of the bill, according to a contemporary economist, was to create "self-sustaining" markets funded entirely by commissions earned on sales rather than tax dollars. The market's purpose was to regulate the "local commission business" in California, reduce the costs of marketing, and enable farmers to bargain more effectively not only with eastern distributors and jobbers but with the local packers and processors who really controlled the prices producers received. See Plehn, "State Market Commission," 9; *Cal. Stats.* (June 10, 1915), chap. 715, §§1–5; Blackford, *Politics of Business in California*, 29–31.

11. Larsen and Erdman, "Aaron Sapiro," 245, 251. Weinstock made a fortune with his half-brother, David Lubin, as dry-goods merchants in San Francisco. Both spent their lives in public service, Lubin as the organizer of the International Institute of Agriculture in 1905 and Weinstock as a state official under Progressive governor Hiram Johnson. Sapiro and Weinstock met through social and religious connections in 1905; Sapiro eventually married Weinstock's daughter. Weinstock served as a mentor to the young lawyer for years before hiring him at the Commission Market. The men severed their close friendship over a business quarrel in 1919 (ibid., 243–50; "Aaron Sapiro, 75, Lawyer, Is Dead," 29).

12. Plehn, "State Market Commission," 10 (also quoted in McEvoy, *Fisherman's Problem*, 169 n. 73). Weinstock and Sapiro's activities kicked up a storm in the legislature, which thought it had authorized the creation of commission markets, not a price-fixing agency. After numerous fights during its 1917 session, the California legislature passed a new statute that ratified Weinstock's expansion of his statutory authority. Further, the Fish Exchange Act of 1917 gave Weinstock sole authority to fix the legal price of all fish taken in waters under California jurisdiction. See Plehn, "State Market Commission," 11–18; Larsen and Erdman, "Aaron Sapiro," 245–47; 1917 State Market Commission Law, *Cal. Stats.* (1917), chap. 802; Fish Exchange Act of 1917, *Cal. Stats.* (1917), chap. 803; Black, "The Sapiro Cooperative Philosophy," 2. For an account of Weinstock's regulation of the fishery markets that portrays him as a maverick, see McEvoy, *Fisherman's Problem*, 169–70.

13. Plehn, "State Market Commission," 24; Harry S. Maddox, "Division of Markets," in California State Department of Agriculture, *Second Report for the Period Ending December 31, 1921*, 736–39.

14. Larsen and Erdman, "Aaron Sapiro," 247; Montgomery, *Cooperative Pattern in Cotton*, 74. See also Wise, *Jews Are Like That!*, 168–69.

15. Quoted in Brooker, *Cooperative Marketing Associations in Business*, 69.

16. Sapiro, "True Cooperative Marketing," 7, 6, in *Papers on Cooperative Marketing*; Sapiro, "Cooperative Marketing," 198. See also Knapp, *Advance of American Cooperative Enterprise*, 9.

17. Sapiro, "True Farmer Cooperation," 90.

18. Knapp, *Advance of American Cooperative Enterprise*, 59; Larsen and Erdman, "Aaron Sapiro," 257–58, 260–62; MacLeod, "Canadian Wheat Pools," 8–9; Tobriner, "Legal Aspect of the Provisions of Cooperative Marketing Contracts," 23. According to Tobriner, 90 percent of dried fruits, 70 percent of tobacco, 65 percent of nuts, 25 percent of milk, and 15 percent of cotton was marketed collectively at this time. On the AFBF, see Fite, *George Peek*.

19. F. J. Hatch to Calvin Coolidge, Feb. 14, 1924, Marketing, 1924, Secretary's Correspondence, RG 16.

20. Henry C. Wallace to C. O. Moser, American Cotton Growers Exchange, Apr. 7, 1924, Reel 168, M440, Secretary's Correspondence, RG 16. When asked for a recommendation of Sapiro by a Florida banker apparently sympathetic to farmers, Wallace was noncommittal: "Mr. Sapiro has planned the organization of many of the large commodity marketing associations" (Wallace to E. L. Mack, President, Central State Bank, President, Lakeland [Florida] Chamber of Commerce, Feb. 8, 1924, Marketing, 1924, Secretary's Correspondence, RG 16). See also J. F. Larson to USDA, Mar. 29, 1921, Agricultural Situation, 1921, Secretary's Correspondence, RG 16, noting that the movement's biggest problem was persuading farmers to give "full and loyal support to the cooperative marketing organizations."

21. Wallace to C. F. Bratnober, Oct. 20, 1923, Reel 167, M440, Secretary's Correspondence, RG 16.

22. Jardine to Norris, Jan. 8, 1925, Reel 170, M440, Secretary's Correspondence, RG 16; Cooperative Marketing Act of 1926, 44 Stat. 802. The USDA sponsored the legislation, which reflected the conclusions of meetings with cooperative leaders on the question of how the department could best help them. The statute gave cooperatives the legal authority to acquire, share, and exchange "crop, market, statistical, economic, and other similar information" — powers that existing antitrust law and the Capper-Volstead Act did not explicitly authorize. The Supreme Court initially took a dim view of these information exchanges (see *American Column & Lumber Co. v. U.S.*, 257 U.S. 377 [1921]) but four years later upheld this practice (*Maple Flooring Assn. v. U.S.*, 268 U.S. 563 [1925]) as long as it did not lead to overt price-fixing (*U.S. v. Trenton Potteries Co.*, 273 U.S. 392 [1927]). See Nourse, *Legal Status of Agricultural Co-operation*, 262, and Keller, *Regulating a New Economy*, 37–39.

23. Revenue Act of 1926, 44 Stat. 39, §231; Federal Farm Board Act of 1929, 46 Stat. 11. Only Capper-Volstead cooperatives qualified for the tax exemption (Nourse, *Legal Status of Agricultural Co-operation*, 263). The Farm Board was empowered to create stabilization corporations for the control of surpluses in any commodity, and it could also lend money for the marketing of these surpluses. Its approach to market stabilization was widely regarded as a failure (Hamilton, *From New Day to New Deal*, 66–108).

24. Ellis, "Robert Worth Bingham and the Crisis of Cooperative Marketing"; Larsen and Erdman, "Aaron Sapiro," 265–66.

Sapiro's critics came from a variety of sources. In 1924, Henry Ford's *Dearborn Independent* warned of Sapiro's "effort to control, direct and regulate every farmer in America" (Robert Morgan, "Jewish Exploitation of Farmers' Organizations," *Dearborn Independent* 24 [Apr. 19, 1924]: 4). Prominent Californians debunked Sapiro's self-professed role as "the savior of the farmer" and unhesitatingly informed the USDA that they did not support him ("Mr. Sapiro Speaks for Himself on the Witness Stand," *Dearborn Independent* 24 [July 5, 1924]: 2). The agricultural economist Edwin G. Nourse was one of Sapiro's chief antagonists. Nourse attacked commodity marketing in academic journals; like Henry C. Taylor of the BAE, he preferred the citrus growers' model of economic democracy to Sapiro's monopolistic, autocratic organizations. For a sample of his thought, see "Economic Philosophy of Co-operation." The citrus growers and their allies, following Nourse, also opposed Sapiro. G. Harold Powell, manager of the California Fruit Growers' Exchange, and O. B. Jesness, a USDA specialist on cooperatives, publicly supported Nourse (Knapp, *Advance of American Cooperative Enterprise*, 557 nn. 1–2).

25. Larsen and Erdman, "Aaron Sapiro," 260; McConnell, *Decline of Agrarian Democracy*, 60–61; Knapp, *Advance of American Cooperative Enterprise*, 67–71. On the farm leaders' complaints, see Dallas H. Gray to Henry C. Taylor, Jan. 3, 1923, Document 84377, Numerical Correspondence, RG 83; Robert W. Bingham to Wallace, Feb. 7, 1924; Walton Peteet to Wallace, Feb. 8, 1924, both in Marketing, 1924, Secretary's Correspondence, RG 16; and Charles E. Bassett to William M. Jardine, Mar. 5, 1925, Marketing, 1925, Secretary's Correspondence, RG 16: "Why should the farmers of this country have to pay Aaron Sapiro and other propagandists hundreds of thousands of dollars annually, to tell them how to organize for business, when Congress is making large appropriations of the people's money for that very kind of service?"

26. Hamilton, *From New Day to New Deal*, 20, 39; Larsen and Erdman, "Aaron Sapiro," 267. A good overview of McNary-Haugen is in Shideler, *Farm Crisis*, 267–90. According to Rexford Tugwell, who served as assistant secretary of agriculture during the New Deal, many farmers did not support McNary-Haugen because their markets were primarily domestic ("Reflections on Farm Relief," 483). The attractiveness of using cooperatives to implement marketing controls was reflected in both the Farm Board Act and the Agricultural Agreement Act of 1933 (Nourse, Davis, and Black, *Three Years of the Agricultural Adjustment Administration*, 222–23).

27. Hanna, *Law of Cooperative Marketing Associations*, 44 n. 35. Though Sapiro often claimed sole credit for the enactment of the uniform law, state and federal politicians, the American Farm Bureau Federation, and agricultural leaders played important roles. On Bingham's role in the passage of the act in Kentucky and the rise of the Burley Tobacco Growers' Association, see Ellis, "Robert Worth Bingham and the Crisis of Cooperative Marketing," 102–5; Hanna, *Law of Cooperative Marketing Associations*, 46 n. 37. On Wallace's philosophical approach to cooperation, see Winters, *Henry C. Wallace as Secretary of Agriculture*.

28. The Bingham Cooperative Marketing Act, Acts of Kentucky, 1922, §§5, 1, in Nourse, *Legal Status of Agricultural Co-operation*, 472, 470; Sapiro, "Law of Cooperative Marketing Associations," 7.

29. Hanna, *Law of Cooperative Marketing Associations*, 47; Meyer, "Law of Co-operative Marketing," 95; *Munn v. Illinois*, 94 U.S. 113 (1877); Scheiber, "Road to *Munn*." The Sapiro-Bingham model act resembled a prototype bill that the USDA drafted in 1917. See Nourse, *Legal Status of Agricultural Co-operation*, 73–92, 100–119, and Chapter 5.

30. Chamberlain, "Cooperative Marketing Act," 417; Tobriner, "Constitutionality of Co-operative Marketing Statutes," 28; Note, "Opposition to Cooperative Marketing," 823 (this writer, like almost all other legal observers, relied on the labor union analogy). See also Henderson, "Cooperative Marketing Associations"; and Wilson, "Legal Status of Cooperative Marketing Movement."

31. These contracts were specifically indemnified from antitrust liability. Bingham Act, §17, in Nourse, *Legal Status of Agricultural Co-operation*, 482–83; Hanna, *Law of Cooperative Marketing Associations*, 82–85. Sapiro took from the CARC's 1918 contract the provisions for resale of the crop by the cooperative; paying to members the retail price minus the costs of marketing and overhead; and limiting dividends to 8 percent. The CARC innovations of giving the association a lien on the land and having the contract run with the land were not, however, incorporated into the Bingham Act, perhaps because Sapiro thought they would not hold up in court (Nourse, *Legal Status of Agricultural Co-operation*, 200–202).

32. Bingham Act, §18, in Nourse, *Legal Status of Agricultural Co-operation*, 483–84; Henderson, "Cooperative Marketing Associations," 97–98. Specific performance required the grower to carry out the terms of the marketing agreement despite the breach; injunctions prevented growers from selling elsewhere the crops pledged to the cooperative. In granting the equitable remedies of specific performance and injunction, the CMA went beyond the earlier cooperative incorporation statutes, which had left cooperatives to the remedies obtainable through "fundamental principles of common law" (Nourse, *Legal Status of Agricultural Co-operation*, 198).

The *Reese* and *Burns* cases are discussed in Chapter 3; Chapter 6 treats pre-CMA decisions in Alabama, California, and New York upholding liquidated damages in spite of the adverse decisions in *Reeves* and *Burns*.

33. Sapiro quoted in Wise, *Jews Are Like That!*, 166–67. See also Sapiro, "Law of Cooperative Marketing Associations," 4. On the penalties for third-party breach, see the Bingham Act, §26, in Nourse, *Legal Status of Agricultural Co-operation*, 487. The fourteen states that did not enact this section provided cooperatives with other safeguards, such as criminalizing the act of spreading false rumors about a cooperative's finances, management, or business conduct (Hanna, *Law of Cooperative Marketing Associations*, 102–3).

34. Note, "Equity," 521; Sapiro, "Law of Cooperative Marketing Associations," 7. See also Sapiro, "True Farmer Cooperation"; Bingham Act, sec. 5, in Nourse, *Legal Status of Agricultural Co-operation*, 471–72. The Bingham Act authorized the appointment of directors selected by the governor, marketing commission, or any other public body to represent the public on the governing body of the cooperative. All but six states included this provision. In addition, thirty-six states required cooperatives to report annually on their business operations, including the amount of capital stock, number of shareholders, total expenses, indebtedness, and balance sheets. Seven states authorized periodic inspections of all records. These requirements came from the practices Weinstock and Sapiro developed in the California State Markets Bureau (Bingham Act, §§12 and 20, in Nourse, *Legal Status of Agricultural Co-operation*, 478–79, 484; Hanna, *Law of Cooperative Marketing Associations*, 69–70, 90–92). Twenty-nine states substituted even stiffer requirements for public

supervision of cooperatives, rejecting the idea that the market could adequately restrain cooperatives with monopoly power (Hanna, *Law of Cooperative Marketing Associations*, 52 n. 51).

35. Sapiro, "Law of Cooperative Marketing Associations," 6. On the whole, the CMAs paid homage to the essence of Rochdale cooperation as they embraced Sapiro's emphasis on efficient marketing. Rochdale features of CMA associations included patronage dividends; one person, one vote; and marketing services conducted at cost for the benefit of members rather than to profit stockholders. The newer style of cooperation was reflected in two features derived from the Capper-Volstead Act: an 8 percent limit on dividends and a 50 percent maximum on business conducted for nonmembers. Many states permitted cooperatives to have capital stock up to a certain limit, as long as they did not operate for profit. The evolution from traditional cooperation to Sapiro cooperation was evidenced in Sapiro's preference for the term "association" rather than "cooperative" — the former conveying the idea of an efficient business rather than the simple pools of Rochdale marketing cooperatives (Ballantine, "Co-operative Marketing Associations").

36. Sapiro, "Law of Cooperative Marketing Associations," 14.

37. *Jones*, 185 N.C. 265, 277 (1923); *Connolly v. Union Sewer Pipe Co.*, 184 U.S. 540 (1902). The Texas and Oregon decisions are *Hollingsworth et al. v. Texas Hay Assn.*, 246 S.W. 1068 (Tex. Civ. App. 1922); *Oregon Growers' Co-op. Assn. v. Lentz*, 107 Ore. 561 (1923).

38. The tone of this opinion subjected its author to accusations of injudicious conduct; Edwin G. Nourse quoted an attorney as saying, "This opinion always impressed me as utterly lacking in judicial tone. The late jurist who wrote it was noted as holding that the courts had little or no right to declare statutes unconstitutional" (Nourse, *Legal Status of Agricultural Co-operation*, 367–68 n. 5). The *Jones* judge effusively supported cooperation, his reasoning relied on prevailing economic conditions rather than immutable legal principles, and the court paid no attention to the cooperative's own practices, which may have included price-fixing and production controls, two tactics not authorized by the statute. This criticism hinted at the active conflict between the legal conservatism of the *Lochner* era and the legal progressivism of some state and federal judges who embraced associationalism and a more activist regulatory state during the 1920s.

39. *Jones*, 274. Two months after *Jones*, the Kansas Supreme Court upheld that state's CMA and its provisions for equitable remedies (*Kansas Wheat Growers' Association v. Schulte*, 113 Kan. 672 [1923]). In the following three years, fourteen states followed suit: *Brown v. Staple Cotton Co-op. Assn.*, 132 Miss. 859 (1923); *Texas Farm Bureau Co-op. Assn. v. Stovall*, 113 Tex. 273 (1923); *Potter v. Dark Tobacco Growers' Co-op. Assn.*, 201 Ky. 441 (1923); *Dark Tobacco Growers' Co-op. Assn. v. Mason*, 150 Tenn. 228 (1924); *Dark Tobacco Growers' Co-op. Assn. v. Dunn*, 150 Tenn. 614 (1924); *Arkansas Cotton Growers' Co-op. Assn. v. Brown*, 168 Ark. 504 (1925); *Warren v. Alabama Farm Bureau Cotton Exchange*, 213 Ala. 61 (1925); *Minnesota Wheat Growers' Co-op. Marketing Assn. v. Huggins*, 162 Minn. 471 (1925); *Nebraska Wheat Growers' Assn. v. Norquest*, 113 Neb. 731 (1925); *Rifle Potato Growers' Co-op. Assn. v. Smith*, 78 Colo. 171 (1925); *Clear Lake Co-op. Livestock Shippers' Assn. v. Weir*, 200 Iowa 1293 (1925); *Dark Tobacco Growers' Co-op. Assn. v. Robertson*, 81 Ind. App. 51 (1926); *List v. Burley Tobacco Growers Co-op. Assn.*, 114 Ohio St. 361 (1926); *South Carolina Cotton Growers' Co-op. Assn. v. English*, 135 S.C. 19 (1926). Sapiro personally appeared or filed briefs in almost half of these cases.

40. 182 Wis. 571, 588–89 (1924). Scores of law review articles summarized the new trends

in the law of cooperative marketing and credited Sapiro with effecting, almost single-handedly, the marked transformation in the legal status of cooperatives. Most also agreed with the substantive conclusions reached by the state courts, that cooperatives did not restrain trade and the CMAs effected a valid public policy. See, e.g., Miller, "Farmers' Cooperative Associations as Legal Combinations"; Henderson, "Cooperative Marketing Associations"; Brown, "Cooperative Marketing of Tobacco"; Ballantine, "Co-operative Marketing Associations"; Tobriner, "Cooperative Marketing and the Restraint of Trade"; Meyer, "Law of Co-operative Marketing"; and Goldberg, "Co-operative Marketing and Restraint of Trade." Only a few of these writers suggested a less roseate assessment of the CMAs; see Tobriner, "Constitutionality of Cooperative Marketing Statutes"; and Keegan, "Power of Agricultural Co-operative Associations to Limit Production."

41. *Huggins*, 475. In *Brown v. Staple Cotton Co-op. Assn.*, cited in note 39, the Mississippi court distinguished the precedents as presenting direct evidence of price-fixing (*Ford* and *Milk Exchange*) or as having been overruled by higher courts (*Turnipseed*). These cases are discussed in Chapter 3. Some courts refused to waive precedent so easily. Even though it recognized the CMA section granting remedies and enforcement powers for the contracts, the Iowa court refused to vitiate state antitrust law or its own decision in *Reeves*. Instead, it held that since the CMA contained no repeal clause, the earlier antitrust statute remained valid. As long as a cooperative acted merely as a selling agency — as opposed to a buyers' trust — it was protected by the CMA (*Clear Lake Co-op. Live Stock Shippers' Assn. v. Weir*, cited in note 39, 299–300). Colorado, another state with an antithetic precedent on its books (*Burns*), upheld its CMA on similarly narrow grounds in *Rifle Potato Growers*: "The Act of 1923 changes the public policy of this state and the contract in this case follows the Act" (174).

42. A Texas appeals court made this clear in 1925. There the defendant cooperative was not organized under the CMA; it purchased eggs outright from its members rather than acting as a selling agent, making it culpable under the antitrust law: "Had appellee association perfected an incorporation as the act itself, under which clearly it was organized and as its contracts show its members contemplated it should do, it would have been relieved of the questions presented" (*Fisher v. El Paso Egg Producers' Assn.*, 278 S.W. 262, 264 [Tex. Civ. App. 1925]). Colorado followed this rule in two cases, holding that the CMA did not apply retroactively and contracts antedating the CMA did not come under its provisions: *Atkinson v. Colorado Wheat Growers' Assn.*, 77 Colo. 559 (1925); *Colorado Wheat Growers' Assn. v. Thede*, 80 Colo. 529 (1927).

43. Sapiro, "Cooperative Marketing," 205–6. The narrow rulings meant that while cooperatives won the right to equitable remedies against members who breached their contracts, they did not always enjoy similar success with regard to third parties (private firms and individuals who induced members to breach) or tenants who refused to deliver even though their lessors belonged to the association. See text below at note 49.

44. *California Prune and Apricot Growers' Assn. v. Pomeroy Orchard Company*, 195 Cal. 264 (1925). The early decisions against specific performance were *Anaheim Citrus Fruit Association v. Yeoman*, 51 Cal. App. 759 (1921); and *Poultry Producers of Southern California v. Barlow*, 189 Cal. 278 (1922). In *Barlow*, the court noted that the mutuality requirement was the only thing standing between cooperatives and unreasonably large damage awards. The California CMA was passed directly in response to the *Barlow* case. See Note, "Specific Performance." Sun-Maid's suits for liquidated damages, as noted in Chapter 8, resulted

in a mixed record; Sun-Maid won some suits, settled some, and lost at least three: *Sun-Maid Raisin Growers of California v. Paul A. Mosesian and Son, Inc.*, 90 Cal. App. 1 (1928); *Sun-Maid Raisin Growers of California v. K. Arakelian, Inc.*, 90 Cal. App. 10 (1928); and *Papazian v. Sun-Maid*, 74 Cal. App. 231 (1925). See Chapter 8.

45. After *Pomeroy*, an intermediate appellate court affirmed the remedy of liquidated damages, but the court again avoided construing the statute and adjudicated the dispute on common law rules alone (*California Canning Peach Growers v. Downey et al.*, 76 Cal. App. 1 [1925]). Another case involved a marketing contract that expired before the CMA was enacted; the cooperative sought liquidated damages for past injury, instead of specific performance to compel future compliance with the contract, and prevailed (*California Bean Growers' Assn. v. Rindge Land and Navigation Co.*, 199 Cal. 168 [1926]).

46. Sapiro, "Law of Cooperative Marketing Associations," 11; *Dairymen's League Co-op. Assn. v. Holmes*, 202 N.Y. Supp. 663, 674 (1924). The New York statute involved was not a CMA but a corporation code provision entitling nonprofit associations to liquidated damages; the difference between the two is less relevant than the court's willingness to enforce the reasonability rule in such a way as to call into question the customary method of fixing liquidated damages. California cooperatives imposed a set cost per unit marketed (for example, two cents per pound of raisins). The Dairymen's League damages ran $10 for each cow up to 30 and $3 for each cow above 30; defendant Holmes happened to have 330 cows, enough to run his tab well over the marketing costs for his contribution—the figure the court chose as the proper estimate of damages. The court recognized the difficulty of estimating actual damages but still found a method to do so when the amount of liquidated damages appeared unreasonable.

47. Kentucky's supreme court enthusiastically supported this provision of the Bingham Act, indicating that it thought the remedy proper even if there were no express statutory authorization (*Potter*, 448). The court in *Minnesota Wheat Growers' Co-operative Marketing Association v. Huggins* framed the issue in terms of the parties' intent: "We must conclude that the parties intended that the contract would be enforced, and it will be" (486). Other decisions granting specific performance were *Brown v. Staple Cotton Co-op. Assn.*, *Nebraska Wheat Growers' Assn. v. Norquest*, *Dark Tobacco Growers' Assn. v. Dunn*, and *Texas Farm Bureau Cotton Assn. v. Stovall*, all cited note 39; and *Hollingsworth et al. v. Texas Hay Assn.*, (Tex. Civ. App.), cited note 37.

48. *Kansas Wheat Growers' Association v. Schulte*, 685. Even when courts awarded specific performance, they restricted the remedy to those lands specified in the contract or to the county in which the cooperative conducted its business. The leading case of *Oregon Growers' Co-operative Assn. v. Lentz* refused to extend the injunction to cover all the defendant's lands, in addition to the nineteen acres named in the contract. Even assuming that the contract covered all the defendant's property, the court held that it was inappropriate to award the remedy until an actual breach occurred with respect to those additional lands (*Lentz*, 585–86). In *Pierce County Dairymen's Assn. v. Templin*, 124 Wash. 567 (1923), the trial court awarded an injunction covering all lands owned by the defendant in Pierce County and "adjoining counties." On appeal, the state high court restricted the order to Pierce County alone (*Templin*, 574).

49. The court ruled that the tenant "merely stepped into the shoes" of the landowner and assumed all contractual obligations. To hold otherwise, the court reasoned, would thwart the legislature's purpose (*Feagain v. Dark Tobacco Growers' Cooperative Assn.*, 202 Ky. 801,

803 [1924]). The tobacco associations, supported by Bingham money and Sapiro's law firm, poured resources into litigating the CMAs and secured more legal victories than growers in any other state. Since so much tobacco production was conducted by tenants, the *Feagain* decision was especially important. Kentucky also protected the crop contract from third-party interference. In *Liberty Warehouse Co. v. Burley Tobacco Growers' Cooperative Association*, 208 Ky. 643 (1925), the court held that protecting the integrity of crop contracts was central to the general policy of the statute: "If the contract is one in furtherance of a legislatively declared purpose, then it was no doubt competent for the legislature to throw such restraints around it as were reasonably necessary to protect its integrity" (650). Remedies for third-party interference were also upheld in Texas (*Hollingsworth et al. v. Texas Hay Assn.*, cited note 37).

50. *Louisiana Farm Bureau Cotton Growers' Co-op. Assn. v. Clark*, 160 La. 294, 307 (1926). The court reached the same result in *Louisiana Farm Bureau Cotton Growers' Co-op. Assn. v. Bannister*, 161 La. 957 (1926); and *Louisiana Farm Bureau Cotton Growers' Co-op. Assn. v. Bacon et al.*, 164 La. 126 (1927). The other rulings are *Tobacco Growers' Co-op. Assn. v. Bissett*, 187 N.C. 180 (1924); and *Minnesota Wheat Growers' Co-op. Marketing Assn. v. Radke*, 163 Minn. 403 (1925).

51. *Bekkedal*, cited in note 40, 598–99; *Radke*, 407. Members did attempt to evade the obligations of their contracts by attacking the statutes as unconstitutional class legislation. Scarcely a case was tried where counsel for the grower failed to cite *Connolly v. Union Sewer Pipe Co.* Courts permitted the liberty of contract rationale to defeat the purpose of the CMAs only with regard to third parties. Where the grower members were concerned, there was no infringement of liberty of contract because the courts presumed that all members joined freely. "Farmers are characteristically meticulous," the Minnesota court grandly stated in *Huggins*; "no one . . . is forced to join [but] does so voluntarily. . . . Defendant . . . assumed the obligation of his contract because he wanted to" (475, 483).

52. *Potter*, cited in note 39, 445.

53. *Bekkedal*, 594–95; *Huggins*, 480, 482; see also *Potter*, 447. Although the Supreme Court had not explicitly overruled *Connolly* (and did not do so until *Tigner v. Texas*, 310 U.S. 141 [1940]), signs from the Taft Court pointed toward a relaxation of the rule. The Wisconsin court, among others, looked to Justice Louis Brandeis's dissenting opinion in *Truax v. Corrigan*, 257 U.S. 312 (1921), in which he supported the state's right to legislate on behalf of organized workers. Other state legislative classifications upheld during this time included a Pennsylvania law exempting bituminous coal from a tax on all anthracite coal mined in the state (*Heisler v. Thomas Colliery Co.*, 260 U.S. 245 [1922]); the exemption of sugarcane farmers from the tax on processed sugar in Louisiana (*American Sugar Refining Co. v. Louisiana*, 179 U.S. 89 [1901]); protective laws for women workers (*Muller v. Oregon*, 208 U.S. 412 [1908]); a law prohibiting combinations among makers or sellers of products but not among purchasers (*International Harvester Co. v. Missouri*, 234 U.S. 199 [1914]); a law setting out special rules for the issuance of crop insurance (*National Fire Insurance Co. v. Wanberg*, 260 U.S. 71 [1922]). The Court looked the other way in cases dealing with child labor, the minimum wage, and laws supporting the right of organized workers to strike and to be paid in legal tender rather than company scrip: *Adkins v. Children's Hospital*, 261 U.S. 525 (1923); *Hammer v. Dagenhart*, 247 U.S. 251 (1918); and *Coppage v. Kansas*, 236 U.S. 1 (1915).

54. Walton H. Hamilton, a public law scholar, argued that the CMA cases made it clear that cooperatives could not violate the antitrust laws. The courts, he noted, "insist[ed] that a strict compliance with statutory provisions is a condition essential to the enjoyment of exceptional privileges. But such instances of disapproval and admonition indicate at most that there are limits of tolerance beyond which the cooperatives will not be permitted to go" (Hamilton, "Judicial Tolerance of Farmers' Cooperatives").

55. Tobriner, "Constitutionality of Cooperative Marketing Statutes," 29 n. 34. If cooperatives violated the antitrust laws, Tobriner contended, the CMA furnished no bar to prosecution: "If the legislature has differentiated between two harmful organizations, the court might well incline to hold that the legislature has done so arbitrarily. The statute, then, is unconstitutional" (29). If the cooperative did not restrain trade, the CMA was superfluous, and its constitutionality was irrelevant. Tobriner claimed that 28 percent of eighty cooperatives he surveyed acknowledged that they in effect limited production by refusing to market poorer-grade produce ("Cooperative Marketing and the Restraint of Trade," 836 n. 47).

56. Tobriner, "Constitutionality of Cooperative Marketing Statutes," 34. The treatise is Mears and Tobriner, *Principles and Practices of Cooperative Marketing*.

57. *Beatrice Creamery Co. v. Cline*, 9 F.2d 176, 177 (D.C.D.Colo. 1925).

58. *Liberty Warehouse*, 276 U.S. 71 (1927).

59. Ibid., 91.

60. Ibid., 90, 96–97.

61. *Frost v. Corporation Commission of Oklahoma*, 278 U.S. 515, 517–19 (1929). To regulate entry into the business of cotton ginning, the Oklahoma legislature declared cotton ginning to be "affected with a public interest" under the *Munn* rationale. In 1915 the state required all persons, firms, or corporations wishing to gin cotton to present the commission with a statement of public necessity demonstrating the need for another gin. Anyone wishing to establish a cooperative gin, however, had merely to submit a petition signed by one hundred persons.

62. *Frost*, 524, 536. Justices Holmes and Stone joined Brandeis's dissent. Clearly, Sutherland concluded that the 1925 exemption unconstitutionally tipped the balance of power in the market in favor of the cooperative. Throughout the opinion, he referred to the cooperative gin company as a corporation—as if it were in form and in practice identical to the plaintiff. Justice Sutherland had authored some of the Taft Court's most conservative decisions; see, e.g., *Adkins v. Children's Hospital*, cited note 53.

63. Hamilton, "Judicial Tolerance of Farmers' Cooperatives," 942.

64. *Proceedings of the Second National Cooperative Marketing Conference* (Chicago: National Council of Farmers' Cooperative Marketing Associations, 1924), 3, quoted in Knapp, *Advance of American Cooperative Enterprise*, 62; Steen, *Cooperative Marketing*, ix.


*Chapter 10*

1. See, e.g., Leuchtenburg, *Franklin D. Roosevelt and the New Deal*; Kelly, Harbison, and Belz, *American Constitution*.

2. Hawley, *New Deal and the Problem of Monopoly*; Tomlins, *State and the Unions*; Gordon, *New Deals*; Campbell, *Farm Bureau and the New Deal*; Parrish, *Anxious Decades*; Saloutos, *American Farmer and the New Deal*.

3. Agricultural Adjustment Act of 1933, 48 Stat. 31 (May 12, 1933). See Hawley, *New Deal and the Problem of Monopoly*, 193–94; Gordon, *New Deals*, 85; Mitchell, *Depression Decade*; Nourse, Davis, and Black, *Three Years of the Agricultural Adjustment Administration*.

4. *Schecter*, 295 U.S. 495 (1935); *Butler*, 297 U.S. 1 (1936). On the AAA litigation, see Irons, *New Deal Lawyers*, 133–55. During the 1920s, the Supreme Court forbade states to regulate prices in a myriad of businesses it defined as private. Food and agricultural commodities were included in this category. In addition to *Frost* (discussed in Chapter 9), see *Wolff Packing Co. v. Court of Industrial Relations*, 262 U.S. 522 (1923) (industrial wages); *New State Ice Co. v. Liebman*, 285 US 262 (1931) (entry into the ice-making business); *Williams v. Standard Oil*, 278 U.S. 235 (1928) (gasoline); *Tyson v. Banton*, 273 U.S. 418 (1926) (theater tickets); *Fairmont Creamery Company Co. v. Minnesota*, 274 U.S. 1 (1927) (dairy prices); *Ribnik v. McBride*, 277 U.S. 350 (1928) (employment agents' fees). For a thorough summary of the Court's rulings on the price issue, see Cushman, "Structure of a Constitutional Revolution," 116–228. In view of the Court's conflicting holdings in these cases, it was not always obvious how the states were to regulate prices or production and come within the rules laid down.

5. In passing the Agricultural Marketing Agreement Act, 50 Stat. 246 (1937), Congress reenacted the first AAA without the processing taxes struck down in *Schecter*.

6. This argument applies chiefly to milk, which legal scholars at the time argued should be treated as a natural monopoly and regulated as a public utility; I extend it to California horticulture, which enjoyed geographic monopolies and thus faced little if any competition from other states. For an example of this argument, see Manley, "Constitutionality of Regulating Milk as a Public Utility."

7. In 1932, New York ranked third in the Union in milk production and second in the value of dairy products. Dairy income in New York between 1927 and 1930 averaged about $190 million, which represented half the state's total agricultural income (Comment, "Legislative Regulation of the New York Dairy Industry," 1260 n. 9).

8. Dillon, *Seven Decades of Milk*, 196–203. Borden and the Sheffield Farms Company operated "the two largest distributing systems in the country" in 1932, together controlling half the metropolitan New York City milk shed (Comment, "Legislative Regulation of the New York Dairy Industry," 1260 n. 10). Borden also aspired to expand into Boston, Chicago, and other cities — ambitions that the Justice Department sought to check with a Sherman Act prosecution in 1937. In *U.S. v. Borden*, 308 U.S. 188 (1939), the Supreme Court ruled that neither the Sherman Act nor the Capper-Volstead Act immunized producers, distributors, or others from prosecution for conspiring to "maintain artificial and noncompetitive prices to be paid to all producers" (205). Nor would the statutes immunize cooperatives when they combined with these parties to affect price.

9. Comment, "Legislative Regulation of the New York Dairy Industry," 1264. Fluid milk is the term applied to milk consumed in liquid form. Much of the milk produced in New York State during the 1930s was not consumed as milk but instead was used in the production of butter, cheese, and other dairy products. Because of the perennial surplus of fluid milk, its diversion into these commodities played an essential role in the distribution of income in the industry. Fluid milk brought the highest prices in an unregulated market, but demand was highly inelastic. In contrast, demand for butter was sensitive to changes in price. In selling both fluid milk and processed dairy products, distributors in New York were "usually able to pass all declines in retail prices back to the farmers" (Goldsmith and Winks, "Price-Fixing," 185).

10. Quoted in U.S. Works Progress Administration, *Marketing Laws Survey*, 7; Dillon, *Seven Decades of Milk*, 203–4. The Pitcher Report is New York State Legislature, Joint Committee on the Milk Industry, *Report of the Joint Legislative Committee to Investigate the Milk Industry*. Law review commentators and the state court opinion noted that the legislature acted under pressure from both producers and large distributors. See Comment, "Legislative Regulation of the New York Dairy Industry," 1265; Comment, "Milk Regulation in New York," 1359; *People v. Nebbia*, 262 N.Y. 259, 268 (1933).

11. WPA, *Marketing Laws Survey*, 7; New York Laws of 1933, chap. 158, §§300, 302, 307, 312. Initially, the law was passed as an emergency measure, but in 1934, the state legislature reenacted the provisions relating to the Milk Control Board as permanent legislation. The price-fixing provisions were maintained as temporary, revised in 1934, and repealed in 1937 (WPA, *Marketing Laws Survey*, 7 n. 21). Six other states (Georgia, Indiana, Massachusetts, Michigan, New Jersey, and Wisconsin) also intended for their milk control laws to be temporary. Alabama, Florida, Montana, New Hampshire, Pennsylvania, Rhode Island, and Vermont all reenacted their control laws as permanent legislation before the end of the decade (ibid., 8–9 nn. 24–25). In all, twenty-one states passed milk control acts between 1933 and 1940, all of them permitting price-fixing by the state, but none of them limiting production. Wisconsin was technically first on the books with its milk control statute, which went into effect four days sooner than the New York law. According to the WPA, the New York law was more widely imitated; however, its own summary of the laws belies that claim (ibid., 21, 25, 34, 41).

The significance of legislation treating milk as a public utility was that it signaled the courts that the state considered the milk industry to be sufficiently public as to meet the requirements of the affectation doctrine. See Comment, "Milk Regulation in New York"; Manley, "Constitutionality of Regulating Milk as a Public Utility."

12. For the facts of the case, see *People v. Nebbia*, 261. The Supreme Court's recitation of the facts adds only the detail that the loaf of bread had a retail value of five cents (*Nebbia v. New York*, 291 U.S. 502, 515 [1934]). The essence of the debate over *Nebbia* is whether the decision prefigured the constitutional revolution of 1937, when the Supreme Court finally cast aside the analytical categories of substantive due process jurisprudence and began to uphold the major planks of the second New Deal. The legal realism movement found its intellectual fuel in attacking substantive due process; see, e.g., Cohen, "Property and Sovereignty"; Hamilton, "Affectation with Public Interest"; and Warren, "New Liberty Under the Fourteenth Amendment." Modern expositions and critiques include Scheiber, "Road to *Munn*"; McCurdy, "Justice Field and the Jurisprudence of Government-Business Relations"; McCurdy, "Roots of 'Liberty of Contract' Reconsidered"; and Mensch, "History of Mainstream Legal Thought." For examples of its contemporary defenders, see Feldman, "Legal Aspects of Federal and State Price Control"; Purcell, "Constitutional Law"; Krug and Dickey, "Note."

Recent scholarship has concluded that *Nebbia* toppled the affectation doctrine. The effect of *Nebbia* was to discard the careful distinction between public and private that the Court had used as a yardstick when taking the measure of economic regulations. As a result, the Court's determination of the constitutionality of such laws would rest on the adequacy of legislative findings to justify the regulation and the reasonability of the relationship between the problem identified by the legislature and the means employed by the law to correct the problem. This test, which until recently seemed entirely settled as a matter of constitutional law (see *Planned Parenthood of Southeastern Pennsylvania v. Casey*, 505 U.S. 833 [1992]), presumes the constitutionality of economic regulations until proven otherwise.

*Nebbia* raised the collateral issue of whether the Supreme Court's "switch in time" in 1937 was so abrupt after all. Especially tantalizing to constitutional historians is the happy circumstance that the author of the opinion in *Nebbia* — Justice Owen Roberts — is the same man who wrote the switch-triggering decision in *West Coast Hotel v. Parrish*, 300 U.S. 379 (1937), a case upholding a state minimum wage law. Carefully examining the doctrinal underpinnings not only of *Nebbia* but also of pre–New Deal exceptions to the substantive due process regime, Barry Cushman argues that the seeds of the "revolution" were sown even before the New Deal, when President Hoover's appointments to the Court established the core of the majority that would ultimately decide the landmark cases. See Cushman, "Structure of a Constitutional Revolution." Most commentary on *Nebbia* deals with its place in constitutional history; here, I give its fuller social and economic context.

13. For contemporary comment on the constitutional issues raised by state regulation of prices (see cases cited in note 4), see Culp, "Comment"; Hale, "The Constitution and the Price System"; and Purcell, "Constitutional Law." Before *Nebbia*, emergency powers supplied the only reliable basis for any exercise of legislative authority affecting prices.

14. *People v. Nebbia*, 268, 272; *Nebbia v. New York*, 539. The degree to which the decision departed from existing constitutional norms was debated by a bevy of legal scholars. Critics of the opinion argued that a legislature could not be permitted simply to declare a business to be public to justify regulation that would otherwise be prima facia unconstitutional. "Doesn't this appear to be an inversion of all previously recognized methods of constitutional regulation of private business?" (Krug and Dickey, "Note," 702).

15. Dillon, *Seven Decades of Milk*, 204. "What was originally designed to provide immediate relief to the farmer became as well a vehicle for the protection of the distributor" (Comment, "Legislative Regulation of Milk Industry," 1265). The final version also contained a clause exempting cooperatives, including the Dairymen's League, from the provision allowing the state to fix prices to farmers; according to John Dillon, editor of a farmers' newspaper, this enabled the League and Borden to buy from farmers at any price they pleased. "The co-operative exemption clause had not been in the original bill. I have never been able to find anyone, legislator or layman, willing to admit authorship of it, or who admitted that he knew when the 'exemption' clause was inserted" (*Seven Decades of Milk*, 206). Amendments gave the Milk Control Board the power to hold hearings on producer prices, but the real power of the law had already been ceded to the distributors.

It is not clear how many farmers belonged to the League. According to one source, in 1933 there were eighty-four thousand individual farmers with an average herd of twenty cows; fifteen thousand of these were independent, unaffiliated with any cooperative, including the League (Comment, "Legislative Regulation of the New York Dairy Industry," 1260 n. 10).

16. Dillon, *Seven Decades of Milk*, 206–8.

17. *People v. Nebbia*, 268: "The condition has given rise to scenes of violence and disorder in the attempt to organize so-called milk strikes as a protest against the low prices paid for milk." On appeal, the state carefully confined its discussion of the violence to three paragraphs (Brief for Appellee, *Nebbia v. New York*, Case No. 531, Supreme Court of the United States, October Term, 1933, 12–13). Had the courts wanted to resort to an especially compelling rationalization for regulating price in a business previously considered not to be among those affected with a public interest, they might have turned to the need to pre-

serve social order. Choosing a case that involved no violence enabled the state to focus the courts' attention exclusively on the constitutionality of the regulation.

18. Comment, "Legislative Regulation of the New York Dairy Industry," 1269; Comment, "Milk Regulation in New York," 1366; see also Dillon, *Seven Decades of Milk*, 209–12, 216. The law exempted cooperatives from paying the retail minimum price, enabling the Dairymen's League to undercut other dealers. Furious competition among independent dealers, price-cutting to producers, and outright circumvention of the law resulted as dealers organized sham cooperatives to take advantage of the loophole. Borden brazenly refused to submit to an audit by the Milk Control Board until specifically ordered by the legislature. Even then, the legislature did not appropriate enough money to complete the audits (Comment, "Milk Regulation in New York," 1366–67 n. 54).

19. *Transcript of Record, Nebbia v. New York*, Case No. 531, Supreme Court of the United States, October Term 1933, 13 (hereafter *Nebbia Transcript*). This calculated strategy guided Nebbia's trial, where the state called only two witnesses. One was the customer who purchased the milk and bread; the other's testimony was omitted from the record as "merely cumulative." The trial judge imposed only a nominal fine — $5 (*Nebbia Transcript*, 11–13).

20. Dillon, *Seven Decades of Milk*, 213, 216; Comment, "Milk Regulation in New York," 1362, 1365. The Milk Control Board's attorney was Henry S. Manley, who in 1933 called for the milk industry to be regulated as a public utility (Manley, "Constitutionality of Regulating Milk as a Public Utility").

21. *Baldwin v. Seelig*, 294 U.S. 511 (1935). On the AAA orders, see Nourse, Davis, and Black, *Three Years of the Agricultural Adjustment Administration*, 104–95, 224–27; and Purcell, "Constitutional Law." On the 1936 interstate proposal, see Comment, "Milk Regulation in New York," 1364–65.

22. In its place the state constructed an entirely different administrative mechanism to regulate prices, dividing the state into marketing areas and authorizing the formation of dealers' and producers' bargaining associations to set prices through a process of hearings and negotiations. Membership in both associations was entirely voluntary. The law exempted dealers from the state antitrust laws and reiterated the cooperative marketing act's exemption for producers' cooperatives. The state built a back door into the new process by giving the Commission of Agriculture the power to fix prices upon the petition of a producers' bargaining agency and the consent of 75 percent of the producers in the area (Dillon, *Seven Decades of Milk*, 225).

23. Ibid., 333; Comment, "Milk Regulation in New York," 1369–70 n. 71. The consistent message of the legal order's response to the cooperative movement and, initially, the New Deal, was that monopoly in agriculture was intolerable, whether conducted by private associations of producers or by the state. Chief Justice Taft's dictum in *Wolff v. Industrial Court* asserted that food prices had never been susceptible to public regulation: "There is no monopoly in the preparation of foods" (538). Legal conservatives criticizing the *Nebbia* case used *Wolff* and its progeny to argue that the affectation doctrine was to be invoked only in cases where a monopoly was extorting unreasonably high prices from the public; in the milk industry, they contended, the economic problem was too much competition in the market rather than not enough (Krug and Dickey, "Note," 702). Dillon and other sources suggest that at the point of sales by producers to dealers there was not too much competition but rather too little. Collusion between the dairy cooperative and Borden kept farm prices down, and the law enabled them to maintain control over prices.

24. Wallace, *New Frontiers*, 171, 179–81, 174–75; White and Maze, *Henry A. Wallace*, 49–50.

25. California Agricultural Prorate Act of 1933, *Cal. Stats.* (June 15, 1933), chap. 754, pp. 1969–70; Watson, "Analysis of Raisin Marketing Controls," 11; Stokdyk, "Economic and Legal Aspects of Compulsory Proration," 18.

26. Prorate Law, §2, p. 1970 (composition of commission); §§9, 10, 11, 21, pp. 1973–74, 1977–78 (establishment of proration zones); §18, p. 1976 (administrative process). The 1935 amendments are Prorate Act, *Cal. Stats.* 1935, chap. 471, pp. 1526–43. One important addition was the institution of surplus pools, through which prorate committees could divert surplus crops to relief and charitable organizations and keep them off the market. To achieve the goals of the act and minimize production surpluses, the program committee was also empowered to secure packing facilities, cooperate with similar agencies in California, other states, or the federal government (Prorate Amendments, §§19.1, 21.1, pp. 1537–38).

On the economic thought of New Deal agricultural planners, see Hawley, *New Deal and the Problem of Monopoly*, 188–94; and Gordon, *New Deals*, 166–86.

27. The crops were tomatoes, sweet potatoes, asparagus, lettuce, olives, prunes, figs, apricots, pears, apples, grapes, lemons, artichokes, and celery—all perishable or semiperishable commodities in which surplus usually resulted in waste. The raisin prorate in 1937 made it fifteen. See C. H. Kinsley, "Some Notes on Proration," *PRP* 136 (Dec. 10, 1938): 568; Donald L. Kieffer, "This Prorate Law—What Is It?" *PRP* 136 (July 16, 1938): 46; "The Most Prorates," *PRP* 141 (Jan. 11, 1941): 21.

28. 5 Cal. 2d. 550, 577 (1936). Citing the state court decision in *People v. Nebbia*, the California court ruled that the state held the authority to regulate even such private businesses as the lemon industry because "demoralizing competition" threatened the livelihood of a substantial number of citizens and undermined their capacity to support the state (578). In view of the U.S. Supreme Court's failure in *Nebbia v. New York* to engage with contrary precedents, the California court determined that *Nebbia* was controlling (582).

29. The author of Comment, "Milk Regulation in New York" concluded by advocating, somewhat guardedly, "government ownership and operation of milk distribution systems." Through eminent domain proceedings the government could purchase the existing marketing infrastructure and eliminate duplications; this approach would be most effective, according to the author, at the state or federal level. "The possibility of political machinations seems worth risking in view of the substantial savings which government ownership would undoubtedly effect" (1370).

30. A massive bailout by the Federal Farm Board was required to reduce the cooperative's debt and return control of the cooperative to grower-directors. This was accomplished by 1931 with a combination of commodity loans, loans to retire Sun-Maid's bonded indebtedness, and a federally sponsored raisin marketing pool. In all, Fresno banks wrote off $6 million in Sun-Maid commodity loans. The U.S. Tariff Commission wrote in 1939 that Sun-Maid's significance as a cooperatively organized company lay only in the "check on other packers" that a grower-owned processor offered (U.S. Tariff Commission, *Grapes, Raisins, and Wines*, 156; Hamilton, *From New Day to New Deal*, 96–102).

31. J. M. Eueless, Chairman, California Raisin Growers Committee, to Franklin D. Roosevelt, Nov. 9, 1933; Budd A. Holt to H. R. Wellman, Mar. 15, 1934, both in Raisins, California, 1933–34, Box 75, RG 136.

32. A Control Board, representing Sun-Maid, the other packers, and the growers' com-

mittee, decided when and in what quantities to market the surplus ("Tentatively Approved Marketing Agreement for the California Raisin Industry," May 21, 1934, California, 1933–42, Box 75, RG 136). The agreement became effective May 29 (Press Release, Agricultural Adjustment Administration, May 29, 1934, Raisins, California, 1933–34, Box 75, RG 136). After one year, Congress eliminated the provision for minimum prices, but that in itself did not affect the raisin marketing program (Benedict and Stine, *Agricultural Commodity Programs*, 374; H. R. Wellman to J. W. Tapp, Memorandum, May 5, 1934, Raisins, California, 1933–34, Box 75, RG 136).

33. P. R. Taylor, "Memorandum for the Secretary," Sept. 12, 1935, Raisins, California, 1934–36, Box 75, RG 136; J. M. Leslie to Growers, Feb. 29, 1936, Sun-Maid Archives. On the packers' decision to pull out, see American Seedless Raisin Company et al. to Secretary of Agriculture, Aug. 9, 1935; A. Setrakian, Chairman, Committee of 140, to Henry A. Wallace, Aug. 20, 1935, both in Raisins, California, 1934–36, Box 75, RG 136; on the cancellation of the 1934 agreement, see "Order Terminating the Marketing Agreement for Packers of California Raisins, as Amended," Sept. 9, 1935, Raisins, California, ibid.

34. With the help of federal loans and by diverting grapes into by-products, Sun-Maid believed that it could control any surplus in 1935, but predictions of a short crop did not come true, and Sun-Maid was unable to control the annual surplus, much less the industry-wide carryovers. The 1935 crop came in at 203,000 tons, as opposed to Sun-Maid's figure of 180,000 tons; the 1936 crop of 182,000 tons exceeded Leslie's projection by 32,000 (Leslie to Growers, Sept. 1, 1936, Sun-Maid Archives). Crop forecasts were unreliable in part because of the different uses possible for raisin grapes—table use, wine, and raisins (Watson, "Analysis of Raisin Marketing Controls," 26).

35. Minutes of Meeting, Sun-Maid General Manager Keeler and Prorate Program Organizers, Mar. 1936, Sunland Sales Cooperative Association Notebook, Sun-Maid Archives; "Sun-Maid Reaffirms Stand in Opposition to Prorate Plan for Raisin Industry," *Fresno Bee*, Sept. 29, 1937, p. 8-A; Watson, "Analysis of Raisin Marketing Controls," 17–18; Leslie to Growers, Sept. 1, 1936; Sun-Maid Press Release, Nov. 24, 1936, Sun-Maid Archives; "At Last a Raisin Prorate," *PRP* 133 (Mar. 6, 1937): 331; "Sun-Maid's Position on Prorate," *Fresno Bee*, Nov. 5, 1936, p. 1. The 1936 season produced a crop of 182,000 tons and a surplus of 60,000. Both figures represented decreases from 1935, but Sun-Maid had not disposed of the surplus (Sun-Maid 1937 Crop Advance Offer, Sun-Maid Archives; "Sun-Maid Advance Set at $60 Ton," *Fresno Bee*, Aug. 9, 1937, p. 1).

36. In December, the cooperative ran out of money when its guaranteed advance of $60 led more growers to sell more raisins to Sun-Maid than it expected. Sun-Maid was forced to stop purchasing raisins from growers, and prices dropped to $55. For two months, the industry held its breath while Leslie and Keeler negotiated additional financing from the Central Bank for Cooperatives of the Federal Farm Credit Administration. It required $2.5 million to reopen the annual pool. The cooperative had badly underestimated both the size of the 1937 crop—250,000 tons, an increase of 68,000 tons over 1936—and the growers' desperation for stable prices. The cooperative was not able to market even half of the industry's total production. See Donald L. Kieffer, "They Said, 'Too Few Raisins to Prorate,'" *PRP* 134 (Aug. 28, 1937): 246; Watson, "Analysis of Raisin Marketing Controls," 29; Kieffer, "Sun Maid Stops Raisin Buying; Growers Are Surprised," *PRP* 134 (Dec. 18, 1937): 670; "Raisin Contracts Deferred," Sun-Maid press release, Dec. 6, 1937, Sun-Maid Archives; "Sun-Maid Stops Raisin Buying," *PRP* 134 (Dec. 18, 1937): 670; Kieffer, "Inside Story of Raisin Situa-

tion," *PRP* 135 (Jan. 29, 1938): 134. The 1937 crop figure is from Kieffer, "California's 1937 Farm Income Review," *PRP* 135 (Jan. 15, 1938): 63.

37. Watson, "Analysis of Raisin Marketing Controls," 45–46; Kieffer, "Government Loans More Raisin Money through New Agency," *PRP* 135 (Feb. 26, 1938): 254; Kieffer, "1938 Raisin Advance to Be $50 per Ton," *PRP* 136 (July 16, 1938): 62; "You May Lose the Prorate Law," *PRP* 136 (Oct. 28, 1938): 412; "Selfishness and Politics Can Kill the Prorate," *PRP* 136 (Dec. 3, 1938): 532; Kieffer, "This Prorate Law—What Is It?," *PRP* 136 (July 16, 1938): 46; C. B. Hutchinson, "The California Agricultural Prorate Act," *PRP* 137 (Jan. 7, 1939): 6; "The Prorate Bills," *PRP* 137 (Feb. 25, 1939): 192.

38. The fact that growers were well represented on the program committees helped to counteract opposition to repeal in the legislature. See "The Prorate Problem and the Alternative," *PRP* 137 (Apr. 1, 1939): 500; R. W. Gray, "Experiences and Lessons of Proration," *PRP* 137 (Apr. 15, 1939): 351; Watson, "Analysis of Raisin Marketing Controls," 47.

39. *Cal. Stats.* (1939), chap. 894, §§3, 6, 10, 16, pp. 2488–90, 2491–92, 2493, 2495–96.

40. Ibid., 1939, chap. 894, §§16, 19.1, pp. 2495–96, 2497–98 (effective Sept. 19, 1939); "California Could Have a Good Year," *PRP* 137 (June 3, 1939): 500; Watson, "Analysis of Raisin Marketing Controls," 47.

41. Leslie to Growers, Aug. 25, 1939, Sun-Maid Archives; Kieffer, "Plans for 1939 Raisin Control," *PRP* 138 (July 29, 1939): 79; "Raisin and Grape Affairs Become Tangled," *PRP* 138 (Aug. 26, 1939): 142–43; Watson, "Analysis of Raisin Marketing Controls," 47–56; Kieffer, "California's 1939 Farm Income Review," *PRP* 139 (Jan. 27, 1940): 30; "Raisin Program Is Like Football," *PRP* 138 (Sept. 9, 1939): 174; "Raisin Chaos Robs Growers of Price Boosts," *PRP* 138 (Sept. 23, 1939): 206; Leslie to Growers, Sept. 30, 1939, Sun-Maid Archives.

42. Kieffer, "Government Raisin Purchase Assured," *PRP* 139 (Mar. 9, 1940): 190; *Brief for Appellants, Parker v. Brown*, Case No. 1040, U.S. Supreme Court, October Term 1941, 4–5, 15; Leslie to Growers, Aug. 30, 1940, Sept. 27, 1940, Sun-Maid Archives. See also Kieffer, "FSCC [Federal Surplus Commodity Commission] Raisin Buying Lifts Field Price," *PRP* 139 (June 1, 1940): 430. Raisins delivered to the surplus pool would bring $27.50 and $25.00 per ton, depending on the variety; for stabilization pool raisins, growers would receive $55.00 per ton for all varieties.

43. Given that 95 percent of the raisins produced in California eventually entered interstate commerce, Brown's attorneys reasoned, this amounted to an unconstitutional interference with interstate trade. The strength of this argument lay in the fact that California had a natural monopoly on U.S. raisin production. Consequently, any marketing control program would affect interstate commerce. What the federal court had to decide was whether this effect was permissible under the Constitution (Brief for Appellants, 16–19; Complaint, *Brown v. Parker*, U.S. District Court, Southern District of California, Case No. 78, filed Dec. 28, 1940, in *Transcript of Record, Parker v. Brown*, No. 1040, October Term 1941, 1–5). During the litigation, the program committee prepared to apply for a marketing agreement under the Agricultural Marketing Agreement Act or a government purchase program, in case the district court struck down the prorate act.

44. *Brown v. Parker*, 39 F.Supp. 895, 901 (S.D.Cal. 1941).

45. Kieffer, "1940 Raisin Pool Payments Completed," *PRP* 143 (Apr. 4, 1942): 254; Leslie to Growers, Apr. 25, May 29, June 27, July 25, 1941, Sun-Maid Archives.

46. *Parker v. Brown*, 317 U.S. 341, 350, 358, 361 (1943). The decision established the "state action doctrine" exempting states from the federal antitrust laws. See Fox and Sullivan,

*Cases and Materials on Antitrust,* 534; and Jorde, "Antitrust and the New State Action Doctrine."

47. Sullivan, *Handbook of the Law of Antitrust,* 734.

48. *Wickard v. Filburn,* 317 U.S. 111 (1942), upheld acreage quotas imposed on wheat farmers under the Agricultural Adjustment Act of 1938 even when the quota was applied to wheat not intended for interstate commerce. This decision, along with *U.S. v. Darby Lumber Co.,* 312 U.S. 100 (1942), and *National Labor Relations Board v. Jones & Laughlin Steel Corp.,* 301 U.S. 1 (1937), charted out the reach of the federal commerce power "as broad as the economic needs of the nation" (*American Power & Light v. S.E.C.,* 329 U.S. 90, 104 [1946]). Only recently has the Supreme Court indicated that the power, considered plenary for over sixty years, is too broad. See *U.S. v. Lopez,* 514 U.S. 549 (1995) (holding that a federal law criminalizing the act of knowingly bringing a handgun within a school zone violated the commerce clause).

## Conclusion

1. *Tigner v. Texas,* 310 U.S. 141, 147 (1939). Aside from the unused restrictions placed on cooperatives by Capper-Volstead, few federal antitrust provisions had any force against farmers after the New Deal. A federal district court construed the Clayton Act in farmers' favor in 1943: "When Congress said that cooperatives were not to be punished even if they became monopolistic it would be ill-considered for me to hold to the contrary" (*U.S. v. Dairy Cooperative Association,* 49 F.Supp. 475, 475 [D.C.D. Ore. 1943]). See Jensen, "Integrating Economic and Legal Thought."

2. "It is true that the economists' concept of a 'business firm' as a 'profit maximizing unit' does not fit cooperatives without qualification, but the ['economic concept of a "pure" cooperative'] is likewise not acceptable" (Erdman, "Thought on Cooperation," 916). See also Breimyer, "Capper-Volstead Act."

3. In 1978, the Justice Department convened the National Commission for the Review of Antitrust Laws and Procedures to conduct a broad study of federal law and enforcement problems. The commission recommended that all existing immunities and exemptions "either be repealed or substantially reduced in scope." National Commission on Antitrust Laws and Procedures, *Report,* x. With regard to agriculture in particular, the commission strongly urged that Section 2 of the Capper-Volstead Act be amended to define more precisely the concept of "undue price enhancement" and to remove enforcement responsibility from the USDA, on the grounds that it was too difficult for the USDA to fulfill its statutory responsibilities for the promotion of cooperatives and police their anticompetitive behavior at the same time (ibid., 262–63). Congress took no action on any of the DOJ proposals. I am indebted to Charles English of the law firm of Ober, Kaler, Grimes & Shriver, Washington, D.C., for this source.

4. Jensen, "Integrating Economic and Legal Thought," 899. For the renaissance of federal antitrust enforcement in the Second New Deal, see *U.S. v. Borden,* 308 U.S. 188 (1939). The case was the culmination of nearly twenty years of DOJ investigations into the milk industry, including both corporate milk dealers such as Borden and cooperative milk associations; one of the defendants in the litigation was a cooperative that was named in the suit because of dealings with noncooperatives that the government alleged restrained trade. The cooperative pled to Capper-Volstead as a defense, but the Court held that the statute only

protected the market activities of farmers' associations abiding by the statutory requirements. Cooperatives conspiring with parties not exempt under Capper-Volstead were not immune under the Sherman Act. *Borden* construed the law just as its opponents had in 1923: as a declaration of the right of agricultural producers to unite for the purposes of marketing their products, and nothing more. The Court noted that Capper-Volstead did not suspend the antitrust laws for farmers; the statute did not "authorize any combination or conspiracy with other persons in restraint of trade that these producers may see fit to devise" (204–5). The government did not undertake any sustained or ongoing antitrust investigation of cooperatives as a result of *Borden*. See Odom, "Associated Milk Producers, Incorporated."

5. The current U.S. Code provisions show no changes from the original statute and no reference to any amending statute (7 U.S.C. 291, 292 [1997]). On the lack of findings by secretaries of agriculture, see Barton, "Principles," 3–7; Heflebower, *Cooperatives and Mutuals*, 37; Jensen, "Integrating Economic and Legal Thought," 899; Guth, "Farmer Monopolies," 81; National Commission on Antitrust Laws and Procedures, *Report*, 262–63. On the 1996 DOJ investigation into price-fixing by Archer, Daniels, Midland, see "$100 Million Fine in ADM Guilty Plea," *Chicago Tribune*, Oct. 15, 1996, p. 1.

The limits of Capper-Volstead immunities relate to market practices that would be illegal even for regular corporations. In 1960, the Supreme Court ruled in *Maryland and Virginia Milk Producers Assn. v. U.S.*, 362 U.S. 458, that cooperatives remained liable for predatory behavior under the Sherman Act. In 1967, the Supreme Court held that cooperatives that included nonproducers as members could not claim the Capper-Volstead exemption as a defense to a Sherman Act lawsuit (*Case-Swayne Co. v. Sunkist Growers, Inc.*, 389 U.S. 384 [1967]). This Sherman Act suit was a private action for treble damages, not a government prosecution, but this had no effect on the Supreme Court's ruling. Another decision holding a cooperative liable in antitrust is *North Texas Producers Assn. v. Metzger Dairies, Inc.*, 348 F.2d 189 (1965). See also National Commission on Antitrust Laws and Procedures, *Report*, 253–58.

In 1995, Sun-Maid and the raisin unit of Dole Foods called off a merger proposal rather than wait for the Federal Trade Commission to conclude its review ("Dancing Alone: Sun-Maid Moves on after Deal with Dole Falls Apart," *Fresno Bee*, Oct. 1, 1995, p. C1). I discuss the 1977–81 FTC investigation of Sunkist Growers below; see text at note 9.

6. On the relationship of associationalism to the New Deal, see Hamilton, *From New Day to New Deal*, 237–50; Hawley, *New Deal and the Problem of Monopoly*, 187–204; and Gordon, *New Deals*, 128–65.

7. Hurst, *Law and Markets in United States History*, 67.

8. McCraw, "Rethinking the Trust Question," 17–19 (quote on p. 18).

9. Among other things, the agreement required Sunkist to divest itself of its holdings in Arizona and to secure FTC approval before acquiring any processing plants for five years. See Mueller, Helmberger, and Patterson, *Sunkist Case*, 6, 186. See also U.S. Department of Agriculture, Division of Cooperative Marketing, "Sunkist Adventure."

10. Personal interview, Gary Marshburn, Sun-Maid Raisin Growers, Kingsburg, California, Mar. 21, 1988; "Dancing Alone," p. C1.

11. Personal interview, Dick Markarian, Fowler, California, July 19, 1990; personal interview, Leo Chooljian, Sanger, California, July 20, 1990.

12. "A Quip by Dole Hits Close to Home," *New York Times*, Dec. 21, 1995, p. B16; "Clinton

Signs Farm Bill Ending Subsidies," *New York Times*, Apr. 5, 1996, p. A22. See Shover, *First Majority, Last Minority*, on post–New Deal Farm policy.

13. At a 1992 hearing before a Massachusetts Food and Agricultural commissioner on the state's price support program, one dairy farmer explained, " 'This isn't about money at all. . . . This is about power. The [milk] dealers have traditionally had the power, and they see this pricing order as an erosion of their power. We [producers] need to be together on this, and we have a moral and a legal right in saying that the government should help us' " (B. J. Roche, "Endangered Species," *Boston Globe Magazine*, Sept. 13, 1992, p. 20).

14. On the Census Bureau's decision to exclude "farmer" from the occupations it counts in the census, see "Too Few Farmers Left to Count, Agency Says," *New York Times*, Oct. 10, 1993, sec. 1, p. 23. Although the farm sector has substantially recovered from the massive debts, commodity surpluses, and foreclosures incurred during the 1980s and the farms in existence now are more efficient and productive than ever, the number of farms continues to decrease. The farming population is aging, without a concomitant infusion of young blood: "The consolidation of cropland into fewer and fewer farms has meant that fewer and fewer people are entering into the profession" ("Fewest U.S. Farms since 1850," *New York Times*, Nov. 10, 1994, p. A24).

*Appendix*

1. For my analysis I used the published agricultural census schedules in both hard copy and electronic form. The electronic data were retrieved from computer files obtained from the Interuniversity Consortium for Political and Social Research at the University of Michigan (ICPSR). All tables and graphs in this appendix rely on ICPSR data and data culled from the published agricultural censuses. In a few cases, I generated variables using ICPSR data. The analysis was performed using SAS Statistical Software for the personal computer (SAS Institute, Cary, North Carolina) or by hand, as noted.

Ideally, the manuscript census would have been the preferred source for this analysis, but the manuscript agricultural census schedules for the decennials including and after 1880 are no longer extant. They were apparently destroyed in a fire at the Interior Department in Washington, D.C., in 1921 (telephone interview, Claire Prechtel-Kluskens, National Archives and Records Administration, Washington, D.C., August 25, 1994). Sucheng Chan reports that the 1880 and 1890 schedules were destroyed (*This Bittersweet Soil*, 409); the result of my inquiries indicates that those for 1900 and 1910 are gone as well. Since I am interested in the period after 1880, the surviving early schedules were not relevant.

2. Figures calculated from data presented in Table A-1. In 1910, the percent of farms under fifty acres was 49; in 1920, 56.

3. See, e.g., the illustrations in *Imperial Fresno*; and Kearney, *Fresno County*.

4. As in the fruit-growing counties, one county—San Joaquin—ranks first throughout the entire period. Here the second rank is also fairly stable. Neither San Joaquin nor the second-place county (usually Stanislaus) figures prominently in fruit growing. Most of the field-crop-producing counties swapped ranks every decennial, and each census saw new counties enter the game while others dropped out.

5. Lerwick, *Fresno County, California*.

6. Fluharty and Wilcox, *Enterprise Efficiency Studies on California Farms*; Adams, "Sea-

sonal Labor Requirements for California Crops." Of course, this kind of horticultural economy could not have prospered as it did without migrant workers to supply the seasonal labor demands of small farmers.

7. For the correlation analysis, I ranked all counties for the number of farms in the various size ranges and for the value of fruit and field crop production for each decennial. I then correlated the rank for number of farms in a given size with the rank of crop production. I used SAS's Spearman correlation function to perform these ranked correlations. Correlations among ranks are preferred when the data cannot be assumed to have a normal distribution, as is the case here, where there are only fifty-eight counties and thus only fifty-eight observations.

8. The correlation between fruit production and farms ten to nineteen acres was very high ($r \cong 0.9$).

9. I chose this method to include all the major horticultural centers for each decennial and to ensure consistency throughout the analysis. Including more counties in the analysis did not change the results; in fact, the differences became more statistically significant when additional counties were included. In each decennial, the county with the largest value of fruit sold that was not included as fruit producing accounted for less than 2 percent of the value of the state's total fruit crop. The counties clustered in the lowest 15 percent of the horticultural economy, then, contributed only marginally to its total value. To illustrate, in 1910, the remaining forty-one nonfruit-producing counties individually contributed from less than $1/100$ of 1 percent to just over 1.5 percent. Thus, including or dropping one or two more counties would not change the conclusion.

A different method of selecting fruit-producing counties might have included counties whose fruit production constituted a high proportion of their total agricultural output. I rejected this approach because it would omit such major fruit-producing counties as Los Angeles and Orange, whose economies were so diversified that, for example, in 1910 the total value of their fruit production was less than half of the total value of all the crops they produced.

10. The distinction between fruit-growing and nonfruit-growing was made on the basis of an arbitrary cut-off point. For example, in 1910, counties with fruit crops valued at greater than $1 million were treated as commercially important horticultural counties. Since the value of the fruit crop changed each decennial, the figure was altered to differentiate between fruit-producing counties and counties with a lower-value fruit crop. I checked this figure by averaging the fruit crop value for all counties; in every decennial the average placed the same number of counties on the fruit-growing list as my arbitrary cut-off point. T-tests to probe for the significance of the differences between the two groups were then performed on the numbers of farms in each size category for both groups of counties.

11. These t-tests were significant at the $P < 0.0010$ level—in statistical parlance, a very highly significant difference.

12. A good indicator of the strength of the finding is that the data were not normally distributed, and the data for the fruit counties had a different variance than the data for the nonfruit counties.

# Bibliography

*Manuscripts*

*Berkeley, California*
Bancroft Library, University of California
  Henry E. Erdman Manuscripts and Papers, 1920–65
  M. Theo Kearney Correspondence and Papers, 1896–1906
  Merritt, Ralph Palmer. *"After Me Cometh a Builder": The Recollections of Ralph Palmer Merritt*. Berkeley and Los Angeles: Regional Cultural History Project and Oral History Project, University of California, 1962.
  Franklin P. Nutting Correspondence and Papers, 1890–1925
  Nutting, Franklin P. *An Interview with Franklin P. Nutting*. Berkeley and Los Angeles: Regional Oral History Project, University of California, 1955.
  Chester Rowell Manuscripts and Papers, 1890–1920
  Charles C. Teague Correspondence and Papers, 1905–30
  Harris Weinstock Scrapbooks, 1892–1922
*Fresno, California*
Fresno City and County Historical Society
  California Raisin Growers Association, 1898–1904
  M. Theo Kearney Papers, 1884–1904
Fresno County Superior Court
  Civil Litigation Docketbooks and Case Files, 1890–1930
Office of *Fresno Bee*
  Historical Files
*Kingsburg, California*
  Sun-Maid Raisin Growers of California Corporate Archives
*San Bruno, California*
National Archives, Pacific Sierra Branch
  RG 21, Records of the United States District Court, Northern District of California, Eastern Division
    Equity Files, 1920–22, Case No. B-67
*Suitland, Maryland*
National Archives, Washington National Records Center
  RG 122, Records of the Federal Trade Commission
    Food Investigations, 1916–19
*Washington, D.C.*
National Archives
  RG 16, Records of the Secretary of Agriculture
    Correspondence of the Secretary's Office, 1908–33
    Records of the Solicitor, 1908–33

RG 60, Records of the Department of Justice
  Anti-Trust Division, 1910–38
RG 83, Records of the Bureau of Agricultural Economics, U.S. Department of Agriculture
  Subject Files, Numerical Correspondence, General Correspondence, 1923–26
RG 122, Records of the Federal Trade Commission
  Docket Section, General Investigations Files, 1916–20
RG 136, Records of the Production and Marketing Administration, Fruit and Vegetable
  Branch, Agricultural Adjustment Administration, 1933–35


*Periodicals*

*Associated Grower*, 1920–25
*California Cultivator*, 1905–20
*Country Gentleman*, 1850–1925
*Fresno Bee*, 1922–43
*Fresno Morning Republican*, 1900–1927
*Journal of Farm Economics*, 1920–45
*Journal of Political Economy*, 1920–45
*Los Angeles Times*, 1922–26
*Pacific Rural Press*, 1890–1943
*Review of Reviews*, 1900–1930
*San Francisco Call*, 1902
*San Francisco Chronicle*, 1923–26
*Sun-Maid Business*, 1925–28
*Sun-Maid Herald*, 1915–19


*Government Documents*

California. State Board of Agriculture. "Report of the State Statistician." In *Fifty-Ninth An-nual Report of the California State Board of Agriculture*. Sacramento: Friend W. Richar-don, Superintendent of State Printing, 1913.
———. State Board of Control. *California and the Oriental: Japanese, Chinese, and Hindus. Report of State Board of Control of California to Gov. William D. Stephens*. Sacramento: California State Printing Office, 1920.
———. State Board of Horticulture. *Annual Report*. Sacramento: State Printing Office, 1892.
———. "Review of the Fruit Season." In *7th Biennial Report of the State Board of Horticul-ture for 1899–1900*. Sacramento: Superintendent of State Printing, 1901.
———. State Commissioner of Horticulture. *First Biennial Report of the Commissioner of Horticulture of the State of California for 1903–04*. Sacramento: W. W. Shannon, Superin-tendent of State Printing, 1905.
———. State Commission of Horticulture. "Benefits of Cooperation." *Monthly Bulletin* 1:9. Sacramento: California State Printing Office, August 1912.
———. *Monthly Bulletin*. Sacramento: California State Printing Office, 1912–28.
———. State Department of Agriculture. *Second Report for the Period Ending December 31, 1921*. Sacramento: California State Printing Office, 1922.

———. *Statistical Report of State Statistician.* Sacramento: State Superintendent of Printing Office, 1911, 1912, 1914, 1916, 1917, 1919, 1921.

———. State Legislature. *Journal of the Fourth Session of the Legislature of the State of California, Proceedings of the Senate.* San Francisco: George Kerr, State Printer, 1853.

New York. State Legislature. Joint Committee on the Milk Industry. *Report of the Joint Legislative Committee to Investigate the Milk Industry.* Albany: J. B. Lyon, 1933.

United Kingdom. Ministry of Agriculture and Fisheries. *Report upon Large Scale Cooperative Marketing in the United States of America.* London: His Majesty's Stationery Office, 1925.

U.S. Bureau of Corporations. *Trust Laws and Unfair Competition.* Washington, D.C.: U.S. Government Printing Office, 1916.

U.S. Commission on Country Life. *Report.* Sen. Doc. 705, 60th Congress, 2d sess., 1909. Reprint. Chapel Hill: University of North Carolina Press, 1944.

U.S. Congress. House of Representatives. Committee on Agriculture. *Hearings on H.R. 6240.* 69th Congress, 1st Sess., Jan. 11–14, 1926. Washington, D.C.: U.S. Government Printing Office, 1927.

———. Committee on the Judiciary. J. J. Speight, ed., *Laws on Trusts and Monopolies.* Washington, D.C.: U.S. Government Printing Office, 1913.

———. *Hearings on Trust Legislation.* 63d Congress, 2d sess., December 16, 1913. Washington, D.C.: U.S. Government Printing Office, 1914.

———. *Hearings on Trust Legislation.* 63d Congress, 2d sess., Dec. 16, 1913. Washington, D.C.: U.S. Government Printing Office, 1914.

U.S. Congress. Senate. Committee on the Judiciary. *Hearings Authorizing Associations of Producers of Agricultural Products.* 67th Cong., 1st sess., June 2, 1921. Washington, D.C.: U.S. Government Printing Office, 1921.

———. *Hearings on H.R. 15657.* 63d Congress, 2d sess., May 13, 1914. Washington, D.C.: U.S. Government Printing Office, 1914.

U.S. Department of Agriculture. *Annual Report for 1913.* Washington, D.C.: U.S. Government Printing Office, 1914.

———. Office of Markets and Rural Organization. *Service and Regulatory Announcements* 20 (Feb. 7, 1917).

———. *Organization and Conduct of a Market Service in the Department of Agriculture, Discussed at a Conference Held at the Department on April 29, 1913.* Washington, D.C.: U.S. Government Printing Office, 1913.

———. "Report of the Secretary of Agriculture." *Yearly Report for 1917,* 9–48. Washington, D.C.: U.S. Government Printing Office, 1918.

———. "The Sunkist Adventure." Farmer Cooperative Service Information *Bulletin* 94. Washington, D.C.: Farmer Cooperative Service, 1975.

———. *Yearbook for 1915.* Washington, D.C.: U.S. Government Printing Office, 1916.

U.S. Department of Justice. National Commission for the Review of Antitrust Laws and Procedures. *Report to the President and the Attorney General.* Vol. 1. Washington, D.C.: U.S. Government Printing Office, 1979.

U.S. Farm Credit Administration. "History of the California Fruit Growers' Exchange." *Circular* C-121. Washington, D.C.: U.S. Government Printing Office, 1940.

U.S. Immigration Commission. *Reports of the Immigration Commission, Immigrants in In-*

*dustries. Part 25: Japanese and Other Immigrant Races in the Pacific Coast and Rocky Mountain States.* Washington, D.C.: U.S. Government Printing Office, 1911.

U.S. Industrial Commission. *Trusts and Industrial Combinations. Reports.* Vol. 2. Washington, D.C.: U.S. Government Printing Office, 1900.

U.S. Tariff Commission. *Grapes, Raisins, and Wines. Report* 134, 2d Series. Washington, D.C.: U.S. Government Printing Office, 1939.

U.S. Works Progress Administration. *Marketing Laws Survey: State Milk and Dairy Legislation.* Washington, D.C.: U.S. Government Printing Office, 1941.


*Books, Articles, and Dissertations*

Adams, Edward F. *The Modern Farmer in His Business Relations.* San Francisco: N. J. Stone, 1899.

Adams, Frank. "The Historical Background of California Agriculture." In *California Agriculture*, edited by Claude B. Hutchison, 17–33. Berkeley: University of California Press, 1946.

Adams, Herbert B., ed. *History of Cooperation in the United States.* Baltimore: Johns Hopkins University Press, 1888.

Adams, R. L. "Seasonal Labor Requirements for California Crops." University of California Agricultural Experiment Station *Bulletin* 623. Berkeley: University of California, 1938.

Arndt, Stanley M. "The Law of California Co-operative Marketing Associations." *California Law Review* 8 (1920): 281–94.

Ayers, Edward L. *The Promise of the New South: Life after Reconstruction.* New York: Oxford University Press, 1992.

Bakalian, Anny. *Armenian-Americans: From Being to Feeling Armenian.* New Brunswick, N.J.: Transaction Publishers, 1993.

Baker, Gladys L., Wayne D. Rasmussen, Vivian Wiser, and Jane M. Porter. *Century of Service: The First 100 Years of the United States Department of Agriculture.* Washington, D.C.: U.S. Government Printing Office, 1963.

Bakker, Elna. *An Island Called California: An Ecological Introduction to Its Natural Communities.* 2d ed. Berkeley: University of California Press, 1971.

Ballantine, Henry W. "Co-operative Marketing Associations." *Minnesota Law Review* 8 (December 1923): 1–27.

Barton, David G. "Principles." In *Cooperatives in Agriculture*, edited by David Cobia, 3–32. Englewood Cliffs, N.J.: Prentice-Hall, 1989.

Bean, Walton, and James J. Rawls. *California: An Interpretive History.* 3d ed. New York: McGraw-Hill, 1978.

Beard, Charles A. *An Economic Interpretation of the Constitution of the United States.* New York: Macmillan, 1925.

Benedict, Murray. "The Economic and Social Structure of California Agriculture." In *California Agriculture*, edited by Claude Hutchinson, 395–435. Berkeley: University of California Press, 1946.

———. *Farm Policies of the U.S., 1790–1900.* New York: Twentieth Century Fund, 1953.

Benedict, Murray, and Oscar C. Stine. *The Agricultural Commodity Programs.* New York: Twentieth Century Fund, 1956.

Benson, Lee. *Merchants, Farmers, and Railroads: Railroad Regulation and New York Politics, 1850–1887.* Cambridge, Mass.: Harvard University Press, 1955.

Berle, Adolf A., Jr., and Gardiner C. Means. *The Modern Corporation and Private Property.* New York: Macmillan, 1933.

Bioletti, Frederic T. *The Phylloxera of the Vine.* California Agricultural Experiment Station *Bulletin* 131. Berkeley: University of California Agricultural Experiment Station, 1901.

Black, William E. "The Sapiro Cooperative Philosophy: Its Relevance to Modern Cooperatives." Paper presented at the Graduate Institute of Cooperative Leadership, University of Missouri, July 17, 1984.

Blackford, Mansel G. *The Politics of Business in California, 1890–1920.* Columbus: Ohio State University Press, 1977.

Blick, James. "Geography and Mission Agriculture." *Pacific Historian* 8 (February 1964): 3–38.

Bragg, James M. "History of Cooperative Marketing in the Raisin Industry to 1923." M.A. thesis, University of California, Berkeley, 1930.

Brand, Charles J. "The Office of Markets of the United States Department of Agriculture." *Annals of the American Academy of Political and Social Science* 50 (November 1913): 252–59.

Breimyer, Harold E. "The Capper-Volstead Act: A Historical and Philosophical Assessment." In *Agricultural Cooperatives and the Public Interest: Proceedings of a North Central Regional Research Committee 117 Sponsored Workshop Held in St. Louis, Missouri, 6–8 June 1977,* North Central Project 117, Monograph 4, 3–10. Madison: University of Wisconsin, College of Agricultural and Life Sciences, Research Division, 1978.

Brinkley, Alan. "Richard Hofstadter's *The Age of Reform*: A Reconsideration." *Reviews in American History* 13 (September 1985): 462–80.

Brody, David. *Steelworkers in America: The Nonunion Era.* New York: Harper & Row, 1969.

Brooker, William C. *Cooperative Marketing Associations in Business.* New York: Privately published, 1935.

Brown, Edmund, Jr. "Cooperative Marketing of Tobacco." *North Carolina Law Review* 1 (1923): 216–22.

Bruchey, Stuart. *Enterprise: The Dynamic Economy of a Free People.* Cambridge, Mass.: Harvard University Press, 1990.

Bryce, James. *The American Commonwealth.* Vol. 2. Rev. ed. London: Macmillan, 1891.

Buck, Solon Justus. *The Granger Movement: A Study of Agricultural Organization and Its Political, Economic, and Social Manifestations, 1870–1880.* Cambridge, Mass.: Harvard University Press, 1913.

Camp, William Roswell. "The Organization of Agriculture in Relation to the Problem of Price Stabilization." *Journal of Political Economy* 32 (1924): 282–314.

Campbell, Christiana McFadden. *The Farm Bureau and the New Deal: A Study of the Making of National Farm Policy, 1933–1940.* Urbana: University of Illnois Press, 1962.

Campbell, Tracy. *The Politics of Despair: Power and Resistance in the Tobacco Wars.* Lexington: University Press of Kentucky, 1993.

Cance, Alexander E. "Farmers' Cooperative Exchanges." *Bulletin of the Extension Service,* Massachusetts Agricultural College, Amherst 1914, reprinted in John Phelan, *Readings in Rural Sociology,* 120–31. New York: Macmillan, 1920.

Capper, Arthur. *The Agricultural Bloc*. New York: Harcourt, Brace, 1922.

Carstensen, Peter. "Dubious Dichotomies and Blurred Vistas: The Corporate Reconstruction of American Capitalism." *Reviews in American History* 17 (1989): 404–11.

Carver, Thomas Nixon. *Principles of Rural Economics*. Boston, Mass.: Ginn, 1911.

Cerny, George. "Cooperation in the Midwest in the Granger Era, 1869–1875." *Agricultural History* 37 (1963): 187–205.

Chamberlain, J. P. "The Cooperative Marketing Act." *American Bar Association Journal* 8 (1922): 416–17.

Chambers, Clarke A. *California Farm Organizations: A Historical Study of the Grange, the Farm Bureau, and the Associated Farmers, 1929–1941*. Berkeley and Los Angeles: University of California Press, 1952.

Chambers, John Whiteclay, II. *The Tyranny of Change: America in the Progressive Era, 1900–1917*. New York: St. Martin's Press, 1980.

Chan, Sucheng. *This Bittersweet Soil: The Chinese in California Agriculture, 1860–1910*. Berkeley: University of California Press, 1986.

Chandler, Alfred D., Jr. "Business History as Institutional History." In *Approaches to American Economic History*, edited by George Rogers Taylor and Lucius F. Ellsworth, 17–24. Charlottesville: University Press of Virginia, 1970.

———. *Scale and Scope: The Dynamics of Industrial Capitalism*. Cambridge, Mass.: Harvard University Press, 1990.

———. *The Visible Hand: The Managerial Revolution in American Business*. Cambridge, Mass.: Harvard University Press, 1977.

Cherny, Robert. "Lawrence Goodwyn and Nebraska Populism: A Review Essay." *Great Plains Quarterly* 1 (1981): 181–94.

Clark, Elizabeth. "'The Sacred Rights of the Weak': Pain, Sympathy and the Culture of Individual Rights in America." *Journal of American History* 82 (1995): 463–93.

Clarke, Sally. "Farmers as Entrepreneurs: Regulation and Innovation in American Agriculture during the Twentieth Century." Ph.D. dissertation, Brown University, 1987.

Cleland, Robert Glass. *The Cattle on a Thousand Hills: Southern California, 1850–1880*. San Marino: Huntington Library, 1951.

Cleland, Robert Glass, and Osgood Harvey. *March of Industry*. Los Angeles: Powell, 1929.

Coben, Stanley. *A. Mitchell Palmer: Politician*. New York: Columbia University Press, 1963.

Cobia, David, ed. *Cooperatives in Agriculture*. Englewood Cliffs, N.J.: Prentice-Hall, 1989.

Cochran, Harrington W. "The Results of Co-operation in the Raisin Industry of California." M.A. thesis, University of California, Berkeley, 1915.

Cohen, Morris R. "Property and Sovereignty." *Cornell Law Quarterly* 13 (1927): 8–30.

Colby, Charles C. "The California Raisin Industry — A Study in Geographic Interpretation." *Annals of the Association of American Geographers* 14 (1924): 49–108.

Comment. "Legislative Regulation of the New York Dairy Industry." *Yale Law Journal* 42 (June 1933): 1259–70.

Comment. "Milk Regulation in New York." *Yale Law Journal* 46 (1937): 1359–70.

Coulter, John Lee. *Cooperation among Farmers: The Keystone of Rural Prosperity*. New York: Sturgis and Walton, 1911.

Croly, Herbert. *The Promise of American Life*. 1909. Reprint. Indianapolis: Bobbs-Merrill, 1965.

Cronon, William. *Nature's Metropolis: Chicago and the Great West*. New York: Norton, 1991.

Crosby, Alfred W. *The Columbian Exchange: Biological and Cultural Consequences of 1492.* Westport, Conn.: Greenwood, 1972.

Cross, Ira. "Cooperation in Agriculture." *American Economic Review* 1 (1911): 535–44.

Crouse, Philip C. "The California Raisin Industry: Its Growth and Financing." Ph.D. dissertation, Stonier Graduate School of Banking, Rutgers University, 1977.

Culp, Maurice S. "Comment: Constitutional Law, Price Fixing, Emergency Legislation." *Michigan Law Review* 32 (1933): 63–71.

Cumberland, William W. *Cooperative Marketing: Its Advantages as Exemplified in the California Fruit Growers Exchange.* Princeton: Princeton University Press, 1917.

Cushman, Barry. "The Structure of a Constitutional Revolution: Nebbia v. New York and the Collapse of Laissez-Faire Constitutionalism." Ph.D. dissertation, University of Virginia, 1995.

Daniel, Cletus. *Bitter Harvest: A History of California Farmworkers, 1870–1941.* Ithaca, N.Y.: Cornell University Press, 1981.

Davidian, Nectar. *The Seropians: First Armenian Settlers in Fresno County, California.* Berkeley: Privately published, 1965.

Davis, Lance E., et al. *American Economic Growth: An Economist's History of the U.S.* New York: Harper & Row, 1972.

Dillon, John J. *Seven Decades of Milk: A History of New York's Dairy Industry.* New York: Orange Judd, 1941.

Dunbar, Robert G. *Forging New Rights in Western Waters.* Lincoln: University of Nebraska Press, 1983.

Durrenberger, Robert W., with Robert B. Johnson. *California: Patterns on the Land.* 5th ed. Palo Alto, Calif.: Mayfield, 1976.

Eaton, Edwin M. *Vintage Fresno: Pictorial Recollections of a Western City.* Fresno: Huntington Press, 1965.

Edwards, Laura F. *Gendered Strife and Confusion: The Political Culture of Reconstruction.* Urbana: University of Illinois Press, 1997.

Eisen, Gustav. *The Raisin Industry: A Practical Treatise on the Raisin Grapes.* San Francisco: H. S. Crocker, 1890.

Elliot, Charles B. *The Law of Corporations.* Indianapolis: Bobbs-Merrill, 1923.

Elliot, Wallace W. *History of Fresno County, California.* 1882. Reprint. Fresno: Valley Publishers, 1973.

Ellis, William E. "Robert Worth Bingham and the Crisis of Cooperative Marketing in the Twenties." *Agricultural History* 56 (1982): 99–121.

Elsworth, R. H. "Statistics of Farmers' Selling and Buying Associations, 1863–91." U.S. Federal Farm Board, Division of Cooperative Marketing *Bulletin* 9 (1932).

Ely, Richard. "Introduction." In *History of Cooperation in the United States*, edited by Herbert B. Adams, 5–9. Baltimore: Johns Hopkins University Press, 1888.

Ensrud, Adolph George. "A History of the Origin and Functions of the Federal Bureau of Markets." M.A. thesis, University of Chicago, 1922.

Erdman, Henry E. *The California Fruit Growers' Exchange: An Example of Cooperation in the Segregation of Conflicting Interests.* New York: American Council, Institute of Pacific Relations, 1933.

———. "The Development and Significance of California Cooperatives, 1900–1915." *Agricultural History* 32 (1958)· 179–84.

———. "Thought on Cooperation—Discussion." *Journal of Farm Economics* 31 (1949): 915–16.

Erdman, Henry E., and Grace H. Larsen. "The Development of Agricultural Cooperatives in California." California Agricultural Experiment Station, Giannini Foundation of Agricultural Economics, *Miscellaneous Publications*, February 20, 1964.

Ernst, Daniel R. *Lawyers against Labor: From Individual Rights to Corporate Liberalism.* Urbana: University of Illinois Press, 1995.

Evans, Frank, and E. A. Stokdyk. *The Law of Agricultural Co-operative Marketing.* Rochester, N.Y.: Lawyers Co-operative Publishing Company, 1937.

Evans, Peter B., Dietrich Rueschemeyer, and Theda Skocpol, eds. *Bringing the State Back In.* Cambridge: Cambridge University Press, 1985.

Everitt, J. A. *The Third Power: Farmers to the Front.* Indianapolis: J. A. Everitt, 1907.

Fay, Charles Ryle. *Cooperation at Home and Abroad.* 2 vols. London: P. S. King & Son, 1908–39.

Feldman, George, J. "Legal Aspects of Federal and State Price Control." *Boston University Law Review* 16 (1936): 570–94.

Filley, H. Clyde. *Cooperation in Agriculture.* New York: Wiley, 1929.

Fisk, Catherine. "Still 'Learning Something of Legislation': The Judiciary in the History of Labor Law." *Law and Social Inquiry* 19 (Winter 1994): 151–86.

Fite, Gilbert C. *The Farmers' Frontier, 1865–1900.* New York: Holt, Rinehart & Winston, 1966.

———. *George N. Peek and the Fight for Farm Parity.* Norman: University of Oklahoma Press, 1954.

———. Review of Lawrence Goodwyn, *Democratic Promise. Pacific Northwest Quarterly* 69 (1978): 137–38.

Fluharty, Lee W., and F. R. Wilcox. *Enterprise Efficiency Studies on California Farms: A Progress Report.* University of California Agricultural Extension Service *Circular* 24. Berkeley: University of California Printing Office, 1929.

Forbath, William E. *Law and the Shaping of the American Labor Movement.* Cambridge, Mass.: Harvard University Press, 1991.

Forrester, Robert. *Report upon Large Scale Co-operative Marketing in the United States of America.* London: H.M. Stationery Office, 1925.

Fox, Eleanor M., and Lawrence A. Sullivan. *Cases and Materials on Antitrust.* St. Paul, Minn.: West, 1989.

Fox, Feramorz Y. "Co-operation in the Raisin Industry of California." M.A. thesis, University of California, Berkeley, 1912.

*Fresno: Where Half the World's Raisins Come From.* Fresno: Fresno County Chamber of Commerce, September 2, 1922.

Fresno County Centennial Committee. *Fresno County Centennial Almanac.* Fresno, Calif.: Privately printed, 1956.

Freyer, Tony. *Regulating Big Business: Antitrust in Great Britain and America.* New York: Cambridge University Press, 1992.

Furner, Mary. *Advocacy and Objectivity: A Crisis in the Professionalization of American Social Science, 1865–1905.* Lexington: University Press of Kentucky, 1975.

Gabriel, Ralph H. "The Farmer in the Commonwealth." *North American Review* 213 (May 1921): 577–86.

Galambos, Louis. "The Emerging Organizational Synthesis in Modern American History." *Business History Review* 64 (1970): 279–90.

———. "Parsonian Sociology and Post-Progressive History." *Social Science Quarterly* 50 (1969): 22–45.

Gates, Paul W. "Adjudication of Spanish-Mexican Land Claims in California." *Huntington Library Quarterly* 21 (1958): 213–36.

———. *The Farmer's Age: Agriculture, 1815–1860*. New York: Holt, Rinehart, & Winston, 1960.

———. *History of Public Land Law Development*. Report to the Public Lands Commission. 1968. Reprint. New York: Arno Press, 1978.

———. *Land and Law in California: Essays on Land Policies*. Ames: Iowa State University Press, 1991.

———. "Public Land Disposal in California." *Agricultural History* 49 (1975): 158–78.

———. "The Suscol Principle, Preemption, and California Latifundia." In *Land and Law in California: Essays on Land Policies*, 209–28. Ames: Iowa State University Press, 1991.

Giglio, James N. *H. M. Daugherty and the Politics of Expediency*. Kent, Ohio: Kent State University Press, 1978.

Gill, Charles O. "Social Effects of Cooperation in Europe." Federal Council of the Churches of Christ in America, Missionary Education Movement of the U.S. and Canada, *Report of Commissions* (New York, 1916). In *Readings in Rural Sociology*, edited by John Phelan, 131–37. New York: Macmillan, 1920.

Gillman, Howard. *The Constitution Besieged: The Rise and Fall of Lochner Era Police Powers Jurisprudence*. Durham, N.C.: Duke University Press, 1993.

Goldberg, Charles L. "Co-operative Marketing and Restraint of Trade." *Marquette Law Review* 12 (1928): 270–92.

Goldschmidt, Walter. *As You Sow*. New York: Harcourt, Brace, 1947.

Goldsmith, Irving B., and Gordon W. Winks. "Price-Fixing: From *Nebbia* to *Guffey*." *Illinois Law Review* 31 (1936): 179–201.

Gonzalez, Gilbert G. *Labor and Community: Mexican Citrus Worker Villages in a Southern California County, 1900–1950*. Urbana: University of Illinois Press, 1994.

Goodwyn, Lawrence. *Democratic Promise: The Populist Moment in America*. New York: Oxford University Press, 1976.

Gordon, Colin. *New Deals: Business, Labor, and Politics in America, 1920–1935*. New York: Cambridge University Press, 1994.

Gordon, Linda. *Heroes of Their Own Lives: The Politics and History of Family Violence, 1880–1960*. New York: Viking, 1988.

Gordon, Robert W. "Critical Legal Histories." *Stanford Law Review* 36 (1984): 57–125.

Gordon, Sarah B. "'The Liberty of Self-Degradation': Polygamy, Woman Suffrage, and Consent in Nineteenth Century America." *Journal of American History* 83 (1996): 815–47.

Grandy, Christopher. *New Jersey and the Fiscal Origins of Modern American Corporation Law*. New York: Garland, 1993.

Grantham, Dewey W., Jr. "Black Patch War: The Story of the Kentucky and Tennessee Night Riders, 1905–1909." *South Atlantic Quarterly* 59 (1960): 215–25.

Gregory, James N. *American Exodus: The Dust Bowl Migration and Okie Culture in California*. New York: Oxford University Press, 1989.

Guarneri, Carl J. *The Utopian Alternative: Fourierism in Nineteenth-Century America.* Ithaca, N.Y.: Cornell University Press, 1991.

Guerin-Gonzalez, Camille. *Mexican Workers and American Dreams: Immigration, Repatriation, and California Farm Labor, 1900–1939.* New Brunswick, N.J.: Rutgers University Press, 1994.

Guth, James L. "Farmer Monopolies, Cooperatives, and the Intent of Congress: Origins of the Capper-Volstead Act." *Agricultural History* 56 (1982): 67–82.

Hahn, Steven. *The Roots of Southern Populism: Yeoman Farmers and the Transformation of the Georgia Upcountry, 1850–1890.* New York: Oxford University Press, 1984.

Hale, Robert L. "The Constitution and the Price System: Some Reflections on *Nebbia v. New York.*" *Columbia Law Review* 34 (1934): 401–25.

Hall, Suzanne M. "Breaking Trust: The Black Patch Tobacco Culture of Kentucky and Tennessee, 1900–1940." Ph.D. dissertation, Emory University, 1989.

Hamilton, David E. "Building the Associative State: The Department of Agriculture and American State-Building." *Agricultural History* 64 (1990): 207–18.

———. *From New Day to New Deal: American Farm Policy from Hoover to Roosevelt, 1928–1933.* Chapel Hill: University of North Carolina Press, 1991.

Hamilton, Walton H. "Affectation with Public Interest." *Yale Law Journal* 39 (1930): 1089–1112.

———. "Judicial Tolerance of Farmers' Cooperatives." *Yale Law Journal* 38 (1929): 936–54.

Hamm, Richard. *Shaping the Eighteenth Amendment: Temperance Reform, Legal Culture, and the Polity, 1880–1920.* Chapel Hill: University of North Carolina Press, 1995.

Handlin, Oscar, and Mary Frug Handlin. *Commonwealth: A Study of the Role of Government in the American Economy, Massachusetts, 1774–1861.* Rev. ed. Cambridge, Mass.: Belknap Press of Harvard University Press, 1969.

Hanna, John. "Cooperative Associations and the Public." *Michigan Law Review* 29 (1930): 148–90.

———. *The Law of Cooperative Marketing Associations.* New York: Ronald Press, 1931.

Harbaugh, William H. *The Life and Times of Theodore Roosevelt.* Rev. ed. London: Oxford University Press, 1975.

Hart, James D. *A Companion to California.* Rev. ed. Berkeley: University of California Press, 1987.

Hartz, Louis. *Economic Policy and Democratic Thought: Pennsylvania, 1776–1860.* Chicago: Quadrangle Paperbacks, 1968.

Hasegawa, Yoshino Tajiri, and Keith Boettcher, eds. *Success through Perseverance: Japanese-Americans in the San Joaquin Valley.* Fresno: Japanese-American Project, San Joaquin Valley Library System, 1980.

Haskell, Henry J. "The Great Farm Movement toward Cooperation." *World's Work* 41 (1921): 547–49.

Hattam, Victoria C. *Labor Visions and State Power: The Origins of Business Unionism in the United States.* Princeton: Princeton University Press, 1993.

Hawley, Ellis W., ed. *Herbert Hoover as Secretary of Commerce: Studies in New Era Thought and Practice.* Iowa City: University of Iowa Press, 1981.

———. *The New Deal and the Problem of Monopoly.* Princeton: Princeton University Press, 1966.

———. "Three Facets of Hooverian Associationalism: Lumber, Aviation, and Movies,

1921–1930." In *Regulation in Perspective: Historical Essays*, edited by Thomas K. McCraw, 95–123. Cambridge, Mass.: Harvard University Press, 1981.

Hay, Douglas. *Albion's Fatal Tree: Crime and Society in Eighteenth-Century England*. New York: Pantheon Books, 1975.

Hayes, Michael N., and Alan L. Olmstead. "The Arvin and Dinuba Controversy Reexamined." Working Papers Series, 8, Agricultural History Center, University of California, Davis, 1981.

Hays, Samuel P. *Conservation and the Gospel of Efficiency: The Progressive Conservation Movement, 1890–1920*. Cambridge, Mass.: Harvard University Press, 1959.

———. *The Response to Industrialism*. Chicago: University of Chicago Press, 1957.

Heflebower, Richard B. *Cooperatives and Mutuals in the Market System*. Madison: University of Wisconsin Press, 1980.

Henderson, Gerard C. "Cooperative Marketing Associations." *Columbia Law Review* 23 (1923): 91–112.

Hendry, George W. "The Source Literature of Early Plant Introduction into Spanish America." *Agricultural History* 8 (1934): 64–71.

Hicks, John D. *The Populist Revolt*. Minneapolis: University of Minnesota Press, 1931.

Higgs, Robert. "Landless by Law: Japanese Immigrants in California to 1941." *Journal of Economic History* 38 (1978): 205–25.

Higham, John, Leonard Krieger, and Felix Gilbert. *History*. Englewood Cliffs, N.J: Prentice-Hall, 1965.

Hine, Robert. *California's Utopian Colonies*. 1953. Reprint. Berkeley: University of California Press, 1983.

Hobson, Asher. "Farmers' Co-operative Associations: Their Legal and Legislative Aspects." *American Economic Review* 11 (1921): 221–26.

Hofstadter, Richard. *The Age of Reform: From Bryan to F. D. R.* New York: Random House, 1955.

Holyoake, George Jacob. *History of Cooperation*. Rev. ed. 2 vols. London: T. Fisher Unwin, 1906.

Horner, J. T. "U.S. Government Activities in the Field of Agriculture Prior to 1913." *Journal of Farm Economics* 10 (1928): 429–60.

Horwitz, Morton. *The Transformation of American Law, 1780–1860*. Cambridge, Mass.: Harvard University Press, 1977.

———. *The Transformation of American Law, 1870–1960*. New York: Oxford University Press, 1992.

Hovenkamp, Herbert. *Enterprise and American Law, 1836–1937*. Cambridge, Mass.: Harvard University Press, 1991.

Howard, Fred K. *History of the Sun-Maid Raisin Growers*. Fresno: F. K. Howard, 1923.

Hurst, J. Willard. *Law and Economic Growth: The Legal History of the Lumber Industry in Wisconsin*. 1964. Reprint. Madison: University of Wisconsin Press, 1984.

———. *Law and Markets in United States History: Different Modes of Bargaining among Interests*. Madison: University of Wisconsin Press, 1982.

———. *Law and the Conditions of Freedom in the 19th-Century U.S.* Madison: University of Wisconsin Press, 1956.

———. *The Legitimacy of the Business Corporation in the Law of the United States, 1780–1970*. Charlottesville: University Press of Virginia, 1970.

Husmann, George C. "The Raisin Industry." U.S. Department of Agriculture *Bulletin* 349. Washington, D.C.: U.S. Government Printing Office, 1916.

Hutchinson, Claude B., ed. *California Agriculture*. Berkeley: University of California Press, 1946.

Ichioka, Yuji. "Japanese Immigrant Response to the 1920 California Alien Land Law." *Agricultural History* 58 (1984): 157–78.

*Imperial Fresno: Resources, Industries and Scenery, Illustrated and Described*. Fresno: Fresno Republican Publishing Co., 1897.

Industrial Survey Committee of Fresno County Chamber of Commerce. *Fresno County Facts*. Fresno: Privately Printed, n.d., but 1922.

Irons, Peter. *The New Deal Lawyers*. Princeton: Princeton University Press, 1982.

Iwata, Masakazu. *Planted in Good Soil: The History of the Issei in United States Agriculture*. New York: Peter Lang, 1992.

Jelinek, Lawrence J. *Harvest Empire: A History of California Agriculture*. 2d ed. San Francisco: Boyd and Fraser, 1982.

Jenks, Jeremiah, and Walter Clark. *The Trust Problem*. 5th ed. Garden City, N.Y.: Doubleday, Doran, 1929.

Jensen, A. Ladru. "Integrating Economic and Legal Thought on Cooperatives." *Journal of Farm Economics* 31 (1949): 891–907.

Jesness, O. B. "The Bill of Rights of U.S. Cooperative Agriculture." *Rocky Mountain Law Review* 20 (1948): 181–89.

———. "Cooperative Purchasing and Marketing Organizations among Farmers in the United States." U.S. Department of Agriculture *Bulletin 547*. Washington, D.C.: U.S. Government Printing Office, 1917.

Jones, Franklin D. "The Status of Farmers' Co-operative Associations under Federal Law." *Journal of Political Economy* 29 (1921): 595–603.

Jorde, Thomas M. "Antitrust and the New State Action Doctrine: A Return to Deferential Economic Federalism." *California Law Review* 75 (1987): 227–56.

Kalman, Laura. *Abe Fortas: A Biography*. New Haven, Conn.: Yale University Press, 1990.

Karst, Kenneth. *Belonging to America: Equal Citizenship and the Constitution*. New Haven, Conn.: Yale University Press, 1989.

Kearney, M. Theo. *Fresno County, a Wonderfully Prosperous District in California*. Chicago: Donohue & Henneberry, 1893.

———. *M. Theo Kearney's Fresno County, California, and the Evolution of the Fruit Vale Estate*. 1899. Reprint. 1903. Facsimile Reproduction, Fresno: Fresno City and County Historical Society, 1980.

Keegan, Milton J. "The Power of Agricultural Cooperative Associations to Limit Production." *Michigan Law Review* 26 (1928): 648–73.

Keller, Morton. *Affairs of State: Public Life in Late Nineteenth Century America*. Cambridge, Mass.: Harvard University Press, 1977.

———. "The Pluralist State: American Economic Regulation in Comparative Perspective, 1900–1930." In *Regulation in Perspective: Historical Essays*, edited by Thomas P. McCraw, 56–94. Cambridge, Mass.: Harvard University Press, 1981.

———. *Regulating a New Economy: Public Policy and Economic Change in America, 1900–1933*. Cambridge, Mass.: Harvard University Press, 1990.

Kelley, Robert. *Battling the Inland Sea: American Political Culture, Public Policy, and the Sacramento Valley, 1850–1986.* Berkeley: University of California Press, 1989.

———. *Gold versus Grain: The Hydraulic Mining Controversy in California's Sacramento Valley.* Glendale, Calif.: A. A. Clark, 1959.

Kelly, Alfred H., Winfred A. Harbison, and Herman J. Belz. *The American Constitution: Its Origins and Development.* 6th ed. New York: Norton, 1983.

Kerber, Linda K. *Women of the Republic: Intellect and Ideology in Revolutionary America.* Chapel Hill: University of North Carolina Press, 1980.

Kerr, W. H. *The Farmers' Union and Federation Advocate Guide: The Minimum Price System.* Topeka, Kans.: Crane, 1919.

Kerr, W. H., and G. A. Nahstoll. "Cooperative Organization Business Methods." U.S. Department of Agriculture *Bulletin* 178. Washington, D.C.: U.S. Government Printing Office, 1915.

Kirkendall, Richard S. "Social Science in the Central Valley: A Rejoinder." *Agricultural History* 53 (1979): 494–505.

Kluger, Richard. *Simple Justice: The History of Brown v. Board of Education and Black America's Struggle for Equality.* New York: Knopf, 1975.

Knapp, Joseph G. *The Advance of American Cooperative Enterprise, 1920–1945.* Danville, Ill.: Interstate Printers and Publishers, 1973.

———. *The Rise of American Cooperative Enterprise, 1620–1900.* Danville, Ill.: Interstate Printers and Publishers, 1969.

Kolko, Gabriel. *Railroads and Regulation, 1877–1916.* Princeton: Princeton University Press, 1965.

Kriebel, Barry. "The Sun-Maid Story." *Audacity* 3 (Fall, 1994): 3.

Kroll, Harry H. *Riders in the Night.* Philadelphia: University of Pennsylvania Press, 1965.

Krug, William W., and Robert R. Dickey Jr. "Note: Constitutionality of Statute Fixing Milk Prices." *Boston University Law Review* 13 (1933): 697–704.

Laidler, Harry W. *Concentration of Control in American Industry.* New York: Thomas Y. Crowell, 1931.

Lamoreaux, Naomi R. *The Great Merger Movement in American Business, 1895–1904.* Cambridge: Cambridge University Press, 1985.

Lanier, Henry Wysham. "Cooperation — A Hopeful Tendency." *Review of Reviews* 66 (1922): 194–97.

LaPiere, Richard T. "The Armenian Colony in Fresno County, California: A Study in Social Psychology." Ph.D. dissertation, Stanford University, 1930.

Larsen, Grace H. "A Progressive in Agriculture: Harris Weinstock." *Agricultural History* 32 (1958): 187–93.

Larsen, Grace H., and Henry E. Erdman. "Aaron Sapiro: Genius of Farm Cooperative Promotion." *Mississippi Valley Historical Review* 49 (1962): 242–68.

Leibhardt, Barbara. "Allotment Policy in an Incongruous Legal System: The Yakima Indian Nation as a Case Study, 1887–1934." *Agricultural History* 65 (1991): 78–104.

———. "Law, Environment, and Social Change in the Columbia River Basin: The Yakima Indian Nation as a Case Study, 1840–1933." Ph.D. dissertation, University of California, Berkeley, 1990.

Lerwick, M. B. *Fresno County, California.* San Francisco: Sunset Magazine Homeseekers Bureau, for the Fresno County Board of Supervisors, n.d. but probably 1913.

Letwin, William. *Law and Economic Policy in America: The Evolution of the Sherman Antitrust Act*. New York: Random House, 1965.

Leuchtenburg, William. *Franklin D. Roosevelt and the New Deal*. New York: Harper & Row, 1963.

Levi, Edward. *An Introduction to Legal Reasoning*. Chicago: University of Chicago Press, 1949.

Levy, Leonard. *The Law of the Commonwealth and Chief Justice Shaw*. Cambridge, Mass.: Harvard University Press, 1957.

Levy, Louis. *History of the Cooperative Raisin Industry of California*. Fresno, Calif.: Privately published, 1928.

Lillard, Richard G. "Agricultural Statesman: Charles C. Teague of Santa Paula." *California History* 65 (1986): 2–16.

Lloyd, John W. "Cooperative and Other Methods of Marketing California Horticultural Products." *University of Illinois Studies in Social Science* 8 (1919).

Lothyan, Phillip E. "A Question of Citizenship." *Prologue* 21 (Fall 1989): 267–73.

Lowitt, Richard B. *George W. Norris: The Persistence of a Progressive, 1913–1933*. Urbana: University of Illinois Press, 1971.

Lukes, Timothy J., and Gary Y. Okihiro. *Japanese Legacy: Farming and Community Life in California's Santa Clara Valley*. Cupertino, Calif.: California History Center, 1985.

MacAvoy, Paul W. *The Economic Effects of Regulation: The Trunk-Line Cartels and the Interstate Commerce Commission before 1900*. Cambridge, Mass.: Massachusetts Institute of Technology Press, 1965.

McConnell, Grant. *The Decline of Agrarian Democracy*. New York: Atheneum, 1959.

———. *Private Power and American Democracy*. New York: Knopf, 1966.

McCraw, Thomas K. *Prophets of Regulation: Charles Francis Adams, Louis D. Brandeis, James M. Landis, and Alfred E. Kahn*. Cambridge, Mass.: Harvard University Press, 1984.

———. "Rethinking the Trust Question." In *Regulation in Perspective: Historical Essays*, edited by Thomas K. McCraw, 1–55. Cambridge, Mass.: Harvard University Press, 1981.

———, ed. *Regulation in Perspective: Historical Essays*. Cambridge, Mass.: Published for the Graduate School of Business Administration by Harvard University Press, 1981.

McCurdy, Charles W. "American Law and the Marketing Structure of the Large Corporation, 1857–1890." *Journal of Economic History* 38 (1978): 631–49.

———. "Justice Field and the Jurisprudence of Government-Business Relations: Some Parameters of Laissez Faire Constitutionalism, 1863–1897." *Journal of American History* 61 (1975): 970–1005.

———. "The *Knight* Sugar Decision of 1895 and the Modernization of American Corporation Law, 1869–1903." *Business History Review* 53 (1979): 304–42.

———. "The Roots of 'Liberty of Contract' Reconsidered: Major Premises in the Law of Employment, 1867–1937." *Supreme Court Historical Society Yearbook* (1984): 20–33.

———. "Stephen J. Field and Public Land Law Development in California, 1850–1866: A Case Study of Judicial Resource Allocation in Nineteenth Century America." *Law and Society Review* 10 (1976): 235–66.

MacCurdy, Rahno Mabel. *The History of the California Fruit Growers Exchange*. Los Angeles: George Rice and Sons, 1925.

McEvoy, Arthur F. *The Fisherman's Problem: Ecology and Law in the California Fisheries, 1850–1980*. New York: Cambridge University Press, 1986.

McKay, Andrew W. *Federal Research and Educational Work for Farmer Cooperatives, 1913–1953.* U.S. Department of Agriculture Farmer Cooperative Service *Service Report* 40. Washington, D.C.: U.S. Government Printing Office, 1959.

———. *Organization and Development of a Cooperative Citrus-Fruit Marketing Agency.* U.S. Department of Agriculture *Bulletin* 1237. Washington, D.C.: U.S. Government Printing Office, 1924.

MacLeod, W. A. "The Canadian Wheat Pools." *News for Farmer Cooperatives,* July 1944, 8–9.

McMath, Robert C., Jr. *American Populism: A Social History, 1877–1989.* New York: Hill and Wang, 1993.

———. "Sandy Land and Hogs in the Timber: (Agri)Cultural Origins of the Farmers' Alliance in Texas." In *The Countryside in the Age of Capitalist Transformation,* edited by Jonathan Prude and Steven Hahn, 205–29. Chapel Hill: University of North Carolina Press, 1985.

McWilliams, Carey. *Factories in the Field.* Boston: Little, Brown, 1939.

Mahakian, Charles. "History of Armenians in California." M.A. thesis, University of California, Berkeley, 1935.

Majka, Linda C., and Theo J. Majka. *Farm Workers, Agribusiness, and the State.* Philadelphia: Temple University Press, 1982.

Malin, James C. "The Background of the First Bills to Establish a Bureau of Markets, 1911–1912." *Agricultural History* 6 (1932): 107–29.

Manley, Henry S. "Constitutionality of Regulating Milk as a Public Utility." *Cornell Law Quarterly* 18 (1933): 410–19.

May, James A. "Political and Economic Theory in Antitrust Analysis, 1880–1918." *Ohio State Law Journal* 50 (1989): 257–395.

Mears, Edwin G., and Mathew O. Tobriner. *Principles and Practices of Cooperative Marketing.* Boston: Ginn, 1926.

Mensch, Elizabeth. "The History of Mainstream Legal Thought." In *The Politics of Law: A Progressive Critique,* edited by David Kairys, 13–37. Rev. ed. New York: Pantheon Books, 1990.

Merritt, Ralph P. "California Cooperatives Set an Example." *Review of Reviews* 72 (1925): 149–58.

Meyer, Albert J. "History of the California Fruit Growers' Exchange, 1893–1920." Ph.D. dissertation, Johns Hopkins University, 1951.

Meyer, Edith Catharine. "The Development of the Raisin Industry in Fresno County, California." M.A. thesis, University of California, Berkeley, 1931.

Meyer, Theodore R. "The Law of Co-operative Marketing." *California Law Review* 15 (1927): 85–112.

Miller, Crane S., and Richard Myslop. *California: The Geography of Diversity.* Pomona: California State Polytechnic University, 1983.

Miller, John D. "Farmers' Co-operative Associations as Legal Combinations." *Cornell Law Quarterly* 7 (1923): 293–309.

Miller, M. Catherine. *Flooding the Courtrooms: Law and Water in the Far West.* Lincoln: University of Nebraska Press, 1993.

———. "Riparian Rights and the Control of Water in California, 1879–1928: The Relationship between an Agricultural Enterprise and Legal Change." *Agricultural History* 59 (1985): 1–24.

Mirak, Robert. *Torn between Two Lands: Armenians in America, 1890 to World War I.* Cambridge, Mass.: Harvard University Press, 1983.

Mitchell, Broadus. *Depression Decade: From New Era through New Deal.* New York: Holt, Rinehart & Winston, 1947.

Montgomery, David. *Fall of the House of Labor: The Workplace, the State, and American Labor Activism, 1865–1925.* New York: Cambridge University Press, 1987.

Montgomery, Robert H. *The Cooperative Pattern in Cotton.* New York: Macmillan, 1929.

Moradian, Ralph. "Armenians." In *Fresno County Centennial Almanac,* edited by Fresno County Centennial Committee, 139. Fresno: Fresno County Centennial Committee, 1956.

Morgan, Edmund. *American Slavery, American Freedom: The Ordeal of Colonial Virginia.* New York: Norton, 1975.

Mowry, George. *The California Progressives.* Chicago: Quadrangle Books, 1976.

Mueller, Willard F., Peter G. Helmberger, and Thomas W. Patterson. *The Sunkist Case: A Study in Legal-Economic Analysis.* Lexington, Mass.: D. C. Heath, 1987.

Nash, Gerald. *State Government and Economic Development: A History of Administrative Policies in California, 1849–1933.* Berkeley: Institute of Governmental Studies, University of California, 1964.

National Commission for the Review of Antitrust Laws and Procedures. *Report to the President and the Attorney General.* 2 vols. Washington, D.C.: U.S. Government Printing Office, 1979.

Nelson, William E. *Americanization of the Common Law: The Impact of Legal Change on Massachusetts Society, 1760–1830.* Cambridge, Mass: Harvard University Press, 1975.

Note. "Equity: Specific Performance of Cooperative Distribution Contracts." *California Law Review* 10 (1922): 518–24.

Note. "Opposition to Cooperative Marketing." *Yale Law Journal* 32 (1923): 819–24.

Note. "Specific Performance Contracts Specifically Enforceable: Contracts for the Co-operative System of Distribution of Agricultural Products under Section 653pp, California Civil Code." *California Law Review* 12 (1924): 146–51.

Nourse, Edwin G. *Agricultural Economics.* Chicago: University of Chicago Press, 1916.

———. "The Economic Philosophy of Co-operation." *American Economic Review* 12 (December 1922): 577–97.

———. *The Legal Status of Agricultural Co-operation.* New York: Macmillan, 1927.

———. "The Place of Agriculture in Modern Industrial Society." *Journal of Political Economy* 27 (1919): 466–97, 561–77.

———. "The Revolution in Farming." *Yale Review* n.s. 8 (1918–19): 90–105.

Nourse, Edwin G., Joseph S. Davis, and John D. Black. *Three Years of the Agricultural Adjustment Administration.* Washington, D.C.: Brookings Institution, 1937.

Novak, William J. *The People's Welfare: Law and Regulation in Nineteenth-Century America.* Chapel Hill: University of North Carolina Press, 1996.

———. "Public Economy and the Well-Ordered Market: Law and Economic Regulation in 19th-Century America." *Law and Social Inquiry* 18 (Winter 1993): 1–32.

Odom, E. Dale. "Associated Milk Producers, Incorporated: Testing the Limits of Capper-Volstead." *Agricultural History* 59 (1985): 41–55.

Okoomian, Janice, and Kenneth L. Schneyer. "The Borders of Asian-American Identity: Inclusion and Exclusion in Some Older Naturalization Cases." Paper delivered at the annual meeting of the Law and Society Association, Toronto, Canada, June 3, 1995.

Olson, Mancur. *The Logic of Collective Action: Public Goods and the Theory of Groups*. Harvard Economic Studies, vol. 124. Cambridge, Mass.: Harvard University Press, 1965.

Ordal, Rolf Waldemar. "History of the California Walnut Industry." Ph.D. dissertation, University of California, Berkeley, 1952.

Ostler, Jeffrey. *Prairie Populism: The Fate of Agrarian Radicalism in Kansas, Nebraska, and Iowa, 1880–1892*. Lawrence: University of Kansas Press, 1993.

Palmer, Bruce. *"Man Over Money": The Southern Populist Critique of American Capitalism*. Chapel Hill: University of North Carolina Press, 1980.

Papazian, Dennis. "The Struggle for Personal and Collective Identity: The Ukrainian and Armenian Experience in America." *Journal of American Ethnic History* 14 (Spring 1995): 53–56.

Parrish, Michael E. *Anxious Decades: America in Prosperity and Depression, 1920–1941*. New York: Norton, 1992.

Parsons, Stanley B., Karen Toombs Parsons, Walter Killiae, and Beverly Borgers. "The Role of Cooperatives in the Development of the Movement Culture of Populism." *Journal of American History* 69 (1983): 866–85.

Paul, Arnold. *Conservative Crisis and the Rule of Law: Attitudes of Bar and Bench, 1887–1895*. 1960. Reprint. Gloucester, Mass.: Peter Smith, 1976.

Paul, Rodman. "The Beginnings of Agriculture in California: Innovation *vs.* Continuity." *California Historical Quarterly* 52 (1973): 16–27.

———. *The Far West and the Great Plains in Transition, 1859–1900*. New York: Harper & Row, 1988.

Payne, Will. "Cooperation — The Raisin Baron." *Saturday Evening Post* 182 (April 30, 1910): 14–15, 44–45.

Perkins, Van L. *Crisis in Agriculture: The Agricultural Adjustment Administration, 1933*. Berkeley: University of California Press, 1969.

Phelan, John. *Readings in Rural Sociology*. New York: Macmillan, 1920.

Pisani, Donald J. *From the Family Farm to Agribusiness: The Irrigation Crusade in California and the West, 1850–1931*. Berkeley: University of California Press, 1984.

———. "Land Monopoly in Nineteenth-Century California." *Agricultural History* 65 (1991): 15–37.

Plehn, Carl C. "The State Market Commission of California." *American Economic Review* 8 (1918): 1–26.

Plunkett, Horace. *The Rural Life Problem of the United States: Notes of an Irish Observer*. New York: Macmillan, 1912.

Powell, Clark. *Organization of a Great Industry: The Success of the California Fruit Growers' Exchange*. Pretoria, South Africa: Government Printing and Stationery Office, 1925.

Powell, Fred W. "Cooperative Marketing of California Fresh Fruit." *Quarterly Journal of Economics* 24 (1910): 392–418.

Powell, G. Harold. *Cooperation in Agriculture*. New York: Macmillan, 1913.

———. "Fundamental Principles of Co-operation in Agriculture." University of California College of Agriculture, Agricultural Experiment Station *Circular* 123 (October 1914): 1–15.

———. "Fundamental Principles of Co-operation in Agriculture." University of California College of Agriculture, Agricultural Experiment Station *Circular* 222 (October 1920): 1–24.

————. *Letters from the Orange Empire.* Edited by Richard G. Lillard. Los Angeles: Historical Society of Southern California, 1990.

Purcell, Conly L. "Constitutional Law—Regulation of the Price of Milk in Interstate Commerce." *Missouri Law Review* 1 (January 1936): 64–68.

Purcell, Robert W. "Constitutional Law: Due Process, Regulation of the Price of Milk." *Cornell Law Quarterly* 19 (December 1933): 85–90.

Rasmussen, Wayne D. *Taking the University to the People: Seventy-Five Years of Cooperative Extension.* Ames: Iowa State University Press, 1989.

————, ed. *Agriculture in the United States: A Documentary History.* 4 vols. New York: Random House, 1975.

Rehart, Schyler, and William K. Patterson. *M. Theo Kearney: Prince of Fresno, a Biography of Martin Theodore Kearney.* Fresno: Fresno City and County Historical Society, 1988.

Ridge, Martin. "Populism Redux: John D. Hicks and *The Populist Revolt.*" *Reviews in American History* 13 (1985): 142–54.

Riley, Elizabeth M. "The History of the Almond Industry in California, 1850–1934." M.A. thesis, University of California, Berkeley, 1948.

Ritchie, Robert Welles. "Rescuing a Raisin Maid: Heroic Efforts Won the Great Cooperative's Fight." *Country Gentleman* 88 (July 7, 1923): 3–4.

Robinson, William W. *Land in California.* Berkeley: University of California Press, 1948.

Rollins, Joan H. *Hidden Minorities.* Lanham, Md.: University Press of America, 1981.

Ross, Dorothy. *The Origins of American Social Science.* New York: Cambridge University Press, 1991.

Ross, Lydia. "The Morality of Co-operation." *Arena* 23 (February 1900): 183–93.

Rothstein, Morton. "The California 'Wheat Kings.' " Working Paper Series, Agricultural History Center, University of California, Davis, 1985.

————. "Farmer Movements and Organizations: Numbers, Gains, Losses." *Agricultural History* 62 (1988): 161–81.

————. "Frank Norris and Perceptions of the Market." *Agricultural History* 52 (1982): 51–66.

————. "West Coast Farmers and the Tyranny of Distance: Agriculture on the Fringes of the World Market." *Agricultural History* 49 (1975): 272–80.

Ruiz, Vicki L. *Cannery Women/Cannery Lives: Mexican Women, Unionization and the California Food Processing Industry, 1930–1950.* Albuquerque: University of New Mexico Press, 1987.

Saloutos, Theodore. *The American Farmer and the New Deal.* Ames: Iowa State University Press, 1982.

————. Review of Lawrence Goodwyn, *Democratic Promise. Pacific Historical Review* 47 (1978): 149–50.

Saloutos, Theodore, and John D. Hicks. *Agricultural Discontent in the Middle West, 1900–1939.* Madison: University of Wisconsin Press, 1951.

Salyer, Lucy E. *Laws Harsh as Tigers: Chinese Immigrants and the Shaping of Modern Immigration Law.* Chapel Hill: University of North Carolina Press, 1995.

Sapiro, Aaron. "Cooperative Marketing." *Iowa Law Bulletin* 8 (May 1923): 193–210.

————. "The Law of Cooperative Marketing Associations." *Kentucky Law Journal* 15 (1926): 1–21.

————. *Papers on Cooperative Marketing*. Pamphlets, 1920–23. General Library, University of California at Berkeley.

————. "True Farmer Cooperation." *World's Work* 46 (1923): 84–96.

Saroyan, William. "The Armenian and the American." *Ararat* 25 (Spring 1984): 7.

————. "'Turn Back the Universe and Give Me Yesterday: Memories of Fresno.'" *American Heritage* 31 (October 1980): 49–60.

Sax, Joseph, and Richard Abrams. *Water Law*. St. Paul, Minn.: West, 1985.

Scheiber, Harry N. "At the Borderland of Law and Economic History: The Contributions of Willard Hurst." *American Historical Review* 75 (1970): 744–56.

————. "Federalism and the American Economic Order, 1789–1910." *Law and Society Review* 10 (1976): 57–118.

————. "Government and the Economy: Studies of the 'Commonwealth' Policy in Nineteenth-Century America." *Journal of Interdisciplinary History* 3 (1972): 135–51.

————. "Property Law, Expropriation, and Resource Allocation by Government, 1789–1910." *Journal of Economic History* 33 (1973): 232–51.

————. "Public Rights and the Rule of Law in American Legal History." *California Law Review* 71 (1984): 217–51.

————. "The Road to *Munn*: The Concept of Public Purpose in the State Courts." *Perspectives in American History* 5 (1971): 329–402.

Schlebecker, John T. *Whereby We Thrive: A History of American Farming, 1607–1972*. Ames: Iowa State University Press, 1975.

Scholl, Carl Albert. "An Economic Study of the California Almond Growers Exchange." Ph.D. dissertation, University of California, Berkeley, 1927.

Schruben, Francis W. "An Even Stranger Death of President Harding." *Southern California Quarterly* 48 (1966): 57–84.

Seftel, Howard. "Government Regulation and the Rise of the California Fruit Industry: The Entrepreneurial Attack on Fruit Pests, 1880–1920." *Business History Review* 59 (1985): 369–402.

Seitz, Don C. "The Kind of Cooperation That Will Afford Farm Relief." *Outlook* 141 (December 30, 1925): 668–69.

Shannon, Fred. *The Farmer's Last Frontier: Agriculture, 1860–1897*. New York: Farrar and Rinehart, 1945.

Sherman, Caroline B. "The Legal Basis of the Marketing Work of the United States Department of Agriculture." *Agricultural History* 11 (1937): 289–301.

Shideler, James H. *Farm Crisis, 1919–1923*. Berkeley and Los Angeles: University of California Press, 1957.

Shover, John L. *First Majority, Last Minority: The Transforming of Rural Life in America*. DeKalb: Northern Illinois University Press, 1976.

Sklar, Martin J. *The Corporate Reconstruction of American Capitalism, 1890–1916: The Market, the Law, and Politics*. New York: Cambridge University Press, 1988.

Skocpol, Theda. *Protecting Soldiers and Mothers: The Political Origins of Social Policy in the United States*. Cambridge, Mass.: Harvard University Press, 1992.

Skowronek, Stephen. *Building a New Administrative State: The Expansion of National Administrative Capacities, 1877–1920*. New York: Cambridge University Press, 1982.

Smith, Wallace. *Garden of the Sun*. 1939. Reprint. Fresno: Max Hardison—A-1 Printers, 1960.

Smythe, William E. "A Benefactor of the State." *Out West* 25 (1906): 146–50.

———. *The Conquest of Arid America*. Rev. ed. Seattle: University of Washington Press, 1969.

———. *Constructive Democracy: The Economics of a Square Deal*. New York: Macmillan, 1905.

Socolofsky, Homer E. *Arthur Capper: Publisher, Politician, and Philanthropist*. Lawrence: University of Kansas Press, 1962.

Soifer, Aviam. "Willard Hurst, Consensus History, and the Growth of American Law." *Reviews in American History* 20 (1992): 124–44.

Spence, W. Y. "Success after Twenty Years." *Sun-Maid Herald* 3 (November 1917)–4 (April 1919).

Spencer, Olin C. *California's Prodigal Sons: Hiram Johnson and the Progressives, 1911–1917*. Berkeley: University of California Press, 1968.

Stanley, Robert Howard. *Dimensions of Law in the Service of Order: Origins of the Federal Income Tax, 1861–1913*. New York: Oxford University Press, 1993.

Starr, Kevin. *Americans and the California Dream, 1850–1915*. New York: Oxford University Press, 1973.

———. *Inventing the Dream: California through the Progressive Era*. New York: Oxford University Press, 1985.

Steen, Herman. "The California Raisin Trust." *Wallace's Farmer* 45 (October 8, 1920): 2356.

———. *Cooperative Marketing: The Golden Rule in Agriculture*. New York: Doubleday, Page, 1923.

———. "Story of the California Raisin 'Trust.'" *Hoard's Dairyman* 60 (October 15, 1920): 532–34.

Stegner, Wallace E. *Beyond the Hundreth Meridian: John Wesley Powell and the Opening of the West*. Boston: Houghton Mifflin, 1984.

Stokdyk, Ellis A. "Economic and Legal Aspects of Compulsory Proration in Agricultural Marketing." University of California Agricultural Experiment Station *Bulletin* 565 (December 1933).

Street, Richard S. "Marketing California Crops at the Turn of the Century." *Southern California Quarterly* 61 (1979): 239–53.

Sullivan, Lawrence A. *Handbook of the Law of Antitrust*. St. Paul, Minn.: West, 1977.

Taylor, Alonzo E. "Your Inelastic Appetite: Its Inability to Expand Is the Hardest Problem of the Cooperative Marketing Associations Today." *Sunset Magazine* 52 (February 1924): 20–21, 58–60.

Taylor, Henry C., and Anne Dewees Taylor. *The Story of Agricultural Economics in the United States, 1840–1932*. Ames: Iowa State College Press, 1952.

Taylor, Paul Schuster. "Immigrant Groups in Western Culture." *Agricultural History* 49 (1975): 179–81.

———. "Walter Goldschmidt's Baptism in Fire: Central Valley Water Politics." In *Paths to the Symbolic Self: Essays in Honor of Walter Goldschmidt*, edited by James P. Louckey and Jeffrey R. Jones, 129–40. Los Angeles: University of California Department of Anthropology, 1976.

Teague, Charles Collins. *Fifty Years a Rancher: Half a Century Devoted to the Citrus and Walnut Industries of California and to Furthering the Cooperative Movement in Agriculture*. Los Angeles: California Fruit Growers' Exchange, 1944.

Teilman, I., and W. H. Shafer. *The Historical Story of Irrigation in Central California.* Fresno: Williams and Son, 1943.

Tevis, Charles V. "A Ku-Klux Klan of Today." *Harper's Weekly* 52 (1908): 14–16, 32.

Thickens, Virginia M. "Pioneer Agricultural Colonies of Fresno County." *California Historical Society Quarterly* 25 (1946): 21–36, 169–77.

Thomas, John L. *Alternative America: Henry George, Edward Bellamy, Henry Demarest Lloyd, and the Adversary Tradition.* Cambridge, Mass.: Harvard University Press, 1983.

Thompson, E. P. *Whigs and Hunters: The Origins of the Black Act.* London: Allen Lane, 1975.

Thorelli, Hans B. *The Federal Antitrust Policy: Origination of an American Tradition.* Baltimore: Johns Hopkins University Press, 1955.

Thorpe, Carlyle. "California Walnuts and Their Co-operative Marketing." *California Magazine* 1 (1915): 481–84.

Tobriner, Mathew O. "The Constitutionality of Cooperative Marketing Statutes." *California Law Review* 17 (1928): 19–34.

———. "Cooperative Marketing and the Restraint of Trade." *Columbia Law Review* 27 (1927): 826–36.

———. "How Shall Cooperatives Be Organized?" *Journal of Farm Economics* 6 (1924): 367–77.

———. "Legal Aspect of the Provisions of Cooperative Marketing Contracts." *American Bar Association Journal* 12 (1926): 19–34.

Tomlins, Christopher L. *Law, Labor, and Ideology in the Early American Republic.* Cambridge: Cambridge University Press, 1993.

———. *The State and the Unions: Labor Relations, Law, and the Organized Labor Movement in America, 1880–1960.* Cambridge: Cambridge University Press, 1985.

Trujillo, Larry Dean. *Parlier, the Hub of Raisin America: A Local History of Capitalist Development.* Berkeley: University of California Institute for the Study of Social Change, 1978.

Tugwell, Rexford G. "Reflections on Farm Relief." *Political Science Quarterly* 43 (December 1928): 481–97.

Tushnet, Mark. *The NAACP's Legal Strategy against Segregated Education, 1925–1950.* Chapel Hill: University of North Carolina Press, 1987.

Vaught, David. "An Orchardist's Point of View: Harvest Labor Relations on a California Almond Ranch, 1892–1921." *Agricultural History* 69 (Fall 1995): 563–91.

Vincent, C. "Co-operation among Western Farmers." *Arena* 31 (1904): 286–92.

Waldrep, Christopher. *Night Riders: Defending Community in the Black Patch, 1880–1915.* Durham, N.C.: Duke University Press, 1992.

Walker, Ben R. *The Fresno County Blue Book.* Fresno: Arthur H. Cawston, 1941.

Wallace, Henry A. *New Frontiers.* New York: Reynal & Hitchcock, 1934.

Wallis, Wilson. *Fresno Armenians (to 1919).* Edited by Nector Davidian. Lawrence, Kan.: Coronado Press, 1965.

Walsh, Robert M. "Raisin Industry Growth Marked by Violence." *Fresno Bee*, August 19, 1984, p. G4.

———. "Turbulent Years in the Raisin Industry." In *Fresno County in the Twentieth Century from 1900 to the 1980s: An All New History*, edited by Charles W. Clough et al., 178–80. Fresno: Panorama West Books, 1986.

Walton, John. *Western Times and Water Wars: State, Culture, and Rebellion in California.* Berkeley: University of California Press, 1992.

Warbasse, James Peter. "The Cooperative Movement." *Nation* 106 (May 11, 1918): 565–66.

Warren, Charles. "The New Liberty under the Fourteenth Amendment." *Harvard Law Review* 39 (1926): 431–65.

Watkins, Marilyn P. *Rural Democracy: Family Farmers and Politics in Western Washington, 1890–1925.* Ithaca, N.Y.: Cornell University Press, 1995.

Watson, Malcolm H. "An Analysis of Raisin Marketing Controls under the California Agricultural Prorate Act." M.S. thesis, University of California, Berkeley, 1940.

Weber, Devra. *Dark Sweat, White Gold: California Farm Workers, Cotton, and the New Deal.* Berkeley: University of California Press, 1994.

Wellman, Harry R. "An Analysis of the Methods of Pooling Employed by the Cooperative Fruit-Marketing Associations in California." Ph.D. dissertation, University of California, Berkeley, 1926.

Weseen, Maurice H. "The Co-operative Movement in Nebraska." *Journal of Political Economics* 28 (1920): 477–98.

White, Graham, and John Maze. *Henry A. Wallace: His Search for a New World Order.* Chapel Hill: University of North Carolina Press, 1995.

White, T. C. "Cooperation among Raisin Growers." In State Board of Horticulture. *Seventh Biennial Report,* 160–61. Sacramento: A. J. Johnston, Superintendent State Printing, 1901.

Wickson, Edward J. *California Fruits and How to Grow Them.* 6th ed. San Francisco: Pacific Rural Press, 1912.

———. "California Mission Fruits." *Overland Monthly* 2d ser., 11 (January 1888): 501–5.

———. "The History of California Fruit Growing." *Pacific Rural Press* 71 (April 14, 1906): 228.

———. *Rural California.* San Francisco: Rural State and Province Series, 1922.

Wiebe, Robert H. *Businessmen and Reform: A Study of the Progressive Movement* Cambridge, Mass.: Harvard University Press, 1959.

Wiley, Jean Giffen. "Three Generations of Giffen Ranchers in California's San Joaquin Valley." Unpublished manuscript, 1989. Photocopy in author's possession.

Williams, Nathan B. *Laws on Trusts and Monopolies.* Washington, D.C.: U.S. Government Printing Office, 1913.

Wilson, Howard B. "Legal Status of Cooperative Marketing Movement." *American Bar Association Journal* 14 (1928): 576–81.

Winchell, Lilbourne Alsip. *History of Fresno County and the San Joaquin Valley.* Fresno: Arthur H. Cawston, 1933.

Winters, Donald L. *Henry C. Wallace as Secretary of Agriculture, 1921–1924.* Urbana: University of Illinois Press, 1970.

Wise, James Waterman. *Jews Are Like That!* New York: Macmillan, 1928.

Woehlke, Walter V. "Raising the Price of the Raisin: Get Together Pays the Growers." *Country Gentleman* 79 (1914): 2039, 2058.

Worster, Donald. *Rivers of Empire: Water, Aridity and the Growth of the American West.* New York: Pantheon Books, 1985.

Youngman, Anna. "The Tobacco Pools of Kentucky and Tennessee." *Journal of Political Economy* 18 (1910): 34–49.

Zavella, Patricia. *Women's Work and Chicano Families: Cannery Workers of the Santa Clara Valley.* Ithaca, N.Y.: Cornell University Press, 1987.

# Index

Capper-Hersman bill, 157

Capper-Volstead Act, 203, 214, 229; embodies benevolent trust model of cooperation, 14, 162, 164; manifests paradox of farmers' cultural image, 138–39; impact on *United States v. CARC* of, 139, 158, 161–62, 165; proposals for, 147; provisions of, 157; Senate hearings on, 159–60; passage of, 160–61; Sun-Maid's compliance with, 167, 181, 182, 183, 185, 186–87, 189; ambiguity of, 195–96, 199; USDA administration of, 195–96, 201; historical importance of, 232–33; implementation of, 234; judicial interpretation of, 332 (n. 5)

Carryovers. *See* Overproduction

Carver, Thomas Nixon, 32

Center firms, 236

Central Pacific Railroad, 44

Central Valley, 48, 53

Chaddock, E. L., 135

Chan, Sucheng, 10, 11, 40, 257–58 (n. 26), 264 (n. 5)

Chandler, Alfred, 10

Chapman, William S., 44, 45

Chinese, 41, 46, 51, 54; exclusion of, 129. *See also* Asians

Citrus industry, 25, 28, 75, 79, 120, 236

Civil War, 21, 65, 67

Clayton Act, 111, 157, 161, 204, 229; provisions, 97–98; effect on legal status of cooperatives, 98–99, 100–101; impact on capital stock cooperatives, 105, 106–7; impact on CARC, 126, 131, 144, 150, 151, 152, 154. *See also* Antitrust law; Capper-Volstead Act; Congress; Sherman Act

Cleland, Robert Glass, 43

Collett, Loraine, 120

Colorado: antitrust law, 72, 211

Commodity Credit Corporation, 227, 229

Commodity marketing. *See* Cooperative marketing

Competition, 193, 213, 235

Congress, 41, 142, 144, 147, 158; and antitrust law reform, 94, 97–100; House, 99, 147; Senate, 100, 148, 159, 194; constrains

administrative authority of USDA, 102; and policy on cooperation, 106, 107, 138, 194, 233

*Connolly v. Union Sewer Pipe Co.*, 70–71, 72, 97, 131, 206, 209–10, 211, 212, 214, 232

Consent decree, 161, 162, 164, 165–66, 179, 180, 181, 182, 196, 199, 206; termed no longer in force against Sun-Maid, 186; contributes to Sun-Maid's collapse, 190; terms of, 300 (n. 49), 304 (n. 78); as initially proposed, 303 (n. 75). *See also* *United States v. CARC*

Consumers, 96, 193. *See also* Advertising

Coolidge, Calvin, 202

Cooperation: ideology of, 1, 8, 31–32, 78, 115, 195, 216, 217, 229; conservativeness of, 8; European roots of, 8, 259 (n. 7); promise to improve marketing without profiteering, 9; in raisin industry, 11, 12, 77–92 passim, 111–89 passim; federal policy on, 13, 93–94, 101, 143–44, 162–63, 201; ideal of, 14, 31–36, 162, 233, 234; as reflection of traditional political and economic values, 19, 103–4; appeal in United States of, 21; "true," 24, 31, 35, 63, 103, 207, 222; as distinctive form of enterprise, 31, 232; historical link to socialism, 32; legal definition of, 35; legal status of, 59, 63, 68, 100, 235; linked by cooperative marketing act to public policy goals, 203; impact on market, 213–14; accepted as definition of farmers' place in market, 214–15; legal change and similarity to corporation, 232. *See also* Cooperative incorporation laws; Cooperative Marketing Act; Cooperative movement; Rochdale cooperation

Cooperative incorporation laws, 34, 35, 64–66, 73, 93, 103–4; nonstock laws, 34, 35, 65, 167, 272 (n. 6); capital stock laws, 65–66, 72, 98, 104, 272 (n. 6)

Cooperative marketing, 24, 102, 199–200, 201, 202. *See also* California Associated Raisin Company; California cooperation; Cooperatives; Sun-Maid Raisin Growers of California

Cooperative Marketing Act (model state law), 178, 194–95, 214, 216; of 1926 (federal statute), 201; exempts farmers from antitrust laws, 203–4; prohibits farmers from activities prohibited by Clayton Act, 204; criminalizes third-party interference with cooperative contracts, 204–5; impact on existing legal rules, 205; provisions on contracts upheld, 206; courts hold cooperatives pose no threat to competition, 206–7, 210–11; limits to judicial acceptance of, 207; as embodiment of "true" cooperation, 207, 214; specific performance not awarded in California, 207, 310 (n. 45); liquidated damages limited in New York, 207–8; provisions on tenants, 208–9; and equal protection law, 209–10; effect on competition, 213–14; impact on cooperatives' legal status, 214; debates among legal scholars on, 320 (n. 40)

Cooperative movement, 3, 4, 7, 18, 106, 230–31, 237; legal changes in, 4–5; and cultural image of farmer, 4–5, 8–10, 19, 234; reorganizes horticultural marketing, 9, 24–31; relationship to business history, 10, 235–36; and relationship to agribusiness, 10–11; alliance with labor movement, 32–33; impact of *Nebbia v. New York* on, 219; unable to limit production, 222; impact of New Deal on, 234; significance of small farms for, 251

Cooperatives: use of corporate trade practices, 5; and monopoly and restraint of trade, 24, 140–41; relation to farm labor, 52; legal status of, 59, 74, 112; "one-member, one-vote" principle, 64, 65; separate incorporation of, 64–65; legal form of, 68, 72, 75; analogy to labor unions, 69, 74; and antitrust law, 95; legal privileges under federal law, 201; and state New Deal programs, 220, 221, 222, 230–31; present-day, 235–36; types of, 262 (n. 25). *See also* California cooperation; Dairy industry; Horticulture; Milk industry; Monopoly

*Coppage v. Kansas*, 74

Corporation law, 5, 64; farmers' use of, 6; CARC organized under California, 113

Corporations, 7, 33, 63, 95, 217; as model for new form of cooperation, 1, 5–6, 35, 75, 76, 77; corporate features of CARC, 113

Cotton, 198–99, 202

*Country Gentleman*, 121

Courts, 33, 35, 73, 105, 111, 159, 214; attitudes toward cooperatives, 63–76 passim; federal, 69–71, 94–96, 100, 111, 211, 228; reaction against cooperatives, 93; state, 106, 111, 140–41, 178, 181, 211

Croly, Herbert, 32

Crop contracts, 118, 126, 132, 150, 156, 172, 199; not enforced by California courts, 86–87, 178; CRGA contract struck by trial court, 89; guaranteed prices in CARC contract, 117–18; CARC, 126–27, 132–33; breach of, 127; imposes liquidated damages, 132; 1918 CARC contract runs with land, 132, 169; 1921 contract, 155–56, 166, 169; 1923 contract, 169, 172, 173, 178; 1921 contract struck down, 187–88; withdrawal clause in, 176; growers obtain cancellation of, 180; voided by courts in cases of coercion, 180–81. *See also* California Associated Raisin Company: membership campaigns; "Iron-clad contract"; Sun-Maid: 1923 crop contract campaign; Sun-Maid Raisin Growers of California: referendum period; Sun-Maid Raisin Growers of California: withdrawal period

Cross, Ira, 28

Cultural image of farming. *See* Agriculture: cultural image of

Cushman, Barry, 325–26 (n. 12)

Dairy industry, 68, 207–8; antitrust prosecutions of, 144, 151, 157, 297 (n. 18). *See also* Milk producers; *United States v. Borden*

Dairymen's League, 218, 220, 221

*Danbury Hatters'* case (*Loewe v. Lawlor*), 96, 97

CARC crop contract, 117–18, 124; retail, 136, 166; as trigger of antitrust enforcement, 138; set by United States Food Administration in 1918, 145; CARC 1919 price, 146, 150, 299 (n. 36); FTC standard on reasonable price, 151

Reedley (town of), 45, 46

*Reeves v. Decorah Farmers' Cooperative Society*, 73–75, 104, 204, 208. *See also* Maintenance clause

Regulation, 1, 59, 63, 213, 217, 223; constitutional limits of, 7; state, 93, 217–31; state-federal cooperation in, 227. *See also* Antitrust law; Corporation law; New Deal; U.S. Department of Agriculture; U.S. Department of Justice

Rehart, Schyler, 91

Restraint of trade, 140, 141. *See also* Antitrust law

Rice industry, 167

Riparianism, 267 (n. 25)

Riter, William D., 183

Rochdale cooperation, 33, 81, 190, 235; ideology of, 8, 9; origins of, 20–21; popularity of, 22; designed for collective purchasing, 23; criticized as unfit for marketing, 24, 34, 190; equated with "true" cooperation, 64, 96; influence on Clayton Act, 98; rejected by CARC organizers, 113. *See also* Cooperation; Cooperative movement; Rochdale principles

Rochdale principles, 20, 24, 35, 64, 103, 104

Roosevelt, Franklin, 224; administration, 216

Roosevelt, Theodore, 117

Rowell, Chester, 89, 280 (n. 11)

Rule of reason, 97, 126, 283–84 (n. 5); decisions, 95. *See also* Antitrust law; Sherman Act

*Rural New Yorker*, 220

Sacramento, 41

*San Francisco Chronicle*, 78, 83

Sanger (town of), 45, 46

San Joaquin Valley, 45, 46, 51, 55. *See also* Central Valley

Sapiro, Aaron, 160, 194, 205, 208, 234; background in cooperative marketing, 196; early career, 197–98, 315 (nn. 9, 11); plan for organizing farmers, 198–99; emphasis on monopoly, 199; criticisms of, 200–202, 317 (n. 24); forced to resign from American Farm Bureau Federation, 202; drafts model state cooperative marketing act, 203–4

Sargent, John G., 184, 185, 187

Saroyan, William, 57, 173

*Schecter Poultry Co. v. United States*, 217

Schoonover, Albert, 120, 131

Selma (town of), 45, 46

Setrakian, Arzapat, 172

Seymour, Fred A., 145

Sherman Act, 5, 66, 119, 229; amended by Clayton Act, 94; enforcement of, 96, 144, 158; CARC liability under, 126, 131; CARC prosecuted under, 138, 141, 151, 152, 155, 156, 159, 233; farmers' liability under, 159; Sun-Maid's liability under, 182, 214. *See also* Antitrust law; Appropriations rider; Capper-Volstead Act; Clayton Act; Congress; U.S. Department of Justice; *United States v. CARC*

Shideler, James, 140

Silver, Gray, 159

Sklar, Martin, 66, 255 (n. 12), 274 (n. 12)

Smith-Lever Act, 102

Social history of law, 3

Socialism, utopian, 32

Southern California Fruit Exchange, 25–28, 261–62 (n. 22)

Southern Pacific Railroad, 44, 45, 46

Spanish missions, 41

Specific performance, 181, 199, 204, 207, 208. *See also* Crop contracts; "Iron-clad contract;" Liquidated damages

Standard Oil, 9

*Standard Oil v. United States*, 95

Steel trust, 87, 91

Steen, Herman, 28, 34, 215

*Steers v. United States*, 96, 97

Steinbeck, John, 10
Stokdyk, E. A., 223
Stone, Harlan F., 229
Stream of commerce jurisprudence, 95, 283
  (n. 4)
Substantive due process. *See* Lochner era
Sunkist, 28, 236, 237
"Sun Made," 120
*Sun-Maid Business*, 188
*Sun-Maid Herald*, 127, 130, 145
Sun-Maid Raisin Growers Association
  (Delaware Company), 168, 169, 189
Sun-Maid Raisin Growers of California,
  4–5, 12, 200, 205, 230, 234; trademark
  of, 120; 1921 contract campaign, 156; reorganization of, 165–68; name changed
  from CARC, 166; economic losses, 166,
  168, 175, 176, 189; 1923 stock campaign,
  169, 172; 1923 crop contract campaign,
  169, 172–75, 178, 182; monopoly of, 172,
  177, 188, 189; and investigations of night
  riding, 174–75; withdrawal period, 176,
  177–78; bankruptcy of old Sun-Maid,
  176, 184, 188, 308 (n. 30); marketing, 176,
  188; referendum period, 177, 179–81, 184;
  complaints about disloyal members,
  177–78, 179; investigated for violations
  of consent decree, 182–87, 196, 206, 214;
  loses breach of contract suits, 187–88,
  320–21 (n. 44); files for bankruptcy in
  1928, 189; reasons for collapse, 189–90;
  Janus-faced interpretation of cooperation, 193; inability to control market
  during Depression, 224, 229, 329 (n. 36);
  insolvency of, 224, 328 (n. 30); role in
  prorate program of, 225–28; market control since World War II, 236–37. *See also*
  Consent decree; Merritt, Ralph P.; U.S.
  Department of Agriculture: approves of
  Sun-Maid's activities in 1923; U.S. Department of Agriculture: prevails over
  Department of Justice in Sun-Maid case
Surplus. *See* Overproduction
Sutherland, George, 212
Sutherland, William A., 112, 116, 144, 148,
  150, 153, 179

Tariffs, 78
Teague, Charles C., 34, 45
Tenny, Lloyd S., 183, 185, 186
Thompson, E. P., 258 (n. 28)
*Tigner v. Texas*, 232
Tobacco, 96, 199, 202, 263 (n. 2); night
  riding in, 277 (n. 34)
*Tobacco Growers' Co-operative Association
  v. Jones*, 206
Tobriner, Mathew, 204, 210
Tri-State Tobacco Growers' Cooperative
  Marketing Association, 201, 202
Trust(s), 6, 66, 77, 92, 121, 235. *See also* Steel
  trust
Tulare County, 45, 55

U.S. attorney, 120, 131, 182, 184
U.S. census, 50; decision not to count
  farmers in 2000, 238
U.S. Department of Agriculture: prevails
  over Department of Justice in Sun-Maid
  case, 14, 178, 180, 181, 184, 185–87, 189,
  190, 206; agricultural economists in,
  94, 101, 103–4; promotion of marketing, 94, 101–3; interpretation of Clayton
  Act, 100; policy on cooperative marketing, 101–3, 106; solicitor's views on
  cooperation, 104, 105; draft of model
  state incorporation law, 105; enforcement of Capper-Volstead, 164–65, 181,
  194, 195–96, 214; approves of Sun-Maid's
  activities in 1923, 172, 175, 183; legal advice to farmers on Capper-Volstead, 196;
  position on Sapiro cooperatives, 200–
  202; New Deal agricultural programs,
  226–27; support of cooperation during
  Progressive Era, 287–88 (n. 35). *See also*
  Agricultural extension; Bureau of Agricultural Economics; Bureau of Markets;
  Jardine, William M.; Wallace, Henry C.
U.S. Department of Commerce, 187, 286 (n.
  27). *See also* Hoover, Herbert
U.S. Department of Justice, 142, 159, 164,
  190, 233; antitrust enforcement policy,
  14, 120, 126, 131, 138, 156, 158, 233; prosecution of CARC, 14, 136–37, 143, 152,

153–56; antitrust enforcement authority of, 94, 99, 100, 106, 145; investigations of CARC, 119–20, 124–26, 130, 147, 161; declines to prosecute CARC in 1917, 132; prosecutions of dairy cooperatives, 144; prosecutions of milk producers, 151; 1925 investigation of Sun-Maid, 177, 184–87; fights with USDA over Capper-Volstead regulatory authority regarding Sun-Maid, 178–79, 180, 181, 182–83, 185–87; 1923 investigation of Sun-Maid, 182; retains authority under consent decree, 184–85, 206; declines to prosecute night riders, 186. *See also* Appropriations rider; Consent decree; *United States v. CARC*

U.S. Federal Trade Commission, 132, 142–43, 147, 148, 221; 1918 investigation of CARC, 142–43, 152; 1919 hearings, 148–51; findings as to price, 151–52; investigation of Sunkist, 236

U.S. Food Administration, 139, 144; sets 1918 raisin prices, 145

U.S. Immigration Commission, 55

U.S. Supreme Court, 5, 70, 71, 72, 99, 111, 194, 211, 219, 228; view of farmers' place in market, 213; "switch in time," 216; and public regulation of prices, 217, 220

*United States v. Borden*, 324 (n. 8), 331–32 (n. 4)

*United States v. Butler*, 217

*United States v. CARC*, 9, 138–39, 152–56, 161–63, 164; consent decree, 161, 162, 181. *See also* Capper-Volstead Act: impact on *United Sates v. CARC* of; Consent decree

*United States v. E. C. Knight*, 95, 155, 156

*United States v. King*, 302 (n. 63)

*United States v. Swift & Co.*, 155

*United States v. Trans-Missouri Freight Association*, 95

*United States v. U.S. Steel*, 156

Urban society, 19, 195–96, 233, 235

Vertical integration. *See* California Associated Raisin Company: vertical integration of; Kearney, M. Theo: applies vertical integration to CRGA

Voting trust agreement, 113, 115. *See also* California Associated Raisin Company: corporate identity of

Wagner Act, 196

Wallace, Henry A., 223, 225, 227

Wallace, Henry C., 34, 214–15; position on Capper-Volstead Act, 169–70, 195–96; endorses Sun-Maid reorganization plan, 172, 175, 182; response to growers' pleas, 181; succeeded by Jardine, 185; views on Sapiro, 198, 200–201, 202; death of, 200

Walnut industry, 230

Walsh, Thomas J., 159, 160

Weinstock, Harris, 28, 144, 197–98; background of, 315 (nn. 11–12)

Welsh, H. H., 113, 117

Wheat, 39, 199. *See also* Canada

Wheat farming, 42

Wheat kings, 42, 44

Wholesale Grocers' Association, 100

*Wickard v. Filburn*, 230

Wickson, Edward J., 41, 88, 90

Wilson, Woodrow, 96, 99, 100, 101, 144

*Wolff v. Industrial Court*, 324 (n. 4), 327 (n. 23)

Women: image in food advertising, 121; as consumers, 121

World War I, 9, 54, 129, 132, 133, 137, 138, 144; effect on food prices, 132; impact on agriculture of, 139–41; postwar depression, 141. *See also* Farm crisis

World War II, 231, 236–37

Wrightson, H. W., 113

*Yick Wo v. Hopkins*, 70